AF845630

VOLUME SIX HUNDRED AND NINETEEN

METHODS IN ENZYMOLOGY

Ubiquitin-dependent Protein Degradation

METHODS IN ENZYMOLOGY

Editors-in-Chief

ANNA MARIE PYLE
Departments of Molecular, Cellular and Developmental Biology and Department of Chemistry
Investigator, Howard Hughes Medical Institute
Yale University

DAVID W. CHRISTIANSON
Roy and Diana Vagelos Laboratories
Department of Chemistry
University of Pennsylvania
Philadelphia, PA

Founding Editors

SIDNEY P. COLOWICK and NATHAN O. KAPLAN

VOLUME SIX HUNDRED AND NINETEEN

METHODS IN ENZYMOLOGY

Ubiquitin-dependent Protein Degradation

Edited by

MARK HOCHSTRASSER

Department of Molecular Biophysics & Biochemistry, Yale University, New Haven, CT, United States

ELSEVIER

ACADEMIC PRESS

An imprint of Elsevier

Academic Press is an imprint of Elsevier
50 Hampshire Street, 5th Floor, Cambridge, MA 02139, United States
525 B Street, Suite 1650, San Diego, CA 92101, United States
The Boulevard, Langford Lane, Kidlington, Oxford OX5 1GB, United Kingdom
125 London Wall, London, EC2Y 5AS, United Kingdom

First edition 2019

Copyright © 2019 Elsevier Inc. All rights reserved.

No part of this publication may be reproduced or transmitted in any form or by any means, electronic or mechanical, including photocopying, recording, or any information storage and retrieval system, without permission in writing from the publisher. Details on how to seek permission, further information about the Publisher's permissions policies and our arrangements with organizations such as the Copyright Clearance Center and the Copyright Licensing Agency, can be found at our website: www.elsevier.com/permissions.

This book and the individual contributions contained in it are protected under copyright by the Publisher (other than as may be noted herein).

Notices
Knowledge and best practice in this field are constantly changing. As new research and experience broaden our understanding, changes in research methods, professional practices, or medical treatment may become necessary.

Practitioners and researchers must always rely on their own experience and knowledge in evaluating and using any information, methods, compounds, or experiments described herein. In using such information or methods they should be mindful of their own safety and the safety of others, including parties for whom they have a professional responsibility.

To the fullest extent of the law, neither the Publisher nor the authors, contributors, or editors, assume any liability for any injury and/or damage to persons or property as a matter of products liability, negligence or otherwise, or from any use or operation of any methods, products, instructions, or ideas contained in the material herein.

ISBN: 978-0-12-818667-1
ISSN: 0076-6879

For information on all Academic Press publications
visit our website at https://www.elsevier.com/books-and-journals

Working together to grow libraries in developing countries

www.elsevier.com • www.bookaid.org

Publisher: Fiona Pattison
Acquisition Editor: Ashlie Jackman
Editorial Project Manager: Shellie Bryant
Production Project Manager: Divya Krishna Kumar
Cover Designer: Alan Studholme

Typeset by SPi Global, India

Contents

Contributors

Zachary Adams
Department of Microbiology and Cell Science, Institute of Food and Agricultural Sciences, University of Florida, Gainesville, FL, United States

Matthew G. Baile
Weill Institute for Cell and Molecular Biology, Department of Molecular Biology and Genetics, Cornell University, Ithaca, NY, United States

Gili Ben-Nissan
Department of Biomolecular Sciences, Weizmann Institute of Science, Rehovot, Israel

Aaron Ciechanover
Technion Integrated Cancer Center (TICC), The Rappaport Faculty of Medicine and Research Institute, Technion-Israel Institute of Technology, Haifa, Israel; Department of Biomedical Sciences, Protein Metabolism Medical Research Center, College of Medicine, Seoul National University, Seoul, South Korea

Victoria Cohen-Kaplan
Technion Integrated Cancer Center (TICC), The Rappaport Faculty of Medicine and Research Institute, Technion-Israel Institute of Technology, Haifa, Israel

Sascha H. Duttke
Department of Medicine, University of California San Diego, La Jolla, CA, United States

Scott D. Emr
Weill Institute for Cell and Molecular Biology, Department of Molecular Biology and Genetics, Cornell University, Ithaca, NY, United States

Valentina Fajner
IFOM, Fondazione Istituto FIRC di Oncologia Molecolare, Milan, Italy

Judith Frydman
Department of Biology; Department of Genetics, Stanford University, Stanford, CA, United States

Xian Fu
Department of Microbiology and Cell Science, Institute of Food and Agricultural Sciences, University of Florida, Gainesville, FL, United States

Néstor García-Rodríguez
Centro Andaluz de Biología Molecular y Medicina Regenerativa (CABIMER), Universidad de Sevilla-CSIC-Universidad Pablo de Olavide; Departamento de Genética, Universidad de Sevilla, Sevilla, Spain

Evan L. Guiney
Weill Institute for Cell and Molecular Biology, Department of Molecular Biology and Genetics, Cornell University, Ithaca, NY, United States

Randolph Y. Hampton
Division of Biological Sciences, The Section of Cell and Developmental Biology, University of California San Diego, La Jolla, CA, United States

Jason Hon
Department of Systems Biology, Harvard Medical School, Boston, MA, United States

Thorsten Hoppe
Institute for Genetics and CECAD Research Center, University of Cologne, Cologne, Germany

Ron R. Kopito
Department of Biology, Stanford University, Stanford, CA, United States

Yong Tae Kwon
Department of Biomedical Sciences, Protein Metabolism Medical Research Center, College of Medicine, Seoul National University, Seoul, South Korea

Byung-Hoon Lee
Department of New Biology, DGIST, Daegu, Republic of Korea

Dara E. Leto
Department of Biology, Stanford University, Stanford, CA, United States

Ido Livneh
Technion Integrated Cancer Center (TICC), The Rappaport Faculty of Medicine and Research Institute, Technion-Israel Institute of Technology, Haifa, Israel

Ying Lu
Department of Systems Biology, Harvard Medical School, Boston, MA, United States

Bayan Mashahreh
Department of Biological Chemistry, The Institute of Life Sciences, The Hebrew University of Jerusalem, Jerusalem, Israel

Elena Maspero
IFOM, Fondazione Istituto FIRC di Oncologia Molecolare, Milan, Italy

Julie Maupin-Furlow
Department of Microbiology and Cell Science, Institute of Food and Agricultural Sciences; University of Florida, Genetics Institute, Gainesville, FL, United States

Srinivasan Muniyappan
Department of New Biology, DGIST, Daegu, Republic of Korea

Sonya Neal
Division of Biological Sciences, The Section of Cell and Developmental Biology, University of California San Diego, La Jolla, CA, United States

Simona Polo
IFOM, Fondazione Istituto FIRC di Oncologia Molecolare; Dipartimento di oncologia ed emato-oncologia, Universita' degli Studi di Milano, Milan, Italy

Tommer Ravid
Department of Biological Chemistry, The Institute of Life Sciences, The Hebrew University of Jerusalem, Jerusalem, Israel

Randi G. Reed
Department of Biomedical Sciences, Florida State University College of Medicine, Tallahassee, FL, United States

Yuval Reiss
Department of Biological Chemistry, The Institute of Life Sciences, The Hebrew University of Jerusalem, Jerusalem, Israel

Rahul S. Samant
Department of Biology, Stanford University, Stanford, CA, United States

Richa Sardana
Weill Institute for Cell and Molecular Biology, Department of Molecular Biology and Genetics, Cornell University, Ithaca, NY, United States

Christian Schlieker
Department of Molecular Biophysics and Biochemistry, Yale University; Department of Cell Biology, Yale School of Medicine, New Haven, CT, United States

Michal Sharon
Department of Biomolecular Sciences, Weizmann Institute of Science, Rehovot, Israel

Alexander Springhorn
Institute for Genetics and CECAD Research Center, University of Cologne, Cologne, Germany

Mark Tarnavsky
Department of Biomolecular Sciences, Weizmann Institute of Science, Rehovot, Israel

Robert J. Tomko Jr.
Department of Biomedical Sciences, Florida State University College of Medicine, Tallahassee, FL, United States

Pei-Ling Tsai
Department of Molecular Biophysics and Biochemistry, Yale University, New Haven, CT, United States

Helle D. Ulrich
Institute of Molecular Biology (IMB), Mainz, Germany

Shay Vimer
Department of Biomolecular Sciences, Weizmann Institute of Science, Rehovot, Israel

Janine Weber
IFOM, Fondazione Istituto FIRC di Oncologia Molecolare, Milan, Italy

Reuven Wiener
Department of Biochemistry and Molecular Biology, Institute for Medical Research Israel-Canada (IMRIC), Hebrew University School of Medicine, Jerusalem, Israel

Chenguang Zhao
Department of Molecular Biophysics and Biochemistry, Yale University, New Haven, CT, United States

Lu Zhu
Weill Institute for Cell and Molecular Biology, Department of Molecular Biology and Genetics, Cornell University, Ithaca, NY, United States

Preface

The ubiquitin-dependent degradation of intracellular proteins is central to a remarkable assortment of physiological pathways, ranging from cell-cycle control and transcriptional regulation to DNA repair and signal transduction (Ciechanover, 2005; Varshavsky, 2017). Although well appreciated now, such intracellular proteolysis was originally regarded as a relatively uncommon feature of cellular regulation (Hochstrasser, 1996). However, it was also recognized early on that most such degradation requires energy despite the fact that proteolysis is an exergonic process (Simpson, 1953). This energy requirement suggested that the elimination of proteins from cells includes mechanistic processes that go beyond simple peptide-bond hydrolysis. For the ubiquitin-proteasome system, this is now well understood: ATP is expended in both the selective tagging of proteins with ubiquitin and the unfolding and translocation of the tagged proteins within the 26S proteasome. The proteasome is a complex molecular machine, and underlying the aforementioned energy expenditure is the execution of a specific cycle of conformational changes coupled to the movement of the substrate through narrow internal channels to the proteolytic core (de la Pena, Goodall, Gates, Lander, & Martin, 2018; Eisele et al., 2018).

The ability of the lysosome, a membrane-bounded organelle filled with hydrolytic enzymes, to degrade cellular proteins was recognized long ago (de Duve, 2005). The lysosome (called the vacuole in yeast) was thought to be fairly nonspecific in choosing its substrates. In a process known as macroautophagy, double-membrane structures envelope random segments of cytoplasm under starvation conditions, and the resulting autophagosomes fuse with the lysosome to disgorge and degrade their contents. This picture of indiscriminate destruction has turned out to be an oversimplification—macroautophagy is now known also to be able to target specific proteins, and specific organelles, to the lysosome.

Intriguingly, both ubiquitin and ubiquitin-like protein (Ubl) ligation have been found to be important for macroautophagy (Gomez-Diaz & Ikeda, 2018; Ohsumi, 2001). Two Ubls that are very distantly related to ubiquitin, but still display the classic β-grasp fold, play key roles. One (ATG12) is ligated to another autophagy factor, resulting in a protein complex that then promotes the attachment of the second Ubl, ATG8, to phosphatidylethanolamine; the membrane-remodeling activity of ATG8 helps drive autophagosome formation (Nakatogawa, Suzuki, Kamada, & Ohsumi, 2009). This mechanism is fundamental to all macroautophagy. On the other hand, selective autophagic elimination of particular cellular proteins or organelles often depends on their prior conjugation to ubiquitin. Adaptor proteins with binding sites for both ATG8 and ubiquitin target the ubiquitinated substrates to the developing autophagosome (Grumati & Dikic, 2018).

Other routes of substrate trafficking to the lysosome have also been found to depend on ubiquitin. Downregulation of cell-surface proteins usually involves endocytosis of the proteins from the plasma membrane; maturation of the endosomes by repeated rounds of invagination and vesiculation of the limiting membrane; and, finally, trafficking of the resulting multivesicular body (MVB) to the lysosome (MacGurn, Hsu, & Emr, 2012). Ubiquitination of target proteins at the cell surface signals the initiation of endocytosis, and their persistent ubiquitination—as well as that of membrane proteins in many other intracellular organelles—results in their being routed to the lysosome, where they are degraded. The primary exception to this mode of degrading ubiquitinated membrane proteins occurs at the endoplasmic reticulum (ER), where most ER-associated protein degradation occurs through extraction of the targeted ubiquitinated protein into the cytoplasm and degradation by the proteasome (Hampton & Sommer, 2012; Mehrtash & Hochstrasser, 2018).

In this volume, a wide range of in vitro and in vivo methodologies is described for the study of intracellular protein degradation in its many guises. As in the previous volume, the description of protocols in each chapter is preceded by a concise summary of the state of understanding in the field. An important aspect of intracellular proteolysis, which is mostly ubiquitin dependent, is its role in protein quality control. Methods for studying such processes for proteins residing in the cytoplasm, ER, and nuclear envelope are detailed. The proteasome is a large protein complex that serves as a platform for multiple reactions. In a series of chapters, protocols are presented for assaying these different features of the proteasome and how they are integrated into the proteolytic mechanism. Finally, techniques are described for the study of lysosomal/vacuolar protein degradation, initiated either by macroautophagy or ubiquitin-dependent membrane trafficking from the plasma membrane or internal membranes.

Altogether, it is hoped that researchers of all stripes will be able to dip into this two-volume collection for experimental insights and detailed technical guidance for their own studies of the ubiquitin system. No doubt, many more new and exciting discoveries will be realized in these endeavors.

Mark Hochstrasser, Editor
Department of Molecular Biophysics & Biochemistry,
Yale University, New Haven, CT, United States

References

Ciechanover, A. (2005). Intracellular protein degradation: From a vague idea, through the lysosome and the ubiquitin-proteasome system, and onto human diseases and drug targeting (Nobel lecture). *Angewandte Chemie (International Ed. in English)*, *44*, 5944–5967.

de Duve, C. (2005). The lysosome turns fifty. *Nature Cell Biology*, 7, 847–849.

de la Pena, A. H., Goodall, E. A., Gates, S. N., Lander, G. C., & Martin, A. (2018). Substrate-engaged 26S proteasome structures reveal mechanisms for ATP-hydrolysis-driven translocation. *Science*, *362*.

Eisele, M. R., Reed, R. G., Rudack, T., Schweitzer, A., Beck, F., Nagy, I., et al. (2018). Expanded coverage of the 26S proteasome conformational landscape reveals mechanisms of peptidase gating. *Cell Reports*, *24*(1301–1315). e5.

Gomez-Diaz, C., & Ikeda, F. (2018). Roles of ubiquitin in autophagy and cell death. *Seminars in Cell & Developmental Biology*. https://doi.org/10.1016/j.semcdb.2018.09.004 (still only available as epub).

Grumati, P., & Dikic, I. (2018). Ubiquitin signaling and autophagy. *The Journal of Biological Chemistry*, *293*, 5404–5413.

Hampton, R. Y., & Sommer, T. (2012). Finding the will and the way of ERAD substrate retrotranslocation. *Current Opinion in Cell Biology*, *24*, 460–466.

Hochstrasser, M. (1996). Ubiquitin-dependent protein degradation. *Annual Review of Genetics*, *30*, 405–439.

MacGurn, J. A., Hsu, P. C., & Emr, S. D. (2012). Ubiquitin and membrane protein turnover: From cradle to grave. *Annual Review of Biochemistry*, *81*, 231–259.

Mehrtash, A. B., & Hochstrasser, M. (2018). Ubiquitin-dependent protein degradation at the endoplasmic reticulum and nuclear envelope. *Seminars in Cell & Developmental Biology*. https://doi.org/10.1016/j.semcdb.2018.09.013 (still only available as epub).

Nakatogawa, H., Suzuki, K., Kamada, Y., & Ohsumi, Y. (2009). Dynamics and diversity in autophagy mechanisms: Lessons from yeast. *Nature Reviews Molecular Cell Biology*, *10*, 458–467.

Ohsumi, Y. (2001). Molecular dissection of autophagy: Two ubiquitin-like systems. *Nature Reviews Molecular Cell Biology*, *2*, 211–216.

Simpson, M. V. (1953). The release of labeled amino acids from the proteins of rat liver slices. *The Journal of Biological Chemistry*, *201*, 143–154.

Varshavsky, A. (2017). The ubiquitin system, autophagy, and regulated protein degradation. *Annual Review of Biochemistry*, *86*, 123–128.

CHAPTER ONE

Assays for protein retrotranslocation in ERAD

Sonya Neal[a,*,†], Sascha H. Duttke[b], Randolph Y. Hampton[a]
[a]Division of Biological Sciences, The Section of Cell and Developmental Biology, University of California San Diego, La Jolla, CA, United States
[b]Department of Medicine, University of California San Diego, La Jolla, CA, United States
*Corresponding author: e-mail address: seneal@ucsd.edu

Contents

Abstract

Elimination of misfolded proteins by endoplasmic reticulum (ER)-associated protein degradation (ERAD) ensures that proteins proceeding through the secretory pathway are correctly folded and processed, which is critical to minimize ER stress. All ERAD pathways include a protein translocation process termed retrotranslocation, in which ubiquitinated misfolded substrates are extracted from the ER and degraded by the cytosolic 26S proteasome. Despite being integral to ERAD, the retrotranslocation process has been largely obscure. Recently, an explosion of discoveries has provided key mechanistic insights into this novel route of protein transport. These advances were facilitated by the development of in vitro and in vivo assays that utilize components from the yeast

† Lead contact.

Methods in Enzymology, Volume 619
ISSN 0076-6879
https://doi.org/10.1016/bs.mie.2019.01.002
© 2019 Elsevier Inc.
All rights reserved.

Saccharomyces cerevisiae. The assays permit detailed study of the distinct steps in ERAD-linked retrotranslocation, including ubiquitination of selected ERAD substrates, substrate removal from the ER, maintenance of cytosolic substrate solubility in the cytosol, and substrate degradation. Here we provide detailed protocols for these assays that pertain to work on retrotranslocation of integral membrane proteins (ERAD-M substrates), with the expectation that these approaches can be adapted for many related biochemical processes.

1. Introduction

Endoplasmic reticulum-associated protein degradation (ERAD) is a highly conserved quality control pathway that eliminates damaged or unassembled ER proteins (Foresti, Ruggiano, Hannibal-Bach, Ejsing, & Carvalho, 2013; Hampton & Garza, 2009). Defects in ERAD result in a toxic buildup of misfolded proteins and is associated with many pressing human maladies such as aging, neurodegenerative diseases, retinal degeneration, and diabetes (Chiti & Dobson, 2017; Hara et al., 2014; Zhao & Ackerman, 2006). ERAD occurs through the ubiquitin-proteasome pathway. ERAD substrates are poly-ubiquitinated, which marks them for degradation by the 26S proteasome. This ubiquitination is mediated by the sequential action of three enzymes. An E1 ubiquitin-activating enzyme uses ATP to create a high-energy thioester between ubiquitin and the E1 and then transfers the ubiquitin to the active-site cysteine of an E2 ubiquitin-conjugating (UBC) enzyme. Ubiquitin is then transferred from the E2 to the substrate or the growing ubiquitin chain by the action of an E3 ubiquitin ligase, resulting in a substrate-attached ubiquitin chain that is recognized and degraded by the proteasome (Bays, Wilhovsky, Goradia, Hodgkiss-Harlow, & Hampton, 2001; Hiller, Finger, Schweiger, & Wolf, 1996; Jarosch et al., 2002). Substrate specificity for a given pathway is largely determined by the E3 ligase.

In *Saccharomyces cerevisiae*, ERAD is mediated by the HRD (HMG-CoA Reductase Degradation) and DOA (Degradation of alpha2) pathways, both of which are conserved in virtually all eukaryotes (Carvalho, Goder, & Rapoport, 2006; Chen et al., 2006; Foresti et al., 2013; Hampton & Garza, 2009). In the HRD pathway, the Hrd1 E3 ligase targets a variety of substrates, including misfolded substrates with lesions within the membrane bilayer (ERAD-M), substrates with luminal lesions (ERAD-L), and some normal proteins such as the HMGR sterol synthetic enzyme

(Bordallo, Plemper, Finger, & Wolf, 1998; Hampton, Gardner, & Rine, 1996). In the DOA pathway, the E3 ubiquitin ligase Doa10 ubiquitinates misfolded soluble and membrane proteins with lesions in their cytosolic domains (ERAD-C) but is also capable of targeting certain ERAD-M substrates (Wangeline, Vashistha, & Hampton, 2017).

Spatial separation of the various ERAD processes presents a major challenge for cells. In all examples of ERAD, there is a requirement to move substrates from their starting point within the ER lumen or ER membrane to the cytosol for proteasome-mediated destruction. This transport component of ERAD, broadly referred to as dislocation or retrotranslocation, has been known to occur since the earliest studies of ERAD when the cytosolic E2 ligase Ubc7 was identified to mediate the ubiquitin-dependent degradation of the luminal ERAD substrate CPY* (Hampton & Sommer, 2012; Hiller et al., 1996). Furthermore, many studies concluded that the Cdc48 ATPase (p97 in mammals) is required to drive the retrotranslocation process (Bays et al., 2001; Jarosch et al., 2002; Ye, Meyer, & Rapoport, 2003). The general requirement for retrotranslocation in all ERAD pathways is gradually revealing the mechanistic aspects of retrotranslocation, including the components required, the energetics, and the mechanism of transfer through the ER membrane. However, many open questions remain; to address these issues, we had previously developed an in vitro assay using the eight-transmembrane-spanning ER-resident substrate Hmg2, an isozyme of HMG-CoA reductase fused to GFP, which undergoes regulated ERAD mediated by the E3 ligase Hrd1 (Fig. 1A) (Garza, Sato, & Hampton, 2009; Garza, Tran, & Hampton, 2009; Hampton, 2005).

This assay has allowed several insights into ERAD-M retrotranslocation. First, it was demonstrated that the entire integral membrane Hmg2 protein is removed from the ER membrane and remains intact and soluble in the

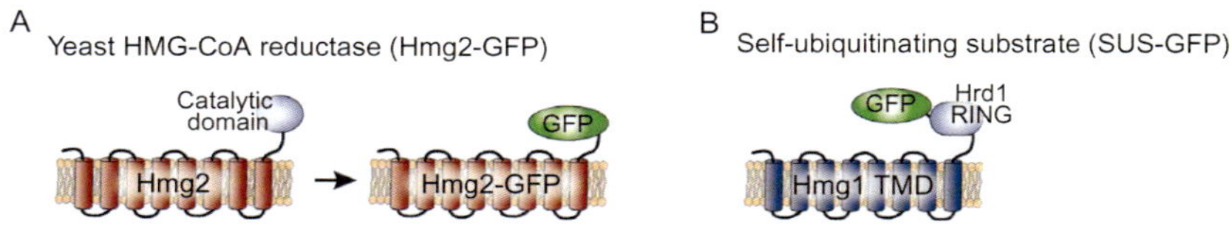

Fig. 1 Fusion substrates used in the in vivo and in vitro retrotranslocation assays. (A) Hmg2 and the fluorescent fusion reporter Hmg2-GFP. Hmg2-GFP has the authentic catalytic domain of Hmg2 replaced by the GFP reporter and is able to undergo regulated degradation. (B) Fluorescent fusion retrotranslocation reporter, SUS-GFP. The transmembrane stable isoform of HMGR, Hmg1, is fused to the Hrd1 cytosolic RING domain along with the GFP at the C-terminus.

cytosol if proteasomal degradation is inhibited. Second, ATP hydrolysis by Cdc48 was found to be required for extraction of the tested integral membrane substrates. Third, the Hrd1 ligase was rate-limiting for retrotranslocation of these substrates (Garza, Sato, & Hampton, 2009; Garza, Tran, & Hampton, 2009).

It is thought that an aqueous protein channel allows the movement of substrates across the ER membrane. Channel candidates include the Derlins (Lilley & Ploegh, 2004; Sun et al., 2006), the Sec61 anterograde channel (Plemper, Böhmler, Bordallo, Sommer, & Wolf, 1997), or the multi-spanning domains of the ER ligases themselves (Baldridge & Rapoport, 2016; Plemper et al., 1997; Ravid, Kreft, & Hochstrasser, 2006). Recently, a strong case has been made for Hrd1 itself serving as an ERAD-L channel by the Rapoport group using in vitro reconstitution studies employing purified components from *S. cerevisiae* (Baldridge & Rapoport, 2016; Stein, Ruggiano, Carvalho, & Rapoport, 2014). This was further supported by a Cryo-EM structure of Hrd1, which revealed that the multispanning E3 ligase includes a channel that at least partially traverses the membrane (Schoebel et al., 2017). Finally, in vivo studies of Hrd1 led to a model in which autoubiquitination of Hrd1 opens the channel to catalyze retrotranslocation (Baldridge & Rapoport, 2016; Stein et al., 2014).

Although the studies above implicate Hrd1 as a channel for ERAD-L retrotranslocation, this factor did not appear to be involved in the retrotranslocation of multispanning integral membrane substrates of the ERAD-M class. The Hampton group used the in vitro retrotranslocation assay to examine the role of Hrd1 as a retrotranslocon for ERAD-M degradation (Garza, Sato, & Hampton, 2009). However, simply removing the ligase to test its role in retrotranslocation was not possible since Hrd1-dependent substrate ubiquitination is also a prerequisite for retrotranslocation. To uncouple E3 ligase activity from any potential channel activity, we devised a self-ubiquitinating substrate (SUS), which was composed of the multispanning membrane region from Hmg1 (the stable HMGR isozyme) attached to the catalytic domain of Hrd1 (Fig. 1B). Using the Hrd1 catalytic domain, SUS catalyzes its own ubiquitination and degradation, with the appropriate dependencies including the attached Hrd1 E3 RING domain, the Ubc7 E2 ligase, and Cdc48 (Garza, Sato, & Hampton, 2009). These studies showed that the integral membrane SUS protein does not require Hrd1 (or Doa10) for its degradation in vivo, despite the fact that it undergoes full retrotranslocation in the in vitro assay in a manner identical to bone fide ERAD-M substrates such as Hmg2. The behavior of the SUS substrate left open the question of how exactly it undergoes degradation and what the mechanism of retrotranslocation might be.

To complement the in vitro assay, we recently developed a physiologically relevant in vivo retrotranslocation assay (Neal, Mak, Bennett, & Hampton, 2017), inspired by earlier work from the Sommer group assaying in vivo ERAD-L retrotranslocation (Mehnert, Sommer, & Jarosch, 2013). The in vivo assay recapitulates two important features of ERAD that, for unknown reasons, were absent in the in vitro assay. First, Hmg2 examined in the in vivo assay undergoes normal, regulated degradation in response to the sterol pathway intermediate geranylgeranyl pyrophosphate (GGPP). Second, the effect of removing the proteasome adaptors Dsk2 and Rad23 in the in vivo assay placed them in the expected position along the HRD pathway: after retrotranslocation and before proteasomal degradation. The ease of the in vivo assay allowed us to quickly score candidate mutants—obtained from a yeast genetic screen—for retrotranslocation function. Use of this assay confirmed that SUS-GFP indeed undergoes retrotranslocation that is independent of Hrd1, leading to the genetic analysis that revealed the Dfm1 retrotranslocation factor for ERAD-M substrates (Neal et al., 2018). The assay is described in detail below.

Due to the unexpected solubility of multispanning membrane ERAD substrates following retrotranslocation, the factors or molecules facilitating the solubility of these hydrophobic proteins in the cytosol are of great interest. A strong suit of the in vivo assay is the ability to isolate the retrotranslocated intermediate from intact cells and to perform biophysical and proteomic analyses on retrotranslocated integral membrane substrates such as Hmg2. Using this assay has led to two important findings: (1) Retrotranslocated Hmg2 is not associated with lipids but is instead associated with proteins and (2) proteomic analysis of retrotranslocated Hmg2 identified Cdc48 as the major "retrochaperone" responsible for solubilizing full-length intact Hmg2 in the cytosol (Neal et al., 2017). Cdc48 was shown to be stochiometrically bound to retrotranslocated, ubiquitinated Hmg2 and is essential for preserving the solubility of this client protein on its way to the proteasome. Protocols for the in vivo assay that led to these discoveries are described in Section 3.

2. An in vitro assay for retrotranslocation of Hmg2-GFP

This section lists the yeast strains required to perform this assay, the detailed descriptions of the methods to monitor ubiquitination of Hmg2-GFP, and in vitro assay used to directly observe Hmg2-GFP retrotranslocation.

In this assay, Hrd1 is expressed from the strong constitutive *TDH3* promoter at levels sufficient to promote ERAD in the absence of other associated ERAD factors, including Hrd3, Usa1, and Yos9 (Carroll & Hampton, 2009; Carvalho et al., 2006; Mehnert et al., 2015; Vashistha, Neal, Singh, Carroll, & Hampton, 2016). (Notably, Hrd1 at native levels does not achieve optimal ubiquitination and degradation of Hmg2-GFP in the in vitro assay.) Accordingly, this approach defines the minimal components sufficient for successful retro-translocation, providing the avenue for complete reconstitution. Hrd1 over-expressed at this level causes sufficient ubiquitination of Hmg2-GFP for the recovery and direct detection of the retrotranslocated substrate (see below). However, ectopic expression of Hrd1 in vivo alters Hmg2 ERAD, abrogating its physiological regulation by GGPP and changing the dependence of the Rad23/Dsk2 adaptors. Nevertheless, it has been very useful for a variety of studies (Bazirgan & Hampton, 2008; Garza, Sato, & Hampton, 2009; Sato, Schulz, Do, & Hampton, 2009).

2.1 Materials

2.1.1 Strains

All yeast strains are derived from the S288C background as described previously (Hampton & Rine, 1994). The initial genotype of the yeast strain used to create the strains used as cytosol and microsome donors is the following:

Matα ura3-52 ade2-101 met2 lys2-801 his3Δ200 trp1::hisG leu2Δ pep4Δ::HIS3 hmg2Δ::1myc-HMG2 hrd1Δ::kanMX ubc7Δ::LEU2 (Strain RHY4288).

The specific genetic insertions added to RHY4288 to create the respective microsome and donor strains are shown below. Mutations in other genes can then be introduced into one or the other strain.

1. Microsome donor strain $(RHY2923)$: $ura3-52::URA3::TDH3_{PROM}-HMG2-GFP$ $trp1::hisG::TRP1::TDH3_{PROM}-HRD1-3HA$
2. Cytosol donor strain $(RHY4295)$: $trp1::hisG::TRP1::TDH3_{PROM}-UBC7-2HA$

2.1.2 Microsome preparation

Microsome preparation from *S. cerevisiae* has been well documented (Garza, Sato, & Hampton, 2009; Neal et al., 2018, 2017) and is described in detail below with additional steps to ensure optimal yields. In most cases, microsome donor strains are grown at 30°C with aeration and vigorous shaking. With temperature-sensitive mutants, cells are grown at permissive temperature at 25°C and then shifted to the restrictive temperature at 37°C. To recapitulate the temperature-sensitive phenotype in the in vitro assay, the duration of growth at the restrictive temperature must be determined empirically.

1. The microsome donor strain is grown at the desired temperature in minimal medium (Garza, Sato, & Hampton, 2009) at an initial optical density at 600 nm (OD_{600}) of 0.1 OD until the culture reaches mid-log phase (OD_{600} of 0.3–0.5). 200 mL of yeast are typically grown for microsome preparation. Note: We have observed that when microsome strains are grown in YPD, this condition induces loss of the overexpressed *HRD1* gene. We do not observe this when strains are grown in minimal media.
2. 30 OD units of cells are collected and prepared for bead lysis in a 50 mL Falcon tube. Cell pellets are resuspended in 600 μL of membrane fractionation (MF) buffer (20 m*M* Tris–HCl pH 7.5, 100 m*M* NaCl, 300 m*M* sorbitol, with the following protease inhibitors (PIs): 1 m*M* phenylmethylsulfonyl fluoride, 260 μ*M* 4-(2-aminoethyl) benzenesulfonyl fluoride hydrochloride, 100 μ*M* leupeptin hemisulfate, 76 μ*M* pepstatin A, 5 m*M* aminocaproic acid, 5 m*M* benzamidine, and 142 μ*M* *N*-tosyl-L-phenylalanine chloromethyl ketone). Note: phenylmethylsulphonyl fluoride is prepared fresh for each experiment.
3. Beads are added until they reach the meniscus of the cell suspension, and cells are lysed at 4°C using hand vortexing at top speed for 6 × 1-min intervals with 1-min intervals on ice between each vortexing. Lysis efficiency can be evaluated under the microscope using 20 × magnification. Lysed cells are clearly distinguishable by their fragmented shape; for optimal yield it is critical to achieve ~80%–90% lysis efficiency. If below this range, continue to vortex for up to three more 1-min cycles at 4°C. We use either 0.5 mm glass-based or silica-based beads for lysis (Biospec Products), but the lysis efficiency with silica-based beads appears to be somewhat higher. It is critical that each of the following steps is performed at 4°C.
4. The crude lysate is collected with a 1 mL pipette and transferred to a 1.5 mL Eppendorf tube. 200 μL of cold MF buffer are used to rinse the beads, and this rinse fraction is added to the original lysate. The extract is centrifuged for 5 s in a microcentrifuge (16,000 × *g*), and the resulting supernatant is transferred to a fresh 2 mL Eppendorf tube.

Centrifugation and transfer to fresh tubes is repeated until no pelleted cellular debris can be seen. This is typically repeated approximately six times.

5. Next, the microsomes are pelleted in a microcentrifuge at 20,000 × *g* for 30 min. The pellets are resuspended in B88 buffer (20 m*M* Hepes pH 6.8, 250 m*M* sorbitol, 150 m*M* KOAc, 5 m*M* MgOAc, freshly made 1 m*M* DTT, and the PIs listed above), to a final concentration of 0.3 OD equivalent units/μL of B88.
6. It is important to immediately use the microsome preparation for in vitro ubiquitination and retrotranslocation assays. In our experience, using fresh microsome preparations works best for in vitro ERAD assays; freezing these preparations does not reliably preserve function, although this avenue has not been intensely pursued.

2.1.3 Cytosol preparation

Cytosol is prepared from strains overexpressing Ubc7 in a *hrd1*Δ *ubc7*Δ double null strain in a manner similar to that described by Spang and Schekman (1998), using rapid freezing of cells with liquid nitrogen followed by mortar and pestle disruption of the frozen cells. Control cytosol is prepared in parallel from an otherwise identical *ubc7*Δ strain (RHY4288). Smaller scale cytosol preparation can be performed with similar activity as described previously (Heck, Cheung, & Hampton, 2010). Again, for temperature-sensitive strains, duration for growth at non-permissive temperature must be validated empirically.

1. Cytosol donor strains are grown at the desired temperature in rich medium to mid-log phase (OD_{600} = ~1) at 30°C. 500 mL of yeast are typically grown for cytosol preparation.
2. 500 OD equivalents of cells are pelleted for 10 min at room temperature at 5000 × *g*. The cells are rinsed once with water, once with B88 buffer, and resuspended in 500 μL of B88 buffer (1 OD/μL B88 buffer) with the same amount of PIs as mentioned for microsomal preparation.
3. For cell lysis, the resuspended cells are transferred to a chilled mortar containing liquid nitrogen. The frozen cells are grounded with a pestle until a fine powder is achieved; this is usually achieved with approximately 15 strokes. During pestle treatment, it can be useful to keep a reservoir of liquid nitrogen to maintain freezing of the pellet during grinding. The frozen powder is transferred to a chilled 50 mL Falcon tube and thawed on ice. It is critical that the each of the following steps is performed at 4°C. Note: the frozen powder may be stored at −80°C for later use.

4. ATP (Sigma) is added to the thawed cytosol from a 500 m*M* stock solution to make a final added concentration of 1 m*M*. Note: For AMP-PNP (Sigma) experiments, which test the role of ATP hydrolysis in ERAD, ATP is not added. ATP stock is typically made with deionized water, adjusted to pH 7.5, aliquoted and frozen at −80°C. An aliquot of ATP is thawed at the start of each in vitro ubiquitination experiment.
5. Once thawed, the crude cytosol lysate is centrifuged for 5 min with a microcentrifuge (3000 × *g*) to remove debris, and then the supernatant is transferred to a new tube and centrifuged for 15 min with a microcentrifuge (20,000 × *g*).
6. The resulting supernatant is then ultracentrifuged at 100,000 × *g* for 1 h using a Beckman TL-1000 tabletop ultracentrifuge along with a TLA 100.4 rotor. After ultracentrifugation, the supernatant is removed carefully with a precision loading tip. The resulting cytosol preparation is measured for protein concentration with a NanoDrop (ThermoFisher) spectrophotometer [Absorbance at 280 nm]. Cytosol protein concentrations are adjusted with B88 to 20–25 mg/mL for ubiquitination and retrotranslocation assays, aliquoted and stored at −80°C. This concentration has worked best in the ubiquitination and retrotranslocation assays in the past as long as samples are not refrozen and thawed again.

2.2 In vitro ubiquitination of Hmg2-GFP

1. Prior to starting the in vitro ubiquitination assay, MG132 (benzyloxycarbonyl-Leu-Leu-aldehyde, Sigma) is added to a final concentration of 300 μ*M* to both microsome and cytosol preparations. Note: We found that the intensity of the ubiquitin signal for a membrane substrate is greater when proteasome inhibitor MG132 is added, suggesting that the 20S core proteasome activity causes a reduction in ubiquitination levels, either due to in vitro degradation of a fraction of the substrate by the proteasome or to increased activity of proteasome-associated ubiquitin-specific proteases that are sensitive to 20S proteasome inhibition (Garza, Sato, & Hampton, 2009; Neal et al., 2017). The schematic for the in vitro ubiquitination assay is depicted in Fig. 2 and described in detail below.
2. A typical in vitro ubiquitination reaction consists of a mixture of 10 μL microsomes and 12 μL cytosol.
3. To initiate the reaction, 1.3 μL ATP (500 m*M* stock) is added to each mixture to a final concentration of 30 m*M* and the reactions are

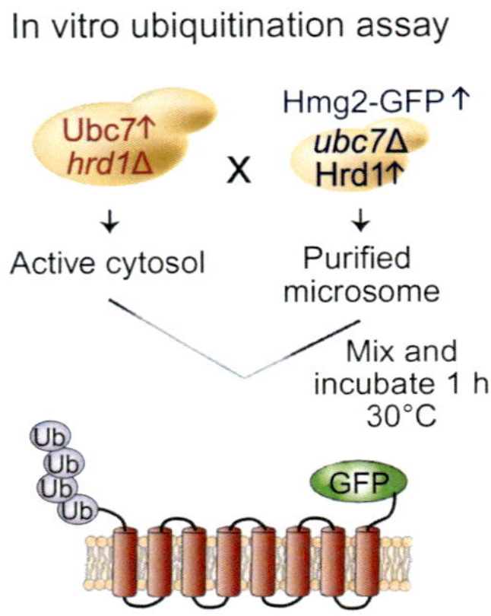

Fig. 2 Schematic of in vitro ubiquitination of Hmg2-GFP. Microsomes were prepared from strains expressing *TDH3*$_{prom}$-Hmg2-GFP with *TDH3*$_{prom}$-Hrd1 and without *ubc7*Δ (RHY2923). Cytosol was prepared from strains expressing *TDH3*$_{prom}$-Ubc7-GFP (RHY4295) or devoid of Ubc7 (*ubc7*Δ) (RHY4288) as a control for no ubiquitination of substrate. To initiate ubiquitination of Hmg2-GFP, microsomes and cytosol preparations were mixed and incubated for 1 h at 30°C as indicated.

incubated for 1 h at 30°C. From our experience, 30 m*M* ATP and a reaction time of 1 h give an optimal signal for Hmg2-GFP ubiquitination. For other substrates of interest, the optimal concentration of ATP and reaction time should be determined empirically. This is especially important since in a number of assays, there is an optimal ATP concentration that if exceed will begin to diminish ubiquitination signal intensity (Heck et al., 2010).

4. The reaction is stopped by addition of 200 μL of SUME buffer (1% SDS, 8 *M* urea, 10 m*M* MOPS pH 6.8, 10 m*M* EDTA) with PIs as above and 5 m*M* *N*-ethylmaleimide (NEM, Sigma) to terminate the reactions and solubilize all proteins. Note: Ubiquitination is reversible and this modification can therefore easily be eliminated by deubiquitinases (DUBs). For this reason, it is essential to include DUB inhibitors such as NEM in the buffers used during the long incubation times used for immunoprecipitation in order to preserve the state of substrate ubiquitination (most DUBs are cysteine proteases that are inhibited by NEM).
5. Ubiquitination of Hmg2-GFP is measured by immunoprecipitation (IP) of Hmg2-GFP, resolution of the IP by SDS-PAGE, and immunoblotting for Hmg2-GFP and ubiquitin. For IP, 600 μL IP buffer (IPB: 15 m*M* Na_2HPO_4, 150 m*M* NaCl, 2% Triton X-100, 0.1% SDS,

0.5% deoxycholate (DOC), 10 mM EDTA, pH 7.5) is added to the 222 μL of the solubilized ubiquitination reaction in step 4; PIs and NEM are added as before.

6. Insoluble material is removed from the IP reaction mix by centrifugation for 5 min with a microcentrifuge (14,000 × g), and the resulting clarified supernatant is transferred to a fresh tube using a 1 mL pipette. 15–25 μL of rabbit polyclonal antiserum is added (the anti-GFP antiserum was prepared in collaboration with Scantibodies, Inc., Santee, CA). The IP incubation is carried out at 4°C for 12–16 h.
7. 100 μL of protein A-Sepharose (GE Healthcare, Piscataway, NJ) equilibrated with IP buffer is added to the IP reactions. The tubes are then incubated for 2 h at 4°C with gentle mixing. Protein A-Sepharose with bound protein is washed once with IPB and once with IPW (50 mM NaCl, 10 mM Tris, pH 7.5).
8. To prepare samples for loading, beads are aspirated to dryness and 55 μL 2 × urea sample buffer (USB: 8 M urea, 4% SDS, 10% ß-mercaptoethanol, 125 mM Tris, pH 6.8) is added. The slurry is incubated at 50°C for 10 min and centrifuged for 5 min with a microcentrifuge (14,000 × g). The eluted proteins are removed to a new tube.
9. Eluted proteins are resolved by SDS-PAGE using 8% gels and are then transferred to nitrocellulose by electroblotting at 0.15 Amps for 2.5 h (4°C).
10. For anti-ubiquitin blots, it is important to note that ubiquitin is small and difficult to denature and the ubiquitin epitopes might not be accessible to antibodies due to insufficient denaturation during SDS-PAGE or renaturation on the membrane (Emmerich & Cohen, 2015). Therefore, after transfer to nitrocellulose membranes, the signal strength of anti-ubiquitin antibodies can frequently be enhanced significantly if the membrane is subjected to a denaturing treatment prior to blocking. Accordingly, the membranes are rinsed with water, sandwiched between sheets of Whatman paper, and placed in a glass dish. Deionized water is added to the dish and the membrane is boiled in a microwave oven at 3 × 1-min intervals; periodically check in between intervals to ensure that the water has not evaporated. This brings about a remarkable increase in signal strength, presumably due to revealing of cryptic epitopes from the heat.
11. After denaturing treatment, the blocking and mouse anti-ubiquitin (Zymed Laboratories, South San Francisco, CA) incubations are

carried out with Tris-buffered saline (TBS-T: 10 m*M* Tris–HCl, pH 8.0, 140 m*M* NaCl, 0.05% Tween 20) containing 20% heat-inactivated bovine serum. First the membrane is blocked by incubation in the serum-containing buffer prior to antibody treatment for 40 min, followed by addition of anti-ubiquitin antibody (1: 4000 dilution) and overnight incubation at 4°C. On the next day, blots are washed three times for 10 min each with TBS-T. Note: From our experience, blocking and primary incubation with 20% inactivated bovine serum in TBS-T produces an enhanced signal for ubiquitin blots.

12. For anti-GFP or anti-HA blots, nitrocellulose membranes are blocked with 5% (w/v) nonfat dried milk in TBS-T and mouse anti-GFP (BD Biosciences, San Jose, CA) or anti-HA (Covance, Princeton, NJ) incubations are carried out in 2% non-fat dried milk in TBS-T. Goat anti-mouse conjugated with HRP (Jackson ImmunoResearch, West Grove, PA) is used to detect primary antibodies. Western Lightning chemiluminescence reagents (Perkin Elmer, Waltham, MA) are used for immunodetection and X-ray film is used for Western blot detection (Fig. 3). Signals on X-ray film are quantified by ImageJ (NIH). Band intensities were measured directly from films scanned in high resolution (600 dpi) in TIFF file format.

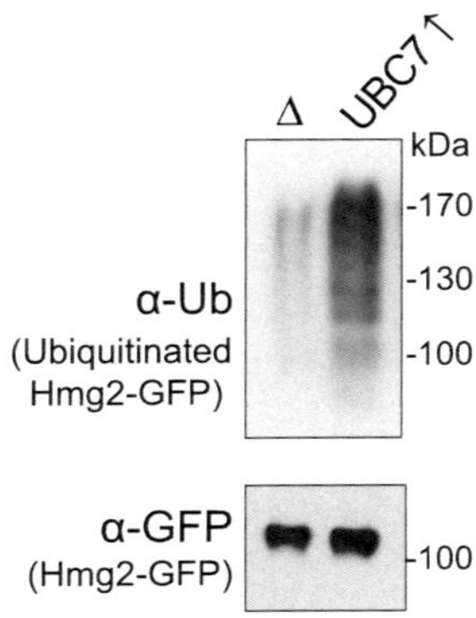

Fig. 3 To discern for ubiquitination of the substrate, Hmg2-GFP is immunoprecipitated with anti-GFP, and immunoblotted for ubiquitin with anti-Ub or unmodified Hmg2-GFP with anti-GFP. *Figure is taken from Neal, S., Mak, R., Bennett, E. J., & Hampton, R. (2017). A Cdc48 "retrochaperone" function is required for the solubility of retrotranslocated, integral membrane endoplasmic reticulum-associated degradation (ERAD-M) substrates.* The Journal of Biological Chemistry 292, *3112–3128.*

2.3 In vitro retrotranslocation assay of Hmg2-GFP

A schematic for the in vitro retrotranslocation assay is shown in Fig. 4, and the procedure is outlined below.

1. For each in vitro retrotranslocation assay, three samples are evaluated: total lysate, supernatant fraction, and pellet fraction. The ubiquitinated membrane protein that is moved from the pellet (membrane) fraction to the supernatants (cytosol) fraction is the portion that has been retrotranslocated. Typically, up to 50% of the ubiquitinated material can be found as solubilized, retrotranslocated supernatant immunoreactivity (Neal et al., 2017). The three fractions are typically derived from a vitro ubiquitination reaction that was scaled up three times. The in vitro ubiquitination reaction is run as described in the preceding section at 30°C for 1 h and terminated by the addition of 1.5 μL of 50 × (NEM) to a final concentration of 5 m*M*.
2. One reaction equivalent (typically 24 μL) is transferred to one tube designated as total (T) and another, equal reaction equivalent is transferred to a tube for centrifugation for 1 h at 25,000 × *g* at 4°C.
3. The resulting supernatant (*S*) is carefully removed, and the resulting pellet (*P*) is resuspended with MF buffer plus PIs (same volume (24 μL) as the supernatant and total fractions). Each fraction is solubilized with

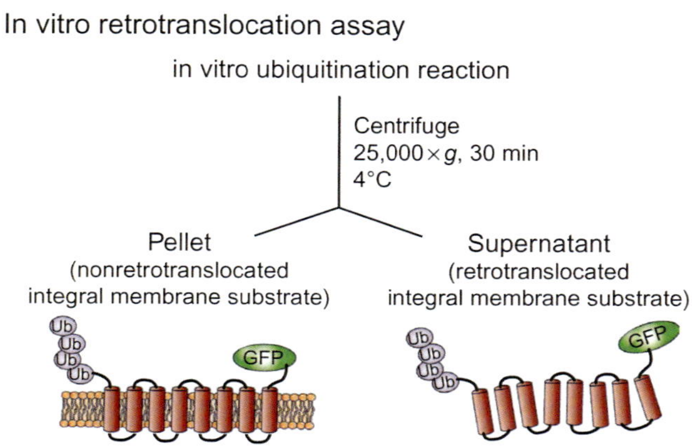

Fig. 4 Schematic of in vitro retrotranslocation of Hmg2-GFP. In vitro ubiquitination reactions were carried out as described in Fig. 2, but with 3 × volume (72 μL). Reaction mixture containing in vitro ubiquitinated Hmg2-GFP is centrifuged at 4°C at 25,000 × *g* for 30 min to discern ubiquitinated Hmg2-GFP that either has been retrotranslocated into the supernatant cytosolic fraction (S) or remained in the microsome pellet fraction (P).

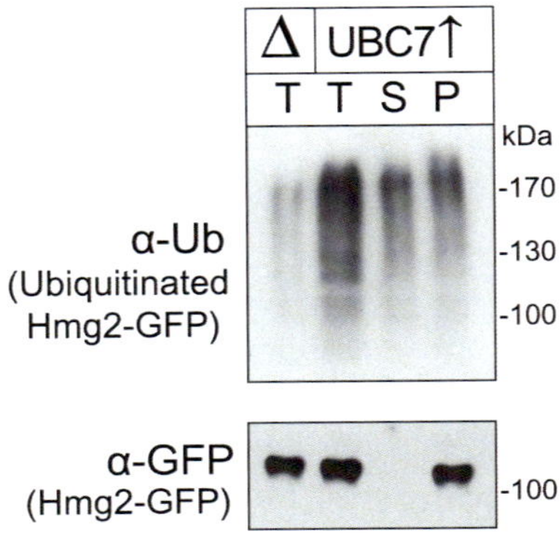

Fig. 5 To discern for nonretrotranslocated and retrotranslocated ubiquitinated Hmg2-GFP, the substrate was immunoprecipitated from a 1× (24μL) aliquot of the non-fractionated mixture (T) or from the supernatant (S) and pellet (P) of an identical volume (24μL) of the same reaction mixture that was centrifuged, as described. *Figure is taken from Neal, S., Mak, R., Bennett, E. J., & Hampton, R. (2017). A Cdc48 "retrochaperone" function is required for the solubility of retrotranslocated, integral membrane endoplasmic reticulum-associated degradation (ERAD-M) substrates.* The Journal of Biological Chemistry 292, *3112–3128.*

SUME plus PIs and immunoprecipitated and detected as described above for the in vitro ubiquitination assay. The use of identical volumes allows for direct visual comparison of the immunoblotted proteins present in each fraction (Fig. 5).

3. An in vivo assay for retrotranslocation of Hmg2-GFP

The in vitro assay is powerful and highly accessible for manipulation of reaction conditions, but the in vivo assay described here allows us to study ERAD in a more physiological context. Because the in vitro assay requires overexpressed Hrd1 to catalyze ERAD, both sterol pathway regulation and the absolute requirement for the Hrd1 partner Hrd3 observed in vivo are lost. Furthermore, the in vivo assay indicates that under normal conditions there is a detectable level of soluble retrotranslocated material en route to the proteasome. Consistent with this idea, addition of proteasome inhibitor MG132 causes a buildup of ubiquitinated Hmg2 in microsomal P and cytosol S fraction. The presence of retrotranslocated material in the absence of proteasome inhibitors supports the idea that full-length extraction and solubilization of multispanning membrane proteins is a part of normal ERAD-M, and amenable to observation with this approach. The in vivo assay also preserves sterol pathway regulation of Hmg2 stability. As shown previously, the 20-carbon sterol pathway molecule geranylgeranyl pyrophosphate (GGPP) is an endogenous regulator of Hmg2-GFP that stimulates increased

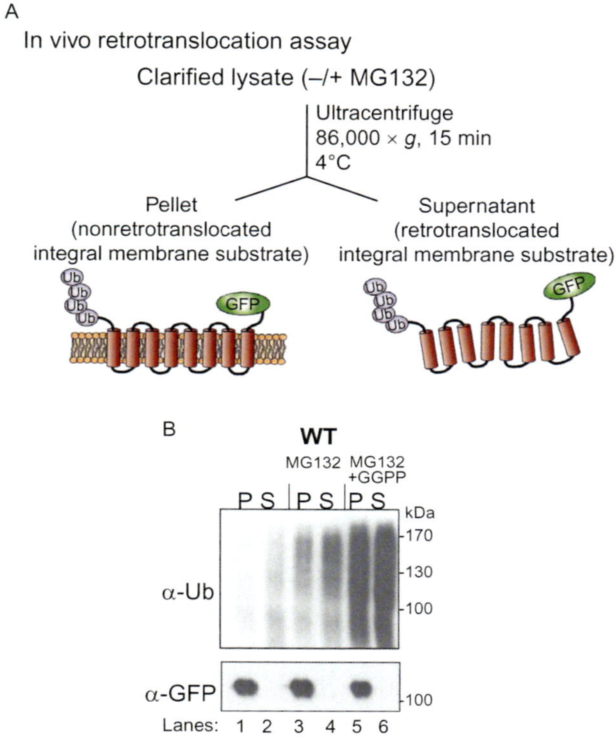

Fig. 6 (A) Schematic of in vivo ubiquitination of Hmg2-GFP. Crude lysate was prepared from WT or various mutants defective in ERAD as indicated and ultracentrifuged to discern ubiquitinated Hmg2-GFP that either has been retrotranslocated into the soluble fraction (S) or remained in the ER membrane (P). (B) GGPP-induced ubiquitination and retrotranslocation of Hmg2-GFP. WT strains were grown to log phase and treated with different combinations of MG132 (25 μg/mL) and GGPP (11 μ*M*). Crude lysate was prepared from each strain and ultracentrifuged to discern ubiquitinated Hmg2-GFP that either has been retrotranslocated into the soluble fraction (S) or remained in the membrane (P). Following fractionation, Hmg2-GFP was immunoprecipitated from both fractions, resolved on 8% SDS-PAGE, and immunoblotted for ubiquitin with anti-Ub or unmodified Hmg2-GFP with anti-GFP. *Figure is taken from Neal, S., Mak, R., Bennett, E. J., & Hampton, R. (2017). A Cdc48 "retrochaperone" function is required for the solubility of retrotranslocated, integral membrane endoplasmic reticulum-associated degradation (ERAD-M) substrates.* The Journal of Biological Chemistry 292, *3112–3128.*

Hmg2 ubiquitination and degradation (Garza, Tran, & Hampton, 2009). We have demonstrated that addition of GGPP to cells along with proteasome inhibitors increases overall ubiquitination of Hmg2-GFP and increases levels of ubiquitinated Hmg2-GFP in the cytosolic fraction (Fig. 6). Together, these findings support the in vivo assay as a powerful tool for studying physiological retrotranslocation.

3.1 Materials

3.1.1 Strains

To generate a suitable strain for in vivo assays involving drug additions such as the proteasome inhibitor MG132, the multidrug resistance gene *PDR5* is disrupted, since *pdr5Δ* null strains allow high potency inhibition of proteasomes of intact yeast with standard inhibitors such MG132 (Neal et al., 2017).

3.2 Assay of in vivo ubiquitination of Hmg2-GFP

1. Strains are grown from a low staring OD_{600} (typically 0.05) at the desired temperature in minimal medium and with vigorous shaking until the cultures reach a mid-log phase OD_{600} of 0.2–0.3. We typically grow 50 mL of yeast culture.
2. For drug treatment, mid-log phase cultures are treated with MG132 at a final concentration of 25 μg/mL, added from a 25 mg/mL stock dissolved in DMSO for 2 h at 30°C and/or final concentration of 11 μ*M* of GGPP, added from a 2.2 m*M* stock of GGPP ammonium salt, for 1 h at 30°C. We typically add GGPP after 1 h of MG132 treatment. From our experience, these concentrations and the duration of drug treatments yield an optimal signal for ubiquitinated substrate. Importantly, only Hmg2 undergoes GGPP-stimulated ubiquitination. Neither unregulated Hmg2 variants, nor any other ERAD-M substrates, show any effect of added GGPP in this assay.
3. 15 OD units of cells are collected and prepared for bead lysis in 50 mL Falcon tubes. Cell pellets are rinsed and resuspended in 400 μL MF buffer with the same amount of PIs as mentioned above for the in vitro assay.
4. Bead-based lysis is performed as described in Section 2.1.2, step 3. It is critical that subsequent steps are performed at 4°C.
5. The lysate is collected into a 2 mL Eppendorf tube, the glass beads are rinsed once with 400 μL of cold MF buffer, and added to the lysate. This combined crude lysate is centrifuged for 5 min at 2500 × *g*. The resulting clarified lysate is transferred to a new chilled 2 mL Eppendorf tube.
6. Ubiquitination of substrates is measured by IP of Hmg2-GFP, SDS-PAGE, and immunoblotting for Hmg2-GFP and ubiquitin. IP and immunoblot analysis were done as described above for the cell-free ubiquitination experiments.

3.3 In vivo retrotranslocation assay of Hmg2-GFP

The in vivo retrotranslocation assay involves ultracentrifugation of the clarified lysate from step 5, prior to addition of any detergents, into microsomal pellet and cytosolic supernatant fractions to assess the extent of retrotranslocation into the supernatant fraction.

1. 200 μL of clarified lysate is ultracentrifuged at 86,000 × *g* for 15 min to yield microsomal pellet (P100) and cytosolic supernatant (S100) fractions.
2. Retrotranslocation of substrates is measured by comparison of immunoprecipitation of Hmg2-GFP from the P100 and S100 fractions, with the S100 material representing retrotranslocation. P100 is resuspended in 200 μL of SUME with PIs and 5 m*M* NEM followed by addition of 600 μL of IPB with PIs and NEM. The S100 supernatant is added directly to 600 μL IPB with PIs and NEM. Subsequent steps for IP and immunoblot analysis of retrotranslocated Hmg2-GFP are carried out as described above for the in vitro ubiquitination assay (Fig. 7).

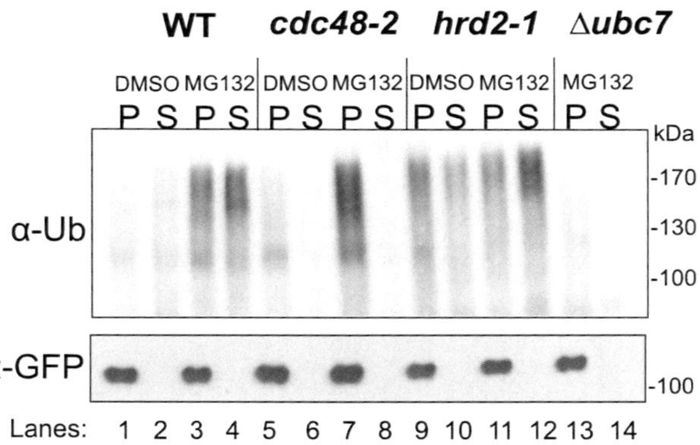

Fig. 7 in vivo retrotranslocation of Hmg2-GFP requires Cdc48. in vivo retrotranslocation assay was performed with WT, *cdc48-2*, proteasome mutant *hrd2-1*, and *ubc7*Δ cells and treated with vehicle or MG132 (25 μg/mL). *cdc48-2* was used as a positive control for block in retrotranslocation (*lanes 5–8*). The ERAD mutant *hrd2-1* is an allele of the 26S proteasome base subunit Rpn1 that is strikingly deficient in Hmg2 degradation. As expected in an *hrd2-1* strain, there was an increase in total ubiquitination of Hmg2-GFP and more accumulation in the soluble fraction compared with WT. Moreover, the signal intensity in each fraction was greater still when MG132 was added to the *hrd2-1* strain due to further inhibition of the proteasome (*lanes 9–12*). In cells lacking Ubc7, the principal HRD pathway E2, Hmg2-GFP was not ubiquitinated, and so no retrotranslocated Hmg2-GFP was generated (*lanes 13* and *14*). *Figure is taken from Neal, S., Mak, R., Bennett, E. J., & Hampton, R. (2017). A Cdc48 "retrochaperone" function is required for the solubility of retrotranslocated, integral membrane endoplasmic reticulum-associated degradation (ERAD-M) substrates.* The Journal of Biological Chemistry 292, *3112-3128.*

Note that employing the same final volumes for the P100 and S100 allows facile visual assessment of the fraction of ubiquitinated substrate that has been moved to the cytosol by retrotranslocation.

4. Characterizing the nature of retrotranslocated Hmg2-GFP

Both in vitro and in vivo retrotranslocation assays allow for the isolation and direct examination of the retrotranslocated intermediate. In this section, the use of biophysical and proteomic analyses to understand the physical state of retrotranslocated Hmg2-GFP is described.

4.1 Ubiquitin cleavage from retrotranslocated Hmg2-GFP

Several experiments indicate that the retrotranslocated protein is soluble, intact and full-length; the most striking result shows that full-length Hmg2 can be recovered from the cytosolic supernatant fraction by cleaving off the attached ubiquitin with Usp2, purified, constitutively active ubiquitin protease (Garza, Sato, & Hampton, 2009; Neal et al., 2017). These studies showed that the intact 8-TMD-containing Hmg2 is in a soluble form after retrotranslocation. The procedure for ubiquitin removal from retrotranslocated Hmg2-GFP is outlined here.

1. As starting material, a supernatant fraction containing either in vivo and in vitro retrotranslocated Hmg2-GFP is used, prepared as above, except that the employed retrotranslocation assay is carried out without added leupeptin or NEM in any of the buffers since these compounds both inhibit ubiquitin protease activity and so prevent removal of the ubiquitin chains from the substrate.
2. For removing ubiquitin from in vitro retrotranslocated Hmg2-GFP, seven separate 24 μL in vitro ubiquitination reactions are incubated for 1 h at 30°C and then centrifuged at 25,000 × *g*. Note: 7 × separate reactions must be performed. Scaling this reaction has been problematic and did not result in optimal ubiquitination of Hmg2-GFP. The resulting supernatant fractions are pooled to yield a sufficient volume for analysis. For in vivo retrotranslocated Hmg2-GFP, supernatant from 15 OD equivalents of MG132 treated cells was enough for analysis. Note: seven separate in vitro ubiquitination reactions are used to recover and detect a robust signal for Hmg2-GFP once the polyubiquitination chain is cleaved.

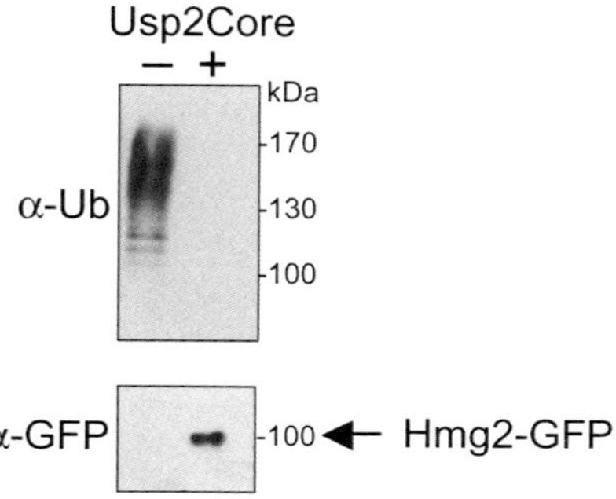

Fig. 8 Full-length Hmg2-GFP retrotranslocates into soluble fraction in vivo. Supernatant fraction containing retrotranslocated Hmg2-GFP was isolated from in vivo retrotranslocation assay and incubated in the presence or absence of Usp2Core for 1 h at 37°C. Full-length Hmg2-GFP was immunoprecipitated and immunoblotted with anti-Ubi to verify that the ubiquitin has been cleaved from the degradation intermediate or with anti-GFP to blot for recovery of full-length retrotranslocated Hmg2-GFP. *Figure is taken from Neal, S., Mak, R., Bennett, E. J., & Hampton, R. (2017). A Cdc48 "retrochaperone" function is required for the solubility of retrotranslocated, integral membrane endoplasmic reticulum-associated degradation (ERAD-M) substrates.* The Journal of Biological Chemistry 292, *3112–3128.*

3. Ubiquitin removal is accomplished with the broadly active Usp2 ubiquitin protease (human recombinant Usp2Core) that either can be purified from *Escherichia coli* that express the recombinant enzyme (a gift from Rohan Baker, Australian National University) or purchased commercially (LifeSensors Inc., Malvern, PA). 50 μL of supernatant containing retrotranslocated Hmg2-GFP is incubated with 10 μL of Usp2Core (5 μg) for 1 h at 37°C. In our experience, this concentration and incubation time is optimal for complete ubiquitin removal.
4. The reaction is quenched with 200 μL of SUME with PIs and retrotranslocated Hmg2-GFP is immunoprecipitated as described above for the in vitro ubiquitination reactions. The final volume of eluted proteins is 55 μL.
5. 20 μL of the immunoprecipitate is used for detection with anti-GFP and anti-ubiquitin antibodies (Fig. 8).

4.2 Sucrose gradient flotation assay of retrotranslocated Hmg2-GFP

We suspected that specific proteins, which we term "retrochaperones" (Neal et al., 2017), were required for the high solubility of retrotranslocated full-length Hmg2-GFP, as has been shown for the Ste6* (also called Ste6-166)

Doa10 pathway substrate (Nakatsukasa et al., 2016). Nevertheless, it remained formally possible that lipids are involved in solubilizing Hmg2 by formation of micelles and/or lipid droplet-like particles (Jo, Hartman, & DeBose-Boyd, 2013). To address this, in vivo retrotranslocated Hmg2-GFP was isolated and analyzed by a sucrose density flotation assay to evaluate its association with lipids: protein complexes "sink" to the lower fractions of a sucrose gradient, while lipid-based complexes such as lipid droplets rise to the top of such gradients due to the buoyancy of such molecular species. This flotation assay is adapted from (Zhang et al., 2001) and described in detail below.

1. Supernatant fraction containing in vivo retrotranslocated Hmg2-GFP is isolated as described above except MSB buffer (50 m*M* HEPES, 150 m*M* NaCl, 5 m*M* EDTA, pH 7.6) was used instead of MF buffer.
2. 100 μL of supernatant fraction containing ubiquitinated soluble Hmg2-GFP is mixed with 300 μL of dense sucrose solution (2.4 *M* sucrose in MSB buffer) in a 1 mL ultracentrifuge tube (Beckman Coulter) on ice.
3. Subsequent steps include the overlaying of sucrose solutions of progressively lower densities. Specifically, MSB supplemented with 1.5 *M* sucrose (500 μL) and then 0.25 *M* sucrose (400 μL) are successively layered into the centrifuge tube containing the dense sucrose solution containing soluble ubiquitinated Hmg2-GFP. The layering must be done without causing any turbulence that could mix the step concentrations and compromise the buoyancy experiment.
4. The resulting sucrose-layered tube is centrifuged in a Beckman SW55 rotor at 100,00 × *g* for 16 h at 4°C.
5. Fractions of 100 μL are gently removed from the top of the gradient and transferred to Eppendorf tubes, allowing analysis of the position of proteins in the gradients.
6. For analysis of retrotranslocated Hmg2-GFP, each fraction is immunoprecipitated with anti-GFP antibodies and immunoblotted as described above in the in vitro ubiquitination assay.
7. Various control proteins are also evaluated in the gradients (Fig. 9) to ensure successful partition based on biophysical properties. The soluble cytoplasmic enzyme phosphoglycerate kinase (PGK) remains in the lower fractions (7–13) closer to the bottom of the tube, whereas the membrane-associated ER-resident integral membrane protein Sec61 floats to higher fractions (4–10). The retrotranslocated Hmg2-GFP shows mobility coincident with PGK, suggesting that the in vivo retrotranslocated Hmg2-GFP is not associated with lipids; rather, it appeared to be fully proteinaceous.

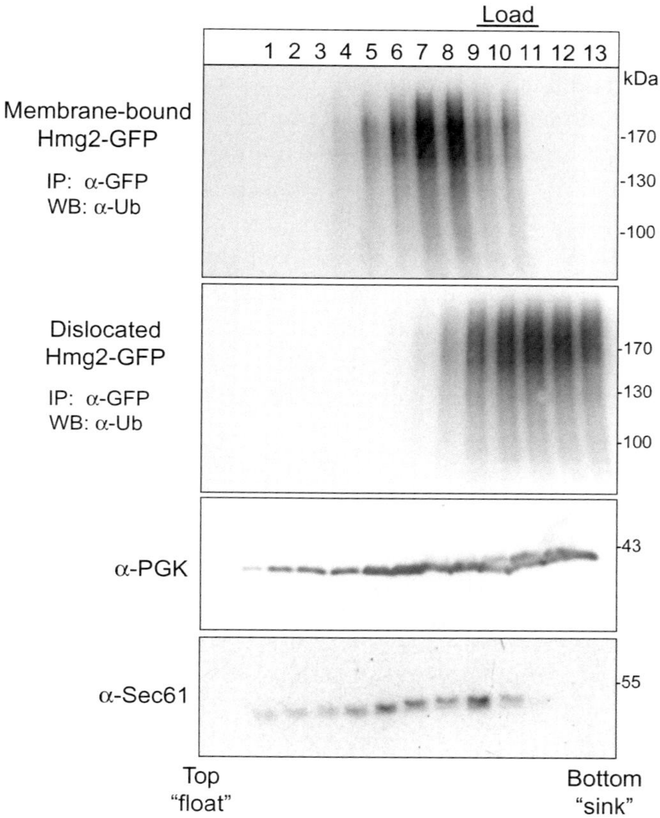

Fig. 9 In vivo retrotranslocated Hmg2 does not associate with lipids. Supernatant fraction containing retrotranslocated Hmg2-GFP was isolated and prepared, layered at the bottom of a centrifuge tube, and subjected to ultracentrifugation as described. Aliquots were removed from the top to the bottom of the sucrose gradient, and each fraction was either directly immunoblotted for Sec61 and PGK with anti-Sec61 or anti-PGK, respectively, or immunoprecipitated with anti-GFP and immunoblotted for ubiquitin and Hmg2-GFP. *Figure is taken from Neal, S., Mak, R., Bennett, E. J., & Hampton, R. (2017). A Cdc48 "retrochaperone" function is required for the solubility of retrotranslocated, integral membrane endoplasmic reticulum-associated degradation (ERAD-M) substrates.* The Journal of Biological Chemistry 292, *3112–3128.*

4.3 Proteomic analysis of retrotranslocated Hmg2-GFP and associated retrochaperones

Our observation that retrotranslocated Hmg2-GFP was proteinacious demanded analysis of the associated proteins that would allow an eight-transmembrane domain ERAD-M substrate to exist in a soluble state.

To identify these retrochaperones, we used an unbiased proteomics approach (Neal et al., 2017). We maximized the amount of retrotranslocated material by modifying and scaling up the in vivo retrotranslocation assay. Accordingly, the in vivo retrotranslocation assay supernatant fraction was derived from simultaneous proteasome inhibitor MG132 and ubiquitin GGPP treatment of a large sample—30 OD units—of cells. Furthermore, the ultracentrifuged S fraction is subjected to preparative anti-GFP co-IP using GFP-Trap® agarose (Chromotek) in the absence of detergent to capture retrotranslocated, ubiquitinated Hmg2-GFP and any accompanying proteins. As controls, Co-IPs are performed on supernatant fractions from otherwise identical strains expressing cytosolic GFP. This Co-IP procedure for proteomic analysis of retrotranslocated Hmg2-GFP is outlined below.

1. Strains are grown from a low starting OD (~0.05) at the desired temperature in minimal medium with vigorous shaking until the cultures reach a mid-log phase OD_{600} of 0.2–0.3. We typically grow 200 mL of yeast culture.
2. For drug treatment, mid-log phase cultures are treated with MG132 at a final concentration of 25 μg/mL (added from a 25 mg/mL stock) for 2 h at 30°C and GGPP ammonium salt at a final concentration of 11 μ*M* (added from a 2.2 m*M* stock) for 1 h at 30°C. GGPP is added after the first hour of MG132 treatment.
3. 30 OD units of cells are collected and prepared for bead lysis in 50 mL Falcon tubes. Cell pellets are rinsed and resuspended in ice cold 800 μL MF buffer with PIs.
4. Bead-based lysis is performed as described above. It is critical that the following steps are performed at 4°C.
5. The lysate is collected and pooled with one bead rinse using 800 μL of cold MF buffer to give crude lysate. The crude lysate is centrifuged for 5 min at 2500 × *g*. The resulting clarified lysate is transferred to an ultracentrifuge tube.
6. The clarified lysate is ultracentrifuged at 86,000 × *g* for 15 min to yield a cytosolic supernatant fraction.
7. While the lysate is being centrifuged, GFP-Trap®-agarose beads are equilibrated for co-IP of retrotranslocated Hmg2-GFP.
   a. GFP-Trap®-agarose stock tube is gently vortexed and 35 μL of agarose slurry is transferred to a chilled Eppendorf tube to which 500 μL of ice cold non-detergent IP buffer (15 m*M* Na_2HPO_4, 150 m*M* NaCl, 10 m*M* EDTA, pH 7.5) with PIs is added.

b. The agarose slurry is centrifuged at 2500 × *g* for 2 min at 4°C and the supernatant is discarded.
c. GFP-Trap®-agarose is resuspended in 600 μL of ice cold non-detergent IP buffer with PIs.
d. The agarose slurry is divided into three fresh chilled Eppendorf tubes (200 μL per tube).
e. All three tube are centrifuged at 2500 × *g* for 2 min at 4°C and the supernatant is discarded from all three tubes.
f. Each agarose pellet is resuspended in 600 μL of ice cold non-detergent IP buffer with PIs.

8. Once ultracentrifugation of the yeast lysate is completed, 600 μL of the supernatant is divided equally into the three tubes (200 μL/tube) containing equilibrated GFP-Trap agarose as prepared above. IP incubation is carried out overnight at 4°C with gentle agitation.
9. Next day: All three IPs are centrifuged at room temperature at 2500 × *g* for 2 min. Supernatants are removed and saved for analysis.
10. GFP-Trap bound protein from all three tubes is resuspended in non-detergent IP buffer. The three IP samples are combined into one Eppendorf tube and washed in 2 × 1 mL with non-detergent IP buffer.
11. For elution of proteins bound to GFP-Trap agarose, beads are resuspended in 50 μL of 0.2 *M* glycine pH 2.5 and incubated for 30 s under constant mixing.
12. Tube is centrifuged at room temperature at 20,000 × *g* for 5 min.
13. Supernatant is transferred to new tube and 5 μL of 1 *M* Tris base pH 10.4 is added for neutralization.
14. Neutralized sample is then subjected for LC–MS/MS mass spectrometry analysis.

Acknowledgments

We thank the Hampton lab members for in depth discussions and technical assistance. These studies were supported by NIH grant 5R37DK051996-18 (to R.Y.H.). S.E.N was funded by the postdoctoral NIH grant 1F32GM111024-01 and Burroughs Wellcome Fund 1013987. S.D. was supported from funding by the CRI Irvington Postdoctoral Fellowship.

References

Baldridge, R. D., & Rapoport, T. A. (2016). Autoubiquitination of the Hrd1 ligase triggers protein Retrotranslocation in ERAD. *Cell*, *166*, 394–407.

Bays, N. W., Wilhovsky, S. K., Goradia, A., Hodgkiss-Harlow, K., & Hampton, R. Y. (2001). HRD4/NPL4 is required for the proteasomal processing of ubiquitinated ER proteins. *Molecular Biology of the Cell*, *12*, 4114–4128.

Bazirgan, O. A., & Hampton, R. Y. (2008). Cue1p is an activator of Ubc7p E2 activity in vitro and in vivo. *The Journal of Biological Chemistry*, *283*, 12797–12810.

Bordallo, J., Plemper, R. K., Finger, A., & Wolf, D. H. (1998). Der3p/Hrd1p is required for endoplasmic reticulum-associated degradation of misfolded lumenal and integral membrane proteins. *Molecular Biology of the Cell*, *9*, 209–222.

Carroll, S. M., & Hampton, R. Y. (2009). Usa1p is required for optimal function and regulation of the Hrd1p endoplasmic reticulum-associated degradation ubiquitin ligase. *The Journal of Biological Chemistry*, *285*(8), 5146–5156.

Carvalho, P., Goder, V., & Rapoport, T. A. (2006). Distinct ubiquitin-ligase complexes define convergent pathways for the degradation of ER proteins. *Cell*, *126*, 361–373.

Chen, B., Mariano, J., Tsai, Y. C., Chan, A. H., Cohen, M., & Weissman, A. M. (2006). The activity of a human endoplasmic reticulum-associated degradation E3, gp78, requires its Cue domain, RING finger, and an E2-binding site. *Proceedings of the National Academy of Sciences of the United States of America*, *103*, 341–346.

Chiti, F., & Dobson, C. M. (2017). Protein Misfolding, amyloid formation, and human disease: A summary of Progress over the last decade. *Annual Review of Biochemistry*, *86*, 27–68.

Emmerich, C. H., & Cohen, P. (2015). Optimising methods for the preservation, capture and identification of ubiquitin chains and ubiquitylated proteins by immunoblotting. *Biochemical and Biophysical Research Communications*, *466*, 1–14.

Foresti, O., Ruggiano, A., Hannibal-Bach, H. K., Ejsing, C. S., & Carvalho, P. (2013). Sterol homeostasis requires regulated degradation of squalene monooxygenase by the ubiquitin ligase Doa10/Teb4. *eLife*, *2*, e00953.

Garza, R. M., Sato, B. K., & Hampton, R. Y. (2009). In vitro analysis of Hrd1p-mediated retrotranslocation of its multispanning membrane substrate 3-hydroxy-3-methylglutaryl (HMG)-CoA reductase. *The Journal of Biological Chemistry*, *284*, 14710–14722.

Garza, R. M., Tran, P. N., & Hampton, R. Y. (2009). Geranylgeranyl pyrophosphate (GGPP) is a potent regulator of HRD-dependent HMG-CoA reductase degradation in yeast. *The Journal of Biological Chemistry*, *284*(51), 35368–35380.

Hampton, R. Y. (2005). Fusion-based strategies to identify genes involved in degradation of a specific substrate. *Methods in Enzymology*, *399*, 310–323.

Hampton, R. Y., Gardner, R. G., & Rine, J. (1996). Role of 26S proteasome and HRD genes in the degradation of 3-hydroxy-3-methylglutaryl-CoA reductase, an integral endoplasmic reticulum membrane protein. *Molecular Biology of the Cell*, *7*, 2029–2044.

Hampton, R. Y., & Garza, R. M. (2009). Protein quality control as a strategy for cellular regulation: Lessons from ubiquitin-mediated regulation of the sterol pathway. *Chemical Reviews*, *109*, 1561–1574.

Hampton, R. Y., & Rine, J. (1994). Regulated degradation of HMG-CoA reductase, an integral membrane protein of the endoplasmic reticulum, in yeast. *The Journal of Cell Biology*, *125*, 299–312.

Hampton, R. Y., & Sommer, T. (2012). Finding the will and the way of ERAD substrate retrotranslocation. *Current Opinion in Cell Biology*, *24*, 460–466.

Hara, T., Hashimoto, Y., Akuzawa, T., Hirai, R., Kobayashi, H., & Sato, K. (2014). Rer1 and calnexin regulate endoplasmic reticulum retention of a peripheral myelin protein 22 mutant that causes type 1A Charcot-Marie-tooth disease. *Scientific Reports*, *4*, 6992.

Heck, J. W., Cheung, S. K., & Hampton, R. Y. (2010). Cytoplasmic protein quality control degradation mediated by parallel actions of the E3 ubiquitin ligases Ubr1 and San1. *Proceedings of the National Academy of Sciences of the United States of America*, *107*, 1106–1111.

Hiller, M. M., Finger, A., Schweiger, M., & Wolf, D. H. (1996). ER degradation of a misfolded luminal protein by the cytosolic ubiquitin-proteasome pathway. *Science*, *273*, 1725–1728.

Jarosch, E., Taxis, C., Volkwein, C., Bordallo, J., Finley, D., Wolf, D. H., et al. (2002). Protein dislocation from the ER requires polyubiquitination and the AAA-ATPase Cdc48. *Nature Cell Biology*, *4*, 134–139.

Jo, Y., Hartman, I. Z., & DeBose-Boyd, R. A. (2013). Ancient ubiquitous protein-1 mediates sterol-induced ubiquitination of 3-hydroxy-3-methylglutaryl CoA reductase in lipid droplet-associated endoplasmic reticulum membranes. *Molecular Biology of the Cell*, *24*, 169–183.

Lilley, B. N., & Ploegh, H. L. (2004). A membrane protein required for dislocation of misfolded proteins from the ER. *Nature*, *429*, 834–840.

Mehnert, M., Sommer, T., & Jarosch, E. (2013). Der1 promotes movement of misfolded proteins through the endoplasmic reticulum membrane. *Nature Cell Biology*, *16*, 77–86.

Mehnert, M., Sommermeyer, F., Berger, M., Kumar Lakshmipathy, S., Gauss, R., Aebi, M., et al. (2015). The interplay of Hrd3 and the molecular chaperone system ensures efficient degradation of malfolded secretory proteins. *Molecular Biology of the Cell*, *26*, 185–194.

Nakatsukasa, K., Kamura, T., Bagola, K., Mehnert, M., Jarosch, E., Sommer, T., et al. (2016). Subcellular Fractionation Analysis of the Extraction of Ubiquitinated Polytopic Membrane Substrate during ER-Associated Degradation. *PLoS One*, *11* e0148327.

Neal, S., Jaeger, P. A., Duttke, S. H., Benner, C. K., Glass, C., Ideker, T., et al. (2018). The Dfm1 Derlin is required for ERAD Retrotranslocation of integral membrane proteins. *Molecular Cell*, *69*, 306–320.

Neal, S., Mak, R., Bennett, E. J., & Hampton, R. (2017). A Cdc48 "retrochaperone" function is required for the solubility of retrotranslocated, integral membrane endoplasmic reticulum-associated degradation (ERAD-M) substrates. *The Journal of Biological Chemistry*, *292*, 3112–3128.

Plemper, R. K., Böhmler, S., Bordallo, J., Sommer, T., & Wolf, D. H. (1997). Mutant analysis links the translocon and BiP to retrograde protein transport for ER degradation. *Nature*, *388*, 891–895.

Ravid, T., Kreft, S. G., & Hochstrasser, M. (2006). Membrane and soluble substrates of the Doa10 ubiquitin ligase are degraded by distinct pathways. *The EMBO Journal*, *25*, 533–543.

Sato, B. K., Schulz, D., Do, P. H., & Hampton, R. Y. (2009). Misfolded membrane proteins are specifically recognized by the transmembrane domain of the Hrd1p ubiquitin ligase. *Molecular Cell*, *34*, 212–222.

Schoebel, S., Mi, W., Stein, A., Ovchinnikov, S., Pavlovicz, R., DiMaio, F., et al. (2017). Cryo-EM structure of the protein-conducting ERAD channel Hrd1 in complex with Hrd3. *Nature*, *548*, 352–355.

Spang, A., & Schekman, R. (1998). Reconstitution of retrograde transport from the Golgi to the ER in vitro. *The Journal of Cell Biology*, *143*, 589–599.

Stein, A., Ruggiano, A., Carvalho, P., & Rapoport, T. A. (2014). Key steps in ERAD of luminal ER proteins reconstituted with purified components. *Cell*, *158*, 1375–1388.

Sun, F., Zhang, R., Gong, X., Geng, X., Drain, P. F., & Frizzell, R. A. (2006). Derlin-1 promotes the efficient degradation of the cystic fibrosis transmembrane conductance regulator (CFTR) and CFTR folding mutants. *The Journal of Biological Chemistry*, *281*, 36856–36863.

Vashistha, N., Neal, S. E., Singh, A., Carroll, S. M., & Hampton, R. Y. (2016). Direct and essential function for Hrd3 in ER-associated degradation. *Proceedings of the National Academy of Sciences of the United States of America*, *113*, 5934–5939.

Wangeline, M. A., Vashistha, N., & Hampton, R. Y. (2017). Proteostatic tactics in the strategy of sterol regulation. *Annual Review of Cell and Developmental Biology*, *33*, 467–489.

Ye, Y., Meyer, H. H., & Rapoport, T. A. (2003). Function of the p97-Ufd1-Npl4 complex in retrotranslocation from the ER to the cytosol: Dual recognition of nonubiquitinated polypeptide segments and polyubiquitin chains. *The Journal of Cell Biology*, *77114*, 21–9525.

Zhang, Y., Nijbroek, G., Sullivan, M. L., McCracken, A. A., Watkins, S. C. Michaelis, S., et al. (2001). NNNN70 molecular chaperone facilitates endoplasmic reticulum-associated protein degradation of cystic fibrosis transmembrane conductance regulator in yeast. *Molecular Biology of the Cell*, *12*, 1303–1314.

Zhao, L., & Ackerman, S. L. (2006). Endoplasmic reticulum stress in health and disease. *Current Opinion in Cell Biology*, *18*, 444–452.

CHAPTER TWO

Methods for measuring misfolded protein clearance in the budding yeast *Saccharomyces cerevisiae*

Rahul S. Samant[a,]*, Judith Frydman[a,b,]*
[a]Department of Biology, Stanford University, Stanford, CA, United States
[b]Department of Genetics, Stanford University, Stanford, CA, United States
*Corresponding authors: e-mail address: rsamant@stanford.edu; jfrydman@stanford.edu

Contents

Abstract

Protein misfolding in the cell is linked to an array of diseases, including cancers, cardiovascular disease, type II diabetes, and numerous neurodegenerative disorders. Therefore, investigating cellular pathways by which misfolded proteins are trafficked and cleared ("protein quality control") is of both mechanistic and therapeutic importance. The clearance of most misfolded proteins involves the covalent attachment of one or more ubiquitin molecules; however, the precise fate of the ubiquitinated protein varies greatly, depending on the linkages present in the ubiquitin chain. Here, we discuss approaches for quantifying linkage-specific ubiquitination and clearance of misfolded proteins in the budding yeast *Saccharomyces cerevisiae*—a model organism used extensively for interrogation of protein quality control pathways, but which presents its own unique challenges for cell and molecular biology experiments. We present a fluorescence microscopy-based assay for monitoring the clearance of misfolded

Methods in Enzymology, Volume 619
ISSN 0076-6879
https://doi.org/10.1016/bs.mie.2018.12.039
© 2019 Elsevier Inc.
All rights reserved.

protein puncta, a cycloheximide-chase assay for calculating misfolded protein half-life, and two antibody-based methods for quantifying specific ubiquitin linkages on tagged misfolded proteins, including a 96-well plate-based ELISA. We hope these methods will be of use to the protein quality control, protein degradation, and ubiquitin biology communities.

1. Introduction

In order for a cell to remain healthy, it must maintain a balanced proteome ("proteostasis") (Balchin, Hayer-Hartl, & Hartl, 2016; Sontag, Samant, & Frydman, 2017). Proteins that do not attain their native three-dimensional structure—e.g., nascent chains from the ribosome, or correctly folded proteins that denature due to various types of stress—must either be refolded or cleared through a process known as protein quality control (PQC). Like any quality-control system, PQC involves different machineries for detection and disposal of nonnative products. For example, molecular chaperones selectively recognize misfolded proteins and target them to refolding or clearance pathways; E3 ubiquitin ligases tag misfolded proteins with one or more molecules of ubiquitin (a posttranslational modification with many roles, one of which is as a proclearance signal); and the proteasome, which recognizes and degrades ubiquitinated proteins (Fig. 1A).

The budding yeast *Saccharomyces cerevisiae* has been used extensively for furthering our mechanistic understanding of protein quality control and clearance pathways (Amm, Sommer, & Wolf, 2014; Schneider, Nyström, & Widlund, 2018; Sontag et al., 2017). Not only is its core proteostasis machinery conserved through to humans, but it offers numerous clear advantages over higher eukaryotes, including an unrivaled genetic toolbox (a fully annotated genome, a complete deletion mutant strain collection, widely available GFP- and TAP-tagged libraries covering the majority of its proteome, as well as high-throughput methods to generate libraries of new genetic crosses within weeks), as well as its scale, i.e., the fact that experiments are performed with entire populations, rather than a small number of organisms representative of a population (Botstein & Fink, 2011). With regard to proteostasis specifically, the reduced complexity and redundancy of PQC systems in yeast (e.g., the human genome encodes 300–350 molecular chaperones and over 600 E3 ubiquitin ligases, compared with 69 and ~60, respectively, in budding yeast) have enabled dissection of protein clearance systems in a way that would not have been possible directly using mammalian models.

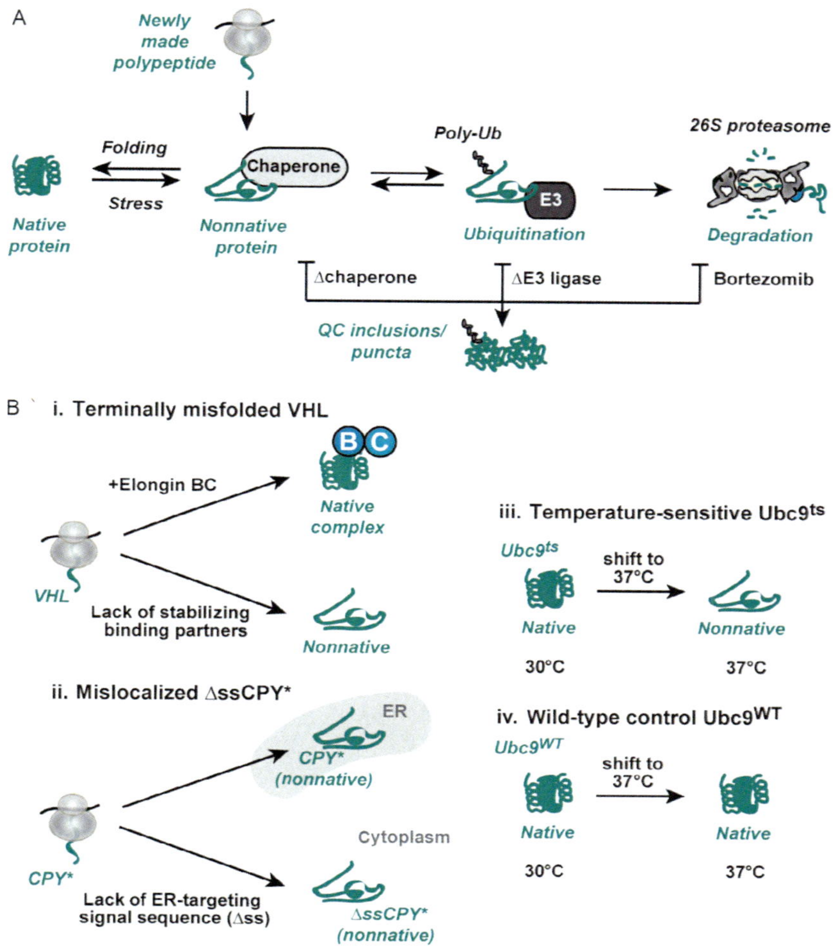

Fig. 1 Misfolded proteins expressed in budding yeast *S. cerevisiae* are cleared through the concerted action of molecular chaperones, E3 ubiquitin ligases, and the proteasome (A). Three model misfolded proteins that misfold for different reasons are shown in (B).

However, standard cell biology protocols generally do not take into account the intricacies of yeast as a model organism, or of ubiquitin as a covalent yet highly dynamic posttranslational modification. This chapter is aimed at discussing some of the unique considerations that must be addressed when studying ubiquitin-mediated clearance of misfolded proteins in budding yeast. Basic yeast maintenance and growth protocols are provided elsewhere (Bergman, 2001; Curran & Bugeja, 2014). We will also present protocols we have used successfully for quantifying relative levels of ubiquitin linkages.

2. Expression and isolation of misfolded proteins in yeast

2.1 Choice of expression system

We routinely use the Gateway® system (Alberti, Gitler, & Lindquist, 2007), which provides a quick and convenient cloning strategy with its suite of plasmids containing various fluorescent proteins or affinity tags (either N- or C-terminal fusions), promoters (GAL, inducible, or GPD, constitutive), selection markers (*HIS3*, *LEU2*, *TRP1*, or *URA3*), and origins of replication (chromosomally integrating, 2μ high-copy number, or *CEN* low-copy number).

The *GAL1* promoter is especially useful for inducible expression. Importantly, this system allows us to switch off misfolded protein expression by replacing galactose with glucose in the growth medium. An inducible system is essential for assaying misfolded protein clearance, as we can track the fate of an existing population of our protein of interest without the confounding effect of newly synthesized species (as would be the case with constitutive expression), or of global PQC alterations triggered by translation inhibitors such as cycloheximide. Note that in order to switch on the *GAL1* promoter, the transcriptional repression circuit present during growth in glucose media must first be inactivated through growth on a nonrepressing sugar (e.g., raffinose), rather than transferring directly from glucose to galactose media.

The galactose system is not suitable for certain aspects of PQC research, such as testing the effects of nutrient deprivation and/or chronological aging. For such applications, other expression systems are available, e.g., the *CUP1* copper-inducible promoter (Macreadie, Jagadish, Azad, & Vaughan, 1989), or the doxycycline-controlled Tet-On or Tet-Off systems (Garí, Piedrafita, Aldea, & Herrero, 1997).

2.2 Choice of misfolded proteins

Studies of PQC in yeast typically involve expression of misfolded proteins from nonnative promoters (see the previous section). The choice of protein will be dictated by the precise question being asked. For the purposes of characterizing general mechanisms of PQC, it is important to include a variety of proteins that misfold for different reasons. In our recent study (Samant, Livingston, Sontag, & Frydman, 2018), we used four different misfolded protein reporters: two thermally unstable proteins (Ubc9^{ts}, luciferase^{ts}); the tumor suppressor VHL, which is terminally misfolded in yeast due to lack of its physiological binding partners Elongin B and Elongin C; and ΔssCPY*, a mutated secretory enzyme that misfolds in the cytoplasm after removal of its ER-targeting signal sequence (Fig. 1B). In this way, we tried to ensure that the machineries and pathways uncovered were general to PQC and not specific only for the protein reporter being expressed.

Of course, comparing and contrasting the PQC requirements for different misfolded proteins could provide insights into context-dependent proteostasis pathways. This approach led us to the discovery that amyloid proteins (such as polyglutamine-expanded mutant Huntingtin) and prions (such as RNQ1) engage different PQC machineries than soluble misfolded proteins (such as VHL and $Ubc9^{ts}$) (Escusa-Toret, Vonk, & Frydman, 2013; Kaganovich, Kopito, & Frydman, 2008). As there is little evidence to suggest that this alternative yeast pathway leads to ubiquitination and/or clearance, amyloidogenic misfolded proteins may not be suitable for probing ubiquitin-mediated clearance mechanisms.

2.3 Choice of lysis and solubilization methods

Cell lysis involves the breakdown of a cell using mechanical, enzymatic and/or osmotic methods to release its contents—the "lysate." Solubilization refers to the extraction of the molecules of interest from the cell into the lysate, usually aided by a detergent or a chaotropic agent. Both processes require methods and conditions that preserve the molecules in a state suitable for, and compatible with, the desired downstream applications. For example, immunoprecipitation techniques for detection of protein–protein interactions must strike a balance between adding enough detergent to solubilize the protein complexes of interest while not adding too much to disrupt the noncovalent interactions in the protein complex itself.

Lysis of higher eukaryote cells is routinely performed by resuspension of cells in a lysis buffer that contains a detergent or urea, which serve both to permeabilize the plasma membrane and to solubilize proteins of interest. In yeast, lysis is complicated by the presence of a cell wall—rendering the aforementioned method unsuitable for releasing the cell's contents in many cases. Most yeast lysis protocols therefore involve some additional form of disruption before or alongside the detergent or urea. Detailed discussions and protocols are available on the European Molecular Biology Laboratory's Protein Expression and Purification Core Facility website (https://www.embl.de/pepcore/). We have used a number of these methods for different protein clearance assays, with each having distinct advantages and drawbacks.

2.3.1 Boiling in SDS sample buffer

Perhaps the simplest and quickest protein extraction methods are boiling pelleted yeast cells in SDS sample buffer (Horvath & Riezman, 1994), as one would do prior to SDS-PAGE analysis. This technique has the added benefit of immediately denaturing the cell extract, thus minimizing opportunities for nonphysiological postlysis alterations (e.g., protein degradation by proteases). For the same reason, however, SDS boiling is unsuitable for many downstream applications (e.g., protein–protein interaction analysis, in vitro activity assays). We have also noticed that SDS is not as effective as 8 *M* urea for solubilization of ubiquitinated proteins from the cell extract. We generally use this method for total protein analysis (e.g., cycloheximide chase assays—see Section 3.2).

2.3.2 Bead beating

One of the most popular mechanical lysis methods for yeast is beating the cells with glass beads to physically break apart the cell walls. This is similar to the classical *Escherichia coli* lysis method, except a bead diameter of 0.5 mm (instead of 0.1 mm) is used, to account for the difference in cell size. One caveat of this approach is that some proteins may aggregate during the bead beating process (Papanayotou, Sun, Roth, & Davis, 2010)—particularly problematic for studying misfolded proteins, which are intrinsically aggregation-prone due to the nonnative surface exposure of hydrophobic sequence elements.

Here, we outline some observations that might be useful when considering bead-beating as the lysis method of choice.

- In our experience, the most important factor in determining lysis efficiency is the ratio of beads:lysate—especially when using cells in stationary phase. We found a ratio of 1:1 (v/v) was most effective. Lower bead volumes considerably decrease the extent of lysis.
- One freeze–thaw step (at −80°C) prior to bead-beating improves lysis efficiency.
- Pretreating glass beads with concentrated acid increases lysis efficiency. Although it is possible to perform this in-house, we routinely use acid-washed glass beads from Sigma (catalog number G-8772).
- The presence of detergent in the lysis buffer is often necessary to solubilize the proteins of interest. However, this results in a considerable degree of "foaming" during the bead-beating process—potentially causing proteins to precipitate out of the lysate. One can add antifoaming agents to the lysis buffer (Wheeler, Jain, Khong, & Parker, 2017) or

add detergent to the lysis buffer only after bead-beating. However, we generally avoid the use of detergents for bead-beating lysis, instead using 8 *M* urea as the solubilizing agent with this method.

- The friction caused by high-frequency homogenization generates heat, which is not desirable for maintaining the proteins in a state suitable for most downstream applications. Therefore, it is necessary to incorporate cooling steps into the lysis protocol. We find that five cycles of bead-beating at 3000 rpm (using a Digital Disruptor Genie from Scientific Industries, catalog number Si-Dd38) for 1 min, with 1 min on ice in between each cycle, is sufficient to lyse and solubilize our model misfolded proteins without overheating the samples. However, this is a common step that could be a source for troubleshooting (e.g., increase time on ice in between cycles). We also perform all steps of this procedure in a cold room set at 4°C.
- We have used various devices for the physical bead-beating process, including different types of vortexer and specialized beading-beating devices—with no clear differences in lysis efficiency.
- Our method of choice for isolating the lysate from the glass beads postbead-beating is to use a 21-gauge needle to punch a hole into the bottom of the 1.5- or 2.0-mL microfuge tube used during the bead-beating, and centrifuging into a new 15-mL conical centrifuge tube (1000 × *g* at 4°C for 2 min, or until all of the lysate has collected into a new tube).

2.3.3 Cryogrinding

As discussed earlier, the friction and heat generated during bead-beating are reported to induce aggregation of certain proteins. One of the best solutions to this problem is mechanical disruption of cells at extremely low temperatures. The process of cryogrinding involves freezing cells (suspended in a minimal volume of lysis buffer) as small pellets in liquid nitrogen, followed by grinding them—in liquid nitrogen-cooled equipment—into a fine powder of broken cells. We use a Retsch Mixer Mill MM301 for the grinding process, following a procedure described previously (Ingolia, 2010), but with five cycles at 30 Hz, 3 min each, with 2 min in liquid nitrogen in between cycles. Note that even one cycle at 30 Hz for 3 min is usually sufficient for lysis of >50% of cells.

One of the largest drawbacks of using cryogrinding for the detection of ubiquitinated proteins is that 8 *M* urea—which is typically required to fully solubilize ubiquitin chains—is close to its saturation point even at room temperature. Therefore, freezing samples containing 8 *M* urea will invariably trigger precipitation out of solution. For this reason, we avoid cryogrinding for lysis when detection of ubiquitination is involved in the downstream application. However, if absolutely necessary, we have found some success with cryogrinding in a minimal volume (1:2 volume buffer:cell pellet) of Tris-containing lysis buffer (50 m*M* Tris–HCl pH 7.5, 150 m*M* NaCl, 2 m*M* EDTA, cOmplete EDTA-free protease inhibitor tablet, 1 m*M* PMSF, 50 m*M* 2-chloroacetamide, 10 μ*M* PR-619 (covalent inhibitor of deubiquitinases, Sigma, catalog number 662141)), followed by solubilization in an equal volume of the same buffer supplemented with 8 *M* urea (resulting in 4 *M* urea final concentration) before clarification of the lysate from the cell debris.

2.3.4 Enzymatic digestion with Zymolyase

We discussed at the start of this section how yeast present an additional challenge over mammalian cells due to the presence of a cell wall. Of course, if the cell wall is removed, standard mammalian cell lysis protocols are applicable. Zymolyase—an endoglucanase from *Arthrobacter luteus* (Kitamura, Kaneko, & Yamamoto, 1971)—digests the cell wall and exposes the yeast cell membrane for lysis without the need for mechanical bead beating or cryogrinding. If cost were no consideration, lysis with the Zymolyase (Zymo Research), followed by solubilization in 8 *M* urea buffer, would be the method of choice for most assays involving ubiquitin detection. For most labs, however, this is not economically feasible, as protein preparations typically involve purification of multiple milligrams of the protein of interest—requiring hundreds of mLs of yeast culture. We reserve the use of Zymolyase for small volumes of culture (e.g., for immunofluorescence detection, which requires the infiltration of antibodies into the cell—a process that is prohibited by cell walls).

3. Assays of protein clearance

3.1 Clearance of misfolded protein puncta

Our lab and others have shown that proteins that misfold for a variety of reasons are all cleared within 1–2 h through the ubiquitin–proteasome system (Escusa-Toret et al., 2013; Kaganovich et al., 2008; Malinovska, Kroschwald, Munder, Richter, & Alberti, 2012; Park et al., 2013). If this process is impaired (e.g., inhibiting the proteasome, or in strains with deletions of genes involved in the clearance pathway), the misfolded proteins accumulate in puncta (Escusa-Toret et al., 2013). One of our standard assays for identifying PQC factors involved in this clearance pathway is expression of the GFP-tagged misfolded protein for 4–6 h, then switching off expression for a set period of time and counting the percentage of cells with puncta at the end of this period using fluorescence microscopy. With this technique, we have identified the role of several molecular chaperones (e.g., Hsp70s, Hsp90, Sse1, Hsp42, Hsp104) and various parts of the ubiquitin–proteasome system (E2 ubiquitin-conjugating enzymes Ubc4 and Ubc5; E3 ubiquitin ligases Ubr1, San1, Doa10, and Hrd1) in misfolded protein clearance pathways (Escusa-Toret et al., 2013; Kaganovich et al., 2008; Samant et al., 2018).

i. Transform yeast strains of interest with a plasmid containing the GFP-tagged misfolded protein under control of a galactose-inducible promoter (see Sections 2.1 and 2.2). Plate cells on SD-glucose selective agar plates—that allow selective growth only of cells carrying the plasmid—and wait for 2–3 days for colonies to form. Plates can then be stored at 4°C for up to 1 month before restreaking onto fresh selective agar plates is required.

ii. On the afternoon/evening before the experiment, select single colonies to grow in 5 mL SD-raffinose selective medium at 30°C with shaking (300 rpm).

iii. The following morning, pellet cells (3000 × *g*, 2 min, room temperature) and resuspend in 2 mL of SD-galactose selective medium to an initial OD_{600} of between 0.05 and 0.1 mL^{-1}; galactose will induce expression of the substrate protein. Allow cells to grow (30°C with shaking) to an OD_{600} 0.6–0.8 mL^{-1} (usually 4–6 h, depending on the strain).

iv. Pellet cells (3000 × *g*, 2 min, room temperature) and resuspend in 2 mL SD-glucose media, to shut off misfolded protein expression. For the WT + bortezomib (Bz) positive control, add the proteasome inhibitor Bz (final concentration 50 μ*M*) to the selective medium 5 min before adding to cells.[a] Allow to grow between 1 and 1.5 h.[b]

v. After the desired amount of time, take 750 μL of each sample and add to 250 μL 16% paraformaldehyde (for final concentration of 4%) in a 1.5-mL microfuge tube.

[a] Bortezomib has poor solubility in water and will precipitate as soon as it is added. Ensure the precipitate has redissolved (5 min shaking at the shut-off temperature) before adding to cells.

[b] The exact amount of time for clearance varies depending on the misfolded protein substrate. It is important to perform an initial time-course for each misfolded protein in the wild-type strain of choice before proceeding with screens of deletion or overexpression strains. In our experience, the majority (>80%) of a wild-type yeast population from the BY4742 background clears misfolded proteasomal substrates in the range of 1–1.5 h.

vi. Incubate for 15 min at room temperature. Don't leave for longer than 20 min—extended formaldehyde exposure diminishes fluorescent protein signal and deforms nuclei.

vii. Pellet cells (8000 × *g*, 0.5 min, 4°C). Resuspend in 1 mL chilled methanol (−20°C) and incubate for 20 min on ice.

viii. Pellet cells (8000 × *g*, 0.5 min, 4°C) and wash 2 × with 1 mL phosphate-buffered saline (PBS).

ix. Resuspend in 1 mL PBS. Mix vigorously to minimize cells clumping together (by sonicating, vortexing, or vigorous pipetting).

x. Add 50 μL of cell suspension onto a concanavalin-A-coated coverslip[c] (prewashed once with 1 mL PBS, to wash off excess concanavalin-A). Allow cells to adhere for 10 min at room temperature.

xi. Wash 3 × with PBS to remove unbound yeast. Any unbound yeast at this stage could interfere with imaging.

xii. Add one drop (~15 μL) of yeast mounting medium (ProLong™ Diamond Antifade Mountant with DAPI, Thermo Fisher Scientific, catalog number P36961) onto coverslip and mount onto slide. Leave to cure[d] for 24 h in dark at room temperature and then seal edges with nail varnish before imaging.

Our standard assay for testing whether deletion strains affect misfolded protein clearance is by calculating the percentage of cells that have puncta in the strain of interest, and comparing that with the percentage obtained for the wild-type strain. We take representative epifluorescence images of the GFP (for counting of misfolded protein puncta) and DAPI (a DNA-binding dye used for counting total number of cells) channels using a 63 × oil objective.[e,f] We then manually count total number of cells (DAPI channel—although

[c] Poly-lysine coated coverslips are not effective for adherence due to the presence of the yeast cell wall. However, if the yeast cells have been spheroplasted (e.g., to allow penetration of antibodies for immunofluorescence experiments), poly-lysine and not concanavalin-A must be used.

[d] ProLong™ mountants harden ("cure") within 24 h, thus forming an optimal optical path for fluorescence microscopy (reflective index = 1.47). For immediate imaging, seal the four corners of the coverslip with nail varnish and allow to dry before microscopy. Leaving part of the coverslip unsealed will still allow the sample to cure. Seal as normal after 24 h.

[e] To obtain a representative sampling of the entire cell population, we ensure at least 300 cells are counted per slide (usually four images from different parts of the slides).

[f] We typically capture images in a single *z*-plane (rather than taking *z*-stacks encompassing the entire cell). Although we may miss some puncta that are out of the focal plane, this is minimized because (1) we are measuring total epifluorescence rather than confocal microscopy; (2) the haploid yeast strains we use are relatively thin (<0.5 μm); and (3) the ProLong™ mountant—a hardening mountant—further flattens cells. Furthermore, the effect of "missed" puncta will be averaged out across the strains, especially as we are counting over 300 cells per sample. However, if this is a concern, it is possible to take maximum intensity projections of *z*-stacks.

the DIC channel is also suitable), and the percentage of those cells with one or more puncta (GFP channel), on ImageJ (https://imagej.nih.gov/ij/) using the Cell Counter plug-in. Automated cell counting macros are also a possibility, although this is complicated by the fact that multiple puncta, of varying sizes, can exist in a single cell.

3.2 Cycloheximide chase assay

The puncta-counting assay provides a means of testing whether the clearance of transient inclusions (e.g., Q-bodies, JUNQ, and/or INQ) that contain misfolded proteins is impaired in different genetic backgrounds. However, it is possible that some genetic backgrounds cause puncta to dissolve (or prevent their formation in the first place) without affecting the total level of the misfolded protein. It is difficult to discern cases where the misfolded protein accumulates diffusely in the cytoplasm from those where it is actually cleared. Therefore, it is important to complement the puncta-counting assays with more direct measures of total substrate protein levels.

The classical method for measuring a protein's half-life is through radiolabeling (e.g., with ^{35}S) a population of newly translated proteins for a short period of time (the "pulse") followed by incubation in unlabeled medium for various times (the "chase") (Simon & Kornitzer, 2014). The clearance of the protein of interest can then be determined by lysing cells at specific timepoints, immunoprecipitating the protein, and detecting radioactivity on a gel by autoradiography. However, this technique is cumbersome due to the use of radioactivity and the requirement for immunoprecipitation at each time point. An alternative is the cycloheximide chase assay, where global protein synthesis is blocked using the translation elongation inhibitor cycloheximide. With this method, the clearance of the protein of interest can be ascertained by direct immunoblot—without requiring radioactivity or the immunoprecipitation steps. However, cycloheximide is a cytotoxic agent and would not be suitable to study clearance pathways taking longer than a few hours.

Detailed protocols for cycloheximide chase assays have been provided elsewhere (Buchanan, Lloyd, Engle, & Rubenstein, 2016). Here, we will provide a brief summary of the method we use for detecting clearance of misfolded proteins under control of a *GAL1* promoter.

i. Grow cells transformed with a plasmid encoding the tagged misfolded protein (as described in Section 3.1) overnight in 5 mL SD-raffinose media. The following morning, pellet cells (3000 × *g*, 2 min, room temperature) and resuspend in 100 mL SD-galactose media (in a 250-mL conical flask, final OD_{600} 0.05–0.1 mL^{-1}) and grow till OD_{600} 0.6–0.8 mL^{-1}.

ii. At this stage, split cells into the number of time-points required for the time-course. For our model misfolded proteins, we have 0, 10, 30, 60, and 90 min. Note that it is also helpful to have a positive control with proteasome inhibition (e.g., bortezomib or MG132).[g] We typically grow 20 mL of culture (in a 50-mL conical flask) for each time-point.

iii. Pellet all samples in 50-mL conical centrifuge tubes (3000 × *g*, 2 min, room temperature). Resuspend the first sample (representing the $t=0$ time-point) with 2 mL PBS containing 15 m*M* sodium azide and 1 × Roche cOmplete™ EDTA-free protease inhibitor tablet. Pellet again (3000 × *g*, 2 min, 4°C) and snap-freeze the pellet in liquid nitrogen. Store at −80°C.

iv. While the cells from the $t=0$ time-point are being frozen, resuspend the remaining samples into SD-glucose selective media with 50 μg mL^{-1} cycloheximide. For the positive control, add 50 μ*M* bortezomib here.

v. Collect samples at the desired time-points by pelleting, resuspending in PBS with sodium azide and protease inhibitors, pelleting again, and snap-freezing, as described for the $t=0$ sample in (iii).

vi. Once all samples are collected, thaw on ice and extract proteins by boiling in an equal volume of 2 × SDS sample buffer for 10 min.

vii. Detect levels of the protein of interest by standard SDS-PAGE and immunoblotting protocols. We use electrochemiluminescence (Pierce ECL Western blotting substrate from Thermo Fisher Scientific) and exposure to GeneMate Blue Ultra Film (BioExpress). Quantitation is performed by densitometry using ImageJ, with values expressed as a percentage of the loading control (e.g., GAPDH, Pgk1) signal for each individual lane.

4. Analysis of ubiquitin linkages

Misfolded protein clearance generally involves the posttranslational attachment of ubiquitin to one or more of the misfolded protein's Lys residues. Although well known as a proteasomal targeting signal, ubiquitination can signal other, diverse fates for the target protein, which affect signal transduction, *endo*-lysosomal trafficking, autophagy, NFκB activation, and DNA

[g] In yeast, the highly active efflux pumps make drug treatments challenging. We find that using a high concentration of bortezomib (50 μ*M*) is effective for a 1.5 h chase. However, some clearance of the misfolded protein is still observed at the later time-points. If this is a concern—or if longer chases are required—one can use strains with increased drug permeability or deleted efflux pumps (e.g., *Δpdr5* or *Δerg6*, respectively). Alternatively, strains with impaired proteasomal activity (e.g., temperature-sensitive *cim3-1* or *cim5-1*) are more effective than pharmacologic inhibitors—however, these can only be used at nonphysiological temperatures, as well as requiring genetic crosses with the deletion strain of interest.

repair, among other processes (Kwon & Ciechanover, 2017; Yau & Rape, 2016). Ubiquitin is often linked to other ubiquitin molecules to form polymers. The precise fate of a ubiquitin-modified protein is dependent on the topology of the ubiquitin chain, with chains linked by any of the seven Lys residues in ubiquitin's 76-residue sequence (as well as the amino-terminal amine to form linear chains) each potentially targeting the modified substrate to a different pathway. Given our evolving appreciation of this diverse role of ubiquitin linkages on critical cellular processes, novel methods and reagents for detecting specific linkages are emerging at a steady pace (e.g., linkage-specific antibodies and affinity reagents; mass spectrometry-based methods for absolute quantification of ubiquitin linkages (Ub-AQUA); in vitro assays making use of the linkage specificity of deubiquitinating enzymes (DUBs) for determination of ubiquitin chain architecture) (Swatek & Komander, 2016). This, in turn, is driving further groundbreaking research. The field of ubiquitin biology is moving past the idea that any ubiquitin "ladder" (by SDS-PAGE and immunoblotting) on a protein of interest must mean that the protein is a proteasomal target.

Here, we outline two antibody-based protocols for quantifying ubiquitin linkages on a tagged misfolded protein of interest.

4.1 Immunoprecipitation of ubiquitinated misfolded proteins

As discussed in Section 2.3, standard immunoprecipitation protocols must also be modified to account for the yeast cell wall. Furthermore, due to the high enzymatic activity of deubiquitinating enzymes (DUBs)—with some residual activity even in 8 *M* urea—typical cell lysates that do not contain DUB inhibitors are not suitable for ubiquitination analysis, as most of the proteome is likely to have been deubiquitinated postlysis. For these reasons, preparing cells for ubiquitin detection is not trivial. We supplement all lysis buffers with 50 m*M* 2-chloroacetamide and 10 μ*M* PR-619 to minimize DUB activity (Rose et al., 2016).

Even after lysis, ubiquitin analysis on specific proteins of interest introduces challenges different from those posed by standard immunoprecipitation protocols (Bloom & Pagano, 2005). Whereas most protocols involve maintaining noncovalent interactions to identify coprecipitating interaction partners, ubiquitination analysis requires the removal of such coprecipitating partners (whose ubiquitination would be indistinguishable from that of the protein of interest) while not preventing interactions between the antibody and the protein of interest. We typically use the previously published approach where cell lysates are initially denatured in 8 *M* urea (to disrupt physiological protein–protein interactions) before diluting the urea in standard Tris-containing immunoprecipitation buffer to enable antibody binding to the protein of interest (Emmerich & Cohen, 2015; Heck, Cheung, & Hampton, 2010; Katzmann & Wendland, 2005). We routinely use Flag-tagged substrate proteins for ubiquitination analysis, for cost and ease of use of the Flag-M2® magnetic beads (Sigma). Note, however, that the same protocol can be used for GFP-tagged proteins, using GFP-Trap nanobodies (ChromoTek) that are resistant to 8 *M* urea and 1% SDS—thereby ensuring fully denaturing conditions throughout the immunoprecipitation protocol.

i. Grow cells transformed with a plasmid encoding the Flag-tagged misfolded protein overnight in 10 mL SD-raffinose selective medium, as described earlier. The following morning, pellet cells in a 15-mL conical centrifuge tube (3000 × *g*, 2 min, room temperature) and switch to 100 mL SD-galactose selective medium (initial OD_{600} 0.05–0.1) in a 250-mL conical flask. Grow to an OD_{600} of 0.6–0.8 mL^{-1}.

ii. Pellet each sample in two 50-mL conical centrifuge tubes (3000 × *g*, 5 min, room temperature) and resuspend in 100 mL SD-glucose selective media with 50 μ*M* bortezomib.[h] Grow at 30°C for 1 h and centrifuge cells as above.

iii. Resuspend cell pellet in an equal volume (approximately 0.5 mL) of fresh urea lysis buffer (50 m*M* Tris–HCl pH 7.5, 8 *M* urea, 150 m*M* NaCl, 2 m*M* EDTA, 1 × cOmplete™ EDTA-free protease inhibitor tablet, 50 m*M* 2-chloroacetamide, 10 μ*M* PR-619, 1 m*M* phenymethylsulfonyl fluoride (PMSF)).[i]

iv. Lyse cells by bead-beating (five cycles at 1 min each, with 1 min on ice in between cycles—see Section 2.3.2).

v. Dilute 10-fold in Triton IP Buffer (same composition as Urea Lysis Buffer, but with 1% Triton X-100 instead of 8 *M* urea), and clarify lysates (16,000 × *g* for 30 min at 4°C). Quantify total protein (e.g., BCA or Bradford assay).

vi. Incubate 2 mg of lysate with Flag-M2® magnetic beads (Sigma catalog number M8823) for 2 h at 4°C to immunoprecipitate the Flag-tagged protein of interest.

---

[h] Note that bortezomib (or some sort of proteasome inhibition) is required to allow accurate detection of ubiquitination, as otherwise ubiquitinated proteins could be degraded by the proteasome within the 1 h of expression shut-off. If it is possible that the protein of interest is cleared through another ubiquitin-mediated pathway (e.g., vacuolar targeting through K63 or mono-Ub linkages), inhibitors for the protease in these pathways (e.g., chloroquine, PMSF) may also be added.

[i] Prepare lysis buffer fresh on day of lysis. For all buffers, add protease inhibitor tablet, 2-chloroacetamide, PR-619 and PMSF immediately before use.

vii. Elute Flag-tagged protein from beads by heating for 30 min at 70°C in NuPAGE™ 1 × nonreducing LDS sample buffer[j] (10% glycerol, 1% lithium dodecyl sulfate, 0.2 *M* triethanolamine–Cl pH 7.6, 1% Ficoll®-400, 0.00625% phenol red, 0.00625% Coomassie G250, 0.5 m*M* EDTA disodium; Thermo Fisher Scientific, catalog number NP0007).

viii. Add DTT (50 m*M* final concentration) and heat again at 70°C for 10 min, to reduce eluted sample prior to SDS-PAGE analysis.

ix. Following SDS-PAGE, transfer proteins to methanol-wetted PVDF membrane, according to standard protocols.[k]

x. Denature proteins on PVDF membrane by heating for 10 min at 95°C.[l]

xi. Immunoblot with the desired antibodies. We have successfully used: pan-Ubiquitin VU-1 (1/1000; Life Sensors, catalog number VU101), pan-conjugated Ubiquitin FK2 (1/1000; Life Sensors, catalog number AB120) K11-Ub (1/100; EMD Millipore, catalog number MABS107-I), HRP-conjugated K48-Ub (1/1000; Cell Signaling Technology, catalog number 12805), HRP-conjugated K63-Ub (1/1000; Cell Signaling Technology, catalog number 12930), and K11/K48-Ub bispecific antibody (1/500; Genentech).

4.2 Ubiquitin linkage ELISA

It is often difficult to compare the degree of ubiquitination of proteins across samples (e.g., different deletion strains), as the ubiquitin "ladder" typical of immunoblots is not readily conducive to densitometric quantification. We therefore developed a ubiquitin enzyme-linked immunosorbent assay (ELISA)—using the same antibodies used for immunoblot—to quantify ubiquitin linkages on Flag- or GFP-tagged proteins of interest (Fig. 2). Here, we describe a protocol to compare K11, K48, and K63 ubiquitin linkages on Flag-tagged proteins of interest between different strains. The same

---

[j] The use of nonreducing LDS sample buffer (without DTT) helps avoid coelution of the Flag-antibody with the protein of interest. This is especially important if the protein of interest runs at a molecular weight similar to the light or heavy immunoglobulin chains. Addition of DTT after this step reduces the eluted samples and helps SDS-PAGE analysis.

[k] We find that PVDF provides much greater sensitivity than nitrocellulose, which is especially important for detection of ubiquitin "ladders."

[l] Ubiquitin itself is a very stable protein, and will not fully denature even after SDS-PAGE. This extra denaturation step is required to fully expose the necessary epitopes for antibody binding by immunoblot. Different protocols for denaturation have been described (e.g., 6 *M* guanidium hydrochloride, autoclaving, pouring boiling water). One of the most effective means we have found is to "stamp" a metal heat-block (set at 95°C) for 10 min onto the PVDF-filter paper sandwich (i.e., just removing the gel after transfer and keeping the rest of the sandwich intact) immersed in water.

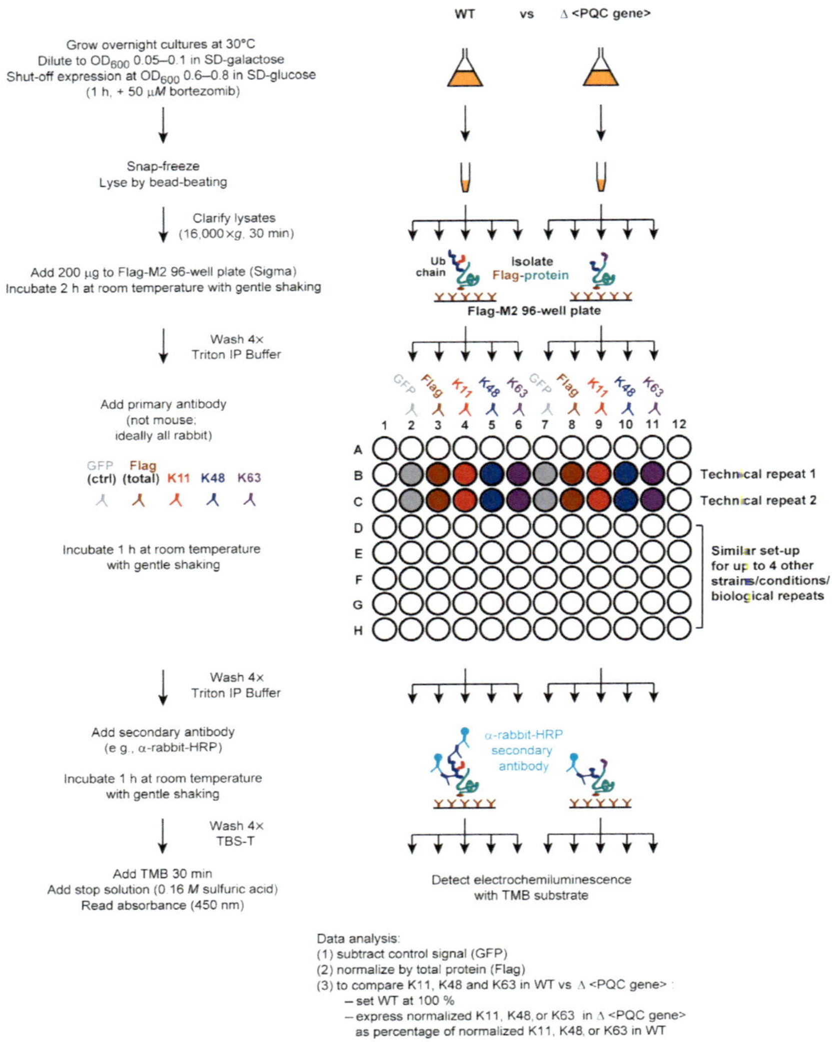

Fig. 2 Schematic for the ubiquitin linkage ELISA to compare relative levels of K11, K48, or K63 ubiquitin linkages in different strains expressing a Flag-tagged protein.

protocol can be followed for quantification of linkages on GFP-tagged proteins, except with the use of GFP-multiTrap 96-well plates (ChromoTek) instead of Flag-M2®-coated plates, and using the GFP and Flag signals as positive and negative controls, respectively.

i. Prepare cell lysates as described for "Immunoprecipitation of ubiquitinated misfolded proteins" (Section 4.1, steps i–v).

ii. Add 200 μg of lysate to each well of an anti-Flag-M2®-coated 96-well plate (Sigma catalog number P2983).[m] Use five wells for each technical replicate.[n] Incubate for 2 h at room temperature with gentle shaking, to allow binding of the Flag-tagged protein of interest to the antibody-coated plate.

iii. Wash 4 × with Triton IP Buffer to remove unbound protein.

iv. To each well, add an antibody to one of: GFP-negative control (1/1000; Cell Signaling Technology, catalog number 2956); Flag-positive control (1/1000; Cell Signaling Technology, catalog number 2368); K11-Ub (1/50; EMD Millipore, catalog number MABS107-I); K48-Ub (1/500; Cell Signaling, catalog number 4289); or K63-Ub (1/500; EMD Millipore, catalog number 05–1308).[o] Each antibody is diluted in 100 μL Triton IP buffer with 0.1% BSA. Incubate for 1 h at room temperature, with gentle shaking.

v. Wash 4 × with Triton IP Buffer to remove unbound primary antibody.

vi. Incubate for 1 h with HRP-conjugated antirabbit secondary antibody (1/2000; Jackson ImmunoResearch, catalog number 711-035-152) diluted in 100 μL Triton IP buffer with 0.1% BSA.

vii. Wash 4 × with Triton IP Buffer to remove unbound secondary antibody.

viii. Incubate for 30 min with 100 μL Pierce tetramethylbenzidine (TMB) substrate (Thermo Fisher Scientific).

ix. Stop the reaction by adding 100 μL 0.16 *M* sulfuric acid, and measure absorbance at 450 nm on a plate reader.

x. Subtract the negative control (GFP) reading from each of the other four readings.

xi. To calculate the K11, K48, and K63 ubiquitin signals for each strain, normalize to the total protein of interest levels by dividing each reading by the Flag reading for that sample. It is now possible to compare the specific ubiquitin linkage levels from one strain to another.[p]

---

[m] The amount of total protein that will saturate the plate needs to be determined prior to ubiquitin analysis. We ensure that the Flag-tagged protein in the lysate is well in excess of the binding capacity of the immobilized antibodies, to ensure that similar amounts of protein are being isolated in all the samples.

[n] There is some degree of variability in this assay; therefore, we ensure there are at least two technical replicates per sample.

[o] As the Flag-M2® antibody immobilized to the 96-well plate is isolated from mouse, it is important to use antibodies at this step that are from a different species. As most current ubiquitin linkage-specific antibodies are produced in rabbit, using rabbit antibodies for all conditions ensures that the same secondary antibody can be used for the next step of the ELISA.

[p] Note that this assay—much like any antibody-based assay—cannot be used to compare relative signals of different Ub linkages, given that antibodies have intrinsically different binding affinities and detection limits. Therefore, we limit this assay to comparison of signals for the same antibodies across different strains (e.g., wild-type vs deletion strains).

Acknowledgments

We thank members of the Frydman Lab past and present for development of many of the described techniques, including S. Escusa-Toret and E. M. Sontag for the puncta clearance assay, and K. C. Stein for advice on optimizing bead-beating lysis. We also thank A. Ordureau from J. W. Harper's lab (Harvard University, USA) for suggesting the use of the DUB inhibitor PR-619. R.S.S. was funded by a Human Frontier Science Program long-term fellowship (LT000695/2015-L). J.F. was funded by grants from the National Institutes of Health.

References

Alberti, S., Gitler, A. D., & Lindquist, S. (2007). A suite of gateway cloning vectors for high-throughput genetic analysis in Saccharomyces cerevisiae. *Yeast*, *24*(10), 913–919. https://doi.org/10.1002/yea.1502.

Amm, I., Sommer, T., & Wolf, D. H. (2014). Protein quality control and elimination of protein waste: The role of the ubiquitin-proteasome system. *Biochimica et Biophysica Acta*, *1843*(1), 182–196. https://doi.org/10.1016/j.bbamcr.2013.06.031.

Balchin, D., Hayer-Hartl, M., & Hartl, F. U. (2016). In vivo aspects of protein folding and quality control. *Science*, *353*(6294), aac4354. https://doi.org/10.1126/science.aac4354.

Bergman, L. W. (2001). Growth and maintenance of yeast. *Methods in Molecular Biology (Clifton, N.J.)*, *177*, 9–14. https://doi.org/10.1385/1-59259-210-4:009.

Bloom, J., & Pagano, M. (2005). Experimental tests to definitively determine ubiquitylation of a substrate. *Methods in Enzymology*, *399*, 249–266. https://doi.org/10.1016/S0076-6879(05)99017-4.

Botstein, D., & Fink, G. R. (2011). Yeast: An experimental organism for 21st century biology. *Genetics*, *189*(3), 695–704. https://doi.org/10.1534/genetics.111.130765.

Buchanan, B. W., Lloyd, M. E., Engle, S. M., & Rubenstein, E. M. (2016). Cycloheximide chase analysis of protein degradation in Saccharomyces cerevisiae. *Journal of Visualized Experiments*, (110), e53975. https://doi.org/10.3791/53975.

Curran, B. P. G., & Bugeja, V. (2014). Basic investigations in Saccharomyces cerevisiae. *Methods in Molecular Biology (Clifton, N.J.)*, *1163*, 1–14. https://doi.org/10.1007/978-1-4939-0799-1_1.

Emmerich, C. H., & Cohen, P. (2015). Optimising methods for the preservation, capture and identification of ubiquitin chains and ubiquitylated proteins by immunoblotting. *Biochemical and Biophysical Research Communications*, *466*(1), 1–14. https://doi.org/10.1016/j.bbrc.2015.08.109.

Escusa-Toret, S., Vonk, W. I. M., & Frydman, J. (2013). Spatial sequestration of misfolded proteins by a dynamic chaperone pathway enhances cellular fitness during stress. *Nature Cell Biology*, *15*(10), 1231–1243. https://doi.org/10.1038/ncb2838.

Garí, E., Piedrafita, L., Aldea, M., & Herrero, E. (1997). A set of vectors with a tetracycline-regulatable promoter system for modulated gene expression in Saccharomyces cerevisiae. *Yeast*, *13*(9), 837–848. https://doi.org/10.1002/(SICI)1097-0061(199707)13:9<837::AID-YEA145>3.0.CO;2-T.

Heck, J. W., Cheung, S. K., & Hampton, R. Y. (2010). Cytoplasmic protein quality control degradation mediated by parallel actions of the E3 ubiquitin ligases Ubr1 and San1. *Proceedings of the National Academy of Sciences of the United States of America*, *107*(3), 1106–1111. https://doi.org/10.1073/pnas.0910591107.

Horvath, A., & Riezman, H. (1994). Rapid protein extraction from Saccharomyces cerevisiae. *Yeast*, *10*(10), 1305–1310. https://doi.org/10.1002/yea.320101007.

Ingolia, N. T. (2010). Genome-wide translational profiling by ribosome footprinting. *Methods in Enzymology*, *470*, 119–142. https://doi.org/10.1016/S0076-6879(10)70006-9.

Kaganovich, D., Kopito, R., & Frydman, J. (2008). Misfolded proteins partition between two distinct quality control compartments. *Nature*, *454*(7208), 1088–1095. https://doi.org/10.1038/nature07195.

Katzmann, D. J., & Wendland, B. (2005). Analysis of ubiquitin-dependent protein sorting within the endocytic pathway in Saccharomyces cerevisiae. *Methods in Enzymology*, *399*, 192–211. https://doi.org/10.1016/S0076-6879(05)99013-7.

Kitamura, K., Kaneko, T., & Yamamoto, Y. (1971). Lysis of viable yeast cells by enzymes of Arthrobacter luteus. *Archives of Biochemistry and Biophysics*, *145*(1), 402–404. Retrieved from http://www.ncbi.nlm.nih.gov/pubmed/5166255.

Kwon, Y. T., & Ciechanover, A. (2017). The ubiquitin code in the ubiquitin-proteasome system and autophagy. *Trends in Biochemical Sciences*, *42*(11), 873–886. https://doi.org/10.1016/j.tibs.2017.09.002.

Macreadie, I. G., Jagadish, M. N., Azad, A. A., & Vaughan, P. R. (1989). Versatile cassettes designed for the copper inducible expression of proteins in yeast. *Plasmid*, *21*(2), 147–150. Retrieved from http://www.ncbi.nlm.nih.gov/pubmed/2500675.

Malinovska, L., Kroschwald, S., Munder, M. C., Richter, D., & Alberti, S. (2012). Molecular chaperones and stress-inducible protein-sorting factors coordinate the spatiotemporal distribution of protein aggregates. *Molecular Biology of the Cell*, *23*(16), 3041–3056. https://doi.org/10.1091/mbc.E12-03-0194.

Papanayotou, I., Sun, B., Roth, A. F., & Davis, N. G. (2010). Protein aggregation induced during glass bead lysis of yeast. *Yeast*, *27*(10), 801–816. https://doi.org/10.1002/yea.1771.

Park, S.-H., Kukushkin, Y., Gupta, R., Chen, T., Konagai, A., Hipp, M. S., et al. (2013). PolyQ proteins interfere with nuclear degradation of cytosolic proteins by sequestering the Sis1p chaperone. *Cell*, *154*(1), 134–145. https://doi.org/10.1016/j.cell.2013.06.003.

Rose, C. M., Isasa, M., Ordureau, A., Prado, M. A., Beausoleil, S. A., Jedrychowski, M. P., et al. (2016). Highly multiplexed quantitative mass spectrometry analysis of ubiquitylomes. *Cell Systems*, *3*(4), 395–403.e4. https://doi.org/10.1016/j.cels.2016.08.009.

Samant, R. S., Livingston, C. M., Sontag, E. M., & Frydman, J. (2018). Distinct proteostasis circuits cooperate in nuclear and cytoplasmic protein quality control. *Nature*. in press https://doi.org/10.1038/s41586-018-0678-x.

Schneider, K. L., Nyström, T., & Widlund, P. O. (2018). Studying spatial protein quality control, proteopathies, and aging using different model misfolding proteins in S. cerevisiae. *Frontiers in Molecular Neuroscience*, *11*(249). https://doi.org/10.3389/fnmol.2018.00249.

Simon, E., & Kornitzer, D. (2014). Pulse-chase analysis to measure protein degradation. *Methods in Enzymology*, *536*, 65–75. https://doi.org/10.1016/B978-0-12-420070-8.00006-4.

Sontag, E. M., Samant, R. S., & Frydman, J. (2017). Mechanisms and functions of spatial protein quality control. *Annual Review of Biochemistry*, *86*, 97–122.

Swatek, K. N., & Komander, D. (2016). Ubiquitin modifications. *Cell Research*, *26*(4), 399–422. https://doi.org/10.1038/cr.2016.39.

Wheeler, J. R., Jain, S., Khong, A., & Parker, R. (2017). Isolation of yeast and mammalian stress granule cores. *Methods*, *126*, 12–17. https://doi.org/10.1016/j.ymeth.2017.04.020.

Yau, R., & Rape, M. (2016). The increasing complexity of the ubiquitin code. *Nature Cell Biology*, *18*(6), 579–586. https://doi.org/10.1038/ncb3358.

CHAPTER THREE

Methodologies to monitor protein turnover at the inner nuclear membrane

Pei-Ling Tsai[a], Chenguang Zhao[a], Christian Schlieker[a,b,*]
[a]Department of Molecular Biophysics and Biochemistry, Yale University, New Haven, CT, United States
[b]Department of Cell Biology, Yale School of Medicine, New Haven, CT, United States
*Corresponding author: e-mail address: christian.schlieker@yale.edu

Contents

Abstract

Lamin B receptor (LBR) is an inner nuclear membrane protein that associates with the nuclear lamina and harbors sterol reductase activity essential for cholesterol biosynthesis. Several LBR mutations implicated in human congenital disorders give rise to C-terminal truncations which render LBR metabolically unstable, resulting in their rapid turnover in the nucleus. These LBR variants serve as model substrates for investigating the poorly understood protein quality control pathways in the mammalian nuclear envelope (NE). Here we describe a split-GFP-based method for tagging these model substrates to enable live cell imaging and flow cytometry for the identification and characterization of NE-resident protein turnover machinery. Furthermore, we describe a facile subcellular fractionation method to isolate a soluble LBR degradation intermediate, allowing the deconvolution of the membrane extraction and proteasomal turnover steps. The combination of imaging-based and biochemical approaches described here facilitates detailed mechanistic studies to dissect protein turnover in the nuclear compartment.

Methods in Enzymology, Volume 619
ISSN 0076-6879
https://doi.org/10.1016/bs.mie.2018.12.033
© 2019 Elsevier Inc.
All rights reserved.

1. Introduction

Lamin B receptor (LBR) is a multifunctional integral membrane protein residing at the inner nuclear membrane (INM) (Olins, Rhodes, Welch, Zwerger, & Olins, 2010). The N-terminal portion of LBR associates with the nuclear lamina (Worman, Yuan, Blobel, & Georgatos, 1988; Ye & Worman, 1994) and contains a Tudor-domain that interacts with heterochromatin (Makatsori et al., 2004; Pyrpasopoulou, Meier, Maison, Simos, & Georgatos, 1996). The polytopic C-terminal domain exhibits sterol C14 reductases activity that catalyzes a reduction step in the cholesterol biosynthesis pathway (Li, Roberti, & Blobel, 2015; Prakash, Sengupta, Aparna, & Kasbekar, 1999; Silve, Dupuy, Ferrara, & Loison, 1998) (for recent reviews, see Kasbekar, 2018; Nikolakaki, Mylonis, & Giannakouros, 2017). Two human congenital diseases, Pelger-Huët anomaly and Greenberg skeletal dysplasia, are known to be associated with mutations in LBR, which cause pathologies ranging from hypolobulated granulocyte nuclei to embryonic lethality depending on the allelic state and nature of the specific mutation (Oosterwijk et al., 2003; Turner & Schlieker, 2016). So far, all LBR mutations implicated in these diseases have been mapped to its C-terminal domain, and we recently found that the disease-causing mutations indeed severely perturb the C14 sterol reductase activity embedded within LBR (Tsai, Zhao, Turner, & Schlieker, 2016).

Notably, two of these mutant LBR variants with C-terminal truncations, LBR1600* and LBR1402TΔ, are still targeted to the INM but are then rapidly degraded via the ubiquitin-proteasome system in the nucleus, suggesting that their degradation is distinct from the canonical endoplasmic reticulum (ER)-associated degradation (ERAD) pathway (Tsai et al., 2016). The degradation of these truncated LBR variants is dependent on the proteasome and p97, an AAA+ (ATPases associated with various cellular activities) ATPase that is also responsible for dislocating misfolded proteins from the ER membrane (Ye, Meyer, & Rapoport, 2001). In the presence of the proteasome inhibitor MG132, LBR1600* is stabilized and accumulates in the nucleoplasm. In contrast, a p97 dominant-negative mutant, p97QQ (Ye et al., 2001), or the p97 inhibitor CB-5083 (Anderson et al., 2015) prevents LBR1600* from dislocating from the nuclear membrane (Fig. 1) (Tsai et al., 2016). These properties provide us with a useful toolkit to probe for activities involved in LBR1600 degradation via assessing the localization changes of LBR1600* upon depleting the components of the degradation pathway.

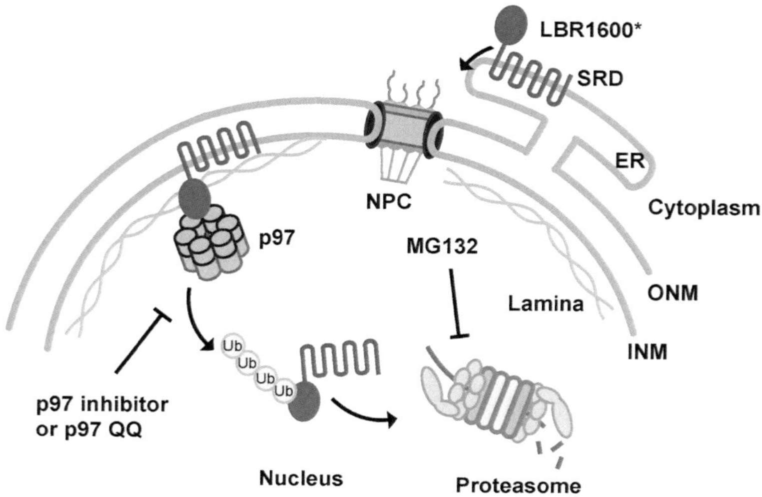

Fig. 1 Proposed inner nuclear membrane protein turnover pathway in mammalian cells. LBR1600* is a C-terminal truncated version of LBR, which is synthesized into the ER, diffuses throughout the ER membrane, and passes through the pore membrane, where it is retained at the INM via interaction with nuclear lamina through its N-terminal domain. The misfolded C-terminal sterol reductase domain leads to its poly-ubiquitylation by a yet-unknown ubiquitin ligase, followed by membrane dislocation/extraction, and ultimately to LBR1600* degradation via proteasome. Protein degradation at the NE can be blocked at the step of membrane dislocation or degradation via a dominant-negative mutant p97QQ or a p97 inhibitor (CB-5083), or after dislocation by the proteasome inhibitor MG132. A small portion of LBR1600* might also be degraded by the canonical ER-associated degradation pathway before reaching the inner nuclear membrane. For clarity, several ERAD components are omitted in this diagram. *ER*, endoplasmic reticulum; *INM*, inner nuclear membrane; *NPC*, nuclear pore complex; *ONM*, outer nuclear membrane; *SRD*, sterol reductase domain.

Currently, the mechanisms responsible for protein homeostasis at the nuclear envelope (NE) of mammalian cells remain poorly understood (Rose & Schlieker, 2012; Shibata & Morimoto, 2014; Ungricht & Kutay, 2017; Webster & Lusk, 2016), and the absence of suitable model substrates in higher eukaryotes has been one of the major limitations toward identifying and characterizing NE-resident turnover machinery. Protein quality control mechanisms at the NE have been identified in yeast (Deng & Hochstrasser, 2006; Foresti, Rodriguez-Vaello, Funaya, & Carvalho, 2014; Khmelinskii et al., 2014) (for a recent review, see Mehrtash & Hochstrasser, 2019); however, many of these components (e.g., the Asi ubiquitin ligase) do not have corresponding orthologs identified in human cells. Given that the truncated LBR variants have very short half-lives and the degradation can be arrested at either the membrane dislocation or degradation steps via dominant-negative constructs or suitable small molecule inhibitors (Fig. 1), we now can utilize these LBR disease variants as model substrates for illuminating the mechanisms of protein quality control at the mammalian NE.

In this chapter, we describe the use of a "split-GFP" system to tag the LBR1600* model substrate, followed by immunofluorescence, pulse-chase analysis, flow cytometry, and biochemical fractionation of cell homogenates as tools to monitor INM protein turnover.

2. Tagging strategy for INM proteins

INM proteins are synthesized at the ER, diffuse through the ER membrane, pass through the pore membrane, and anchor at the INM via association with chromatin and nuclear lamina (Boni et al., 2015; Ungricht, Klann, Horvath, & Kutay, 2015; Ungricht & Kutay, 2015). Tagging INM proteins can be challenging, since the fused fluorescent protein may cause protein misfolding or mislocalization, particularly when the size of the nuclear domain exceeds 60kDa, which causes the protein to become incapable of diffusing through the pore membrane due to constrictions imposed by the nuclear pore complex, resulting in ER retention (Boni et al., 2015; Ohba, Schirmer, Nishimoto, & Gerace, 2004; Ungricht et al., 2015). We previously demonstrated that truncated variants of LBR are degraded rapidly in the nucleus (Tsai et al., 2016). To study the turnover mechanism of LBR1600* without disrupting its localization and the turnover rate, we here employ the split-GFP system that has been intensely used in visualizing subcellular protein localization (Cabantous, Terwilliger, & Waldo, 2005; Feng et al., 2017; Kamiyama et al., 2016; Van Engelenburg & Palmer, 2010). In this system, GFP (green fluorescent protein) is split into two fragments composed of a small fragment of GFP, S11 (β-strand 11, amino acids 215–230) and a large fragment of GFP, GFP1–10 (β-strands 1–10, amino acids 1–214). Neither of these individual fragments exhibit fluorescence, but upon association, these fragments spontaneously mature into the active fluorescent protein (Cabantous et al., 2005).

The advantage of using the split-GFP system is that the small fragment of GFP is only 16 amino acids in length, which does not interfere with INM targeting of LBR1600* (Fig. 2A and B), in contrast to fusion with intact GFP, which results in ER retention (data not shown). The tagging strategy we used was to fuse the small fragment of GFP, S11, to LBR1600* and supply the cell line with the complementary fragment, GFP1–10 (Fig. 2A).

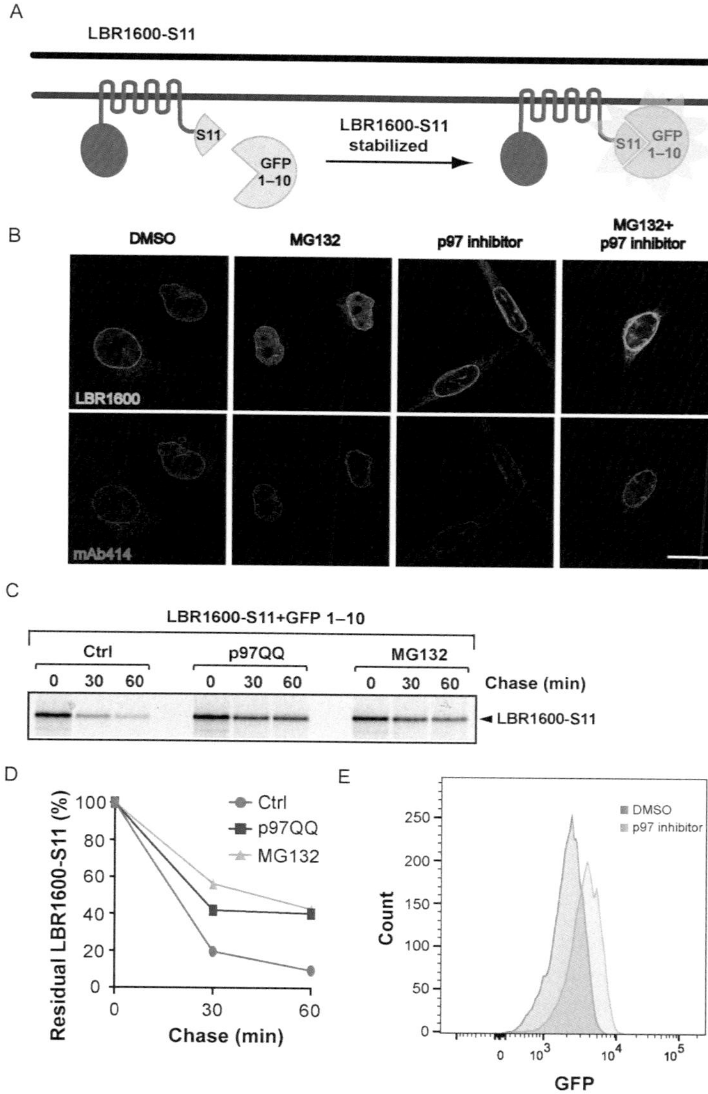

Fig. 2 Imaging approaches to monitor INM/LBR1600* turnover. (A) A split-GFP system to monitor LBR1600* stability. LBR1600-S11 is comprised of LBR1600* and the C-terminal fragment S11 of GFP (amino acids 215–230). A stable LBR KO-derived cell line, termed LBR1600$^{S11/GFP1-10}$, was engineered to uniformly express both LBR1600-S11 and the large GFP fragment GFP1–10 (amino acids 1–214). (B) LBR1600$^{S11/GFP1-10}$ cells were treated with DMSO, 10 μM MG132, or 10 μM p97 inhibitor individually, or a combination of 10 μM MG132 and 10 μM p97 inhibitor for 4 h, stained with anti-LBR and mAb414 (recognizes several FG-nucleoporins and serves as NE marker), and imaged via confocal microscopy. Scale bar = 20 μm. (C) Pulse-chase analysis of LBR1600$^{S11/GFP1-10}$ cells transfected with empty vector or p97QQ, or treated with 10 μM MG132 during the chase period. (D) Densitometric quantification of the pulse-chase experiment depicted in (C). (E) LBR1600$^{S11/GFP1-10}$ cells were treated with DMSO or 10 μM p97 inhibitor for 4 h and analyzed using flow cytometry. Note that a subpopulation of cells exhibits higher GFP signal upon p97 inhibition indicating the LBR1600*S11 was stabilized.

LBR1600*-S11 has a rapid turnover rate (Fig. 2C) that faithfully mimics previously observed turnover kinetics of the naturally occurring disease allele (Tsai et al., 2016), resulting in a low GFP signal; however, if LBR1600*-S11 turnover is inhibited (e.g., by proteasomal inhibition via MG132, cf. Fig. 2C) more GFP molecules are assembled into active forms, leading to an increased signal intensity (Fig. 2A and E).

2.1 Construct design and stable cell line generation

2.1.1 Cloning

To establish a cell line stably expressing the LBR1600*-S11 fusion and the complementary fragment of GFP, GFP1–10, we used a retroviral vector to carry gene expression constructs (Retro-X Tet-On advanced inducible expression system, Takara Bio). The sequence of the small fragment GFP S11 was fused to LBR1600* using PCR. The reverse primer was composed of an EcoRI site, a linker encoding the GDGGSGGGS peptide, and amino acids 215–230 of GFP (RDHMVLHEYVNAAGIT); the forward primer carried a BamHI site. The sequence of GFP1–10 (amino acids 1–214) was cloned from the pCMV-mGFP 1–10 plasmid (Sandia Biotech) by PCR with a forward primer containing a BamHI site and the SV40 nuclear localization sequence PKKKRKV together with a reverse primer carrying an EcoRI site. The resulting PCR products were then subcloned into the pRetroX-Tight-Pur-GOI vector (Takara Bio) after restriction enzyme digestion.

2.1.1.1 Notes

The GFP S11 fragment can be fused to the N- or C-terminus of a protein. When tagging a new protein, one might need to try both ends and optimize the length of the linker sequence in order to maximize accessibility to GFP1–10. In some cases, protein folding or aggregation might cause the inaccessibility of the S11 tag, preventing GFP complementation. Also, GFP S11 appended to a protein terminus that lies in the ER-lumen will be inaccessible to GFP1–10 unless the protein is retrotranslocated into the cytosol or nucleoplasm. In general, we found it to be critical that LBR1600* derivatives are expressed at low to moderate levels since transient transfections with high-expression CMV-based promoters can overburden the INM-resident turnover machinery or lead to an undesired accumulation of the fusion protein at the ER. Therefore, the generation of stable cell lines is generally recommended. In any case, it is critical to confirm the authentic NE localization of the fusion construct via microscopy (cf. Section 2.2).

2.1.2 Virus transduction and single cell selection

The detailed procedures of retrovirus production were performed using the Retro-X Tet-On system (Takara Bio) following the manufacturer's instructions. This system establishes a stable cell line in which target gene expression is regulated by the presence of the antibiotic doxycycline. Below is a summary of the modified procedures.

2.1.2.1 Buffers and reagents

- Growth medium: Dulbecco's Modified Eagle Medium (DMEM) + 10% fetal bovine serum + 100 units/mL penicillin and 100 μg/mL streptomycin (Thermo Fisher Scientific)
- 0.05% Trypsin-EDTA (Thermo Fisher Scientific)
- Phosphate-buffered saline (PBS) (Thermo Fisher Scientific)
- Puromycin dihydrochloride (MilliporeSigma)
- Geneticin selective antibiotic (G418 sulfate) (Thermo Fisher Scientific)

2.1.2.2 Equipment

- CO_2 incubator (37°C, 5% CO_2)
- Class II Laminar Flow Hood designed for BSL-2 (SterilGARD, Baker or similar)

2.1.2.3 Procedure

1. Seed low-passage 293T cells in three 10-cm plates with a density of 3.5×10^6 cells per plate the day before transfection.
2. On the next day, replace the medium with 6 mL fresh medium and transfect each plate with three plasmids encoding: MMLV gag/pol (2 μg), viral envelope protein VSV-G (1 μg), and either pRetroX-Tight-Pur-LBR1600*-S11 (6 μg), pRetroX-Tight-Pur-GFP1–10 (6 μg), or pRetroX-Tet-On (6 μg). Transfection reagent X-tremeGENE 9 (Roche) is used at a 3:1 ratio to DNA, 3 μL transfection reagent to 1 μg plasmid DNA.

   Note: For producing retroviral particles, three plasmid components are required: (1) A packaging vector encoding MMLV proteins gag and pol. Gag is a polyprotein further processed into matrix and other viral core proteins. Pol is the reverse transcriptase. (2) An envelope vector specifying VSV-G. VSV-G mediated viral entry via membrane fusion does not depend on proteinaceous cell surface receptors; therefore, VSVG-enveloped retroviruses can be used to transduce a broad range of cell types. (3) A transfer vector encoding the transgene of interest.

3. After 72 h, harvest supernatants containing the three respective retroviruses carrying genes encoding LBR1600*-S11, GFP1–10, and a mutant Tet repressor. Filter each through a 0.45-μm filter unit, aliquot into screw-cap tubes, and store at −80°C. These cryostocks are stable for at least 12 months.

   Note: Ideally, use a multiplicity of infection (MOI) between 0.5 and 1 for retroviral transduction. Retrovirus titers can be determined using commercial qRT-PCR kits (Takara Bio or similar) or colony formation; however, it takes 7–14 days for colony formation to work. An alternative and faster way to achieve adequate ratios of virus per host cell is to add a serial dilution of viral solution by adding 50–200 μL into a 6-well plate with a cell density of 2×10^5 cells/well (adjust the volume for all wells to 2 mL with growth medium). Forty-eight hours post-transduction, apply 0.75 μg/mL puromycin, incubate for 48 h, and observe the cell death of each well under a light microscope. If there is no observable cell death in a well, this indicates that each cell is infected with at least one or multiple viruses. It is important to include a negative control where no virus is added, which should result in near-complete cell death. For low copy insertions, use at most the smallest virus volume necessary to achieve complete survival, or the next lowest dilution of the series.
4. Seed the targeted cells (e.g., HeLa) in a 6-well plate at a density of 2×10^5 per well 1 day before transduction.
5. On the next day, replace the medium with 2 mL fresh medium supplemented with 4 μg/mL of the cationic polymer polybrene, which is used to neutralize the electrostatic repulsion between the virus and the cell surface (Davis, Rosinski, Morgan, & Yarmush, 2004). Add 100 μL of each retrovirus solution made in step 3 into the same well, shake the plate gently and incubate for 48 h.
6. Aspirate the medium, wash the cells with PBS once, add 0.5 mL of 0.05% trypsin-EDTA to cover the cells, incubate at 37°C for 5 min, and then monitor the cells under a microscope to ensure that cells are rounded up and detached. Add 1 mL of growth medium that includes 10% fetal bovine serum to neutralize the trypsin, pipet the solution five times and transfer them into a 10-cm plate with the medium supplemented with 0.75 μg/mL puromycin and 800 μg/mL G418 for selecting cells expressing the gene of interest and the Tet repressor. Incubate for 3 days.
7. Trypsinize the cells as described in step 6 and seed them into several 10-cm plates at very low densities, such as 25, 50, 100, and 200 cells/plate.

Supplement the medium with 800 μg/mL G418. Replace the medium every 3 days until the single cell-derived colonies form. This usually takes ~10 days. Puromycin usually takes 3–4 days to kill all nontransduced cells, and G418 takes roughly 10–14 days.

8. Mark the colony position at the bottom of the plate with a marker pen and prepare a 96-well plate filled with 20 μL of 0.05% trypsin-EDTA in each well.
9. Gently wash the cells in the plate with 5 mL PBS twice to remove unattached cells and overlay the plate with 2 mL PBS. Set a P20 pipet to 6 μL and directly pick up the target colony with the pipet tip and pipet into a well of the 96-well plate. Mark the well having cells with marker pen and repeat the steps until ~20 colonies are picked up. After a 5-min incubation, add 80 μL of growth medium into each well to neutralize the trypsin and wait for 3 days.
10. When the cells are confluent, trypsinize the cells with 30 μL of 0.05% trypsin-EDTA, neutralize them with 70 μL of growth medium, and transfer 30% of cells from each well into a well of 24-well plate that includes a coverslip and 0.5 mL growth medium supplemented with 0.5 μg/mL doxycycline to induce gene expression.
11. Transfer the remaining cells of each well into a duplicate 24-well plate for passaging.
12. After 24 h doxycycline induction, treat cells with either vehicle (DMSO) or 10 μ*M* MG132. After 4 h, fix cells with 4% paraformaldehyde, mount, and image GFP using a fluorescence microscope. Colonies with enhanced GFP can be propagated from the duplicate plate and used for future experiments.

2.1.2.4 Notes

1. The retrovirus using VSV-G as an envelope protein can infect human cells. This handling therefore requires a BioSafety Level 2 facility (BSL2); researchers need to acquire proper training and equipment before starting these experiments.
2. Different cell types require different concentrations of antibiotics and durations for selection. Therefore, the optimal concentration of puromycin and G418 need to be titrated before step 6. A serial dilution of puromycin ranging from 0.5 to 2 μg/mL and G418 ranging from 200 to 800 μg/mL can be applied to nontransduced cells. The minimum concentration capable of killing all cells in 3 days (puromycin) and 10 days (G418) was selected for use. Avoid applying very high concentrations of antibiotics for selection, which will select only the cells with high copy number integration.

3. An alternative way to select the cells with proper GFP signal intensity is using flow cytometry to directly sort out the single cells into a 96-well plate. For weakly adherent cells (e.g., 293T), washing cells with PBS might disturb the colonies; single cell sorting might be a better option for this type of cell line.
4. It is recommended to prepare permanent cryostocks of stable cell lines as soon as the functionality of the transgene is established as judged by NE localization (cf. Section 2.2, Fig. 2B) and pulse-chase analysis (cf. Section 2.3, Fig. 2C and D).

2.2 Immunofluorescence analysis of tagged INM proteins

Immunofluorescence analysis is a very straightforward way to examine protein localization and to assess expression levels to ensure the validity of the generated cell lines for future experimentation. More importantly, we can exploit this technique in conjunction with transfections of dominant-negative alleles or pharmacological inhibitors to evaluate individual steps of protein turnover at the NE and nucleoplasm. For example, in the presence of the proteasome inhibitor MG132, the degradation intermediate of LBR1600* accumulates in the nucleoplasm after its dislocation from the INM (Tsai et al., 2016) (Fig. 2B). This characteristic provides a useful tool to probe the protein turnover pathways at the nucleus via examining the localization changes of LBR1600* in the presence of MG132 upon depletion or inhibition of other components of the turnover pathway. This "epistasis strategy" also allows the experimentalist to place a newly identified component involved in INM turnover at a specific point in the pathway. For example, both the p97 inhibitor CB-5083 and a dominant-negative p97 variant (delivered by transfection) inhibit INM turnover but also result in a blockage of the membrane dislocation step as judged by the lack of nucleoplasmic accumulation in the presence of MG132 and the concomitant accumulation of the model substrate at the INM. This allowed us to assign p97 to a role in dislocating LBR1600* from the INM prior to proteasomal turnover (Tsai et al., 2016) (Figs. 1 and 2B). On the other hand, any newly identified component acting downstream of the INM dislocation step would be predicted to lead to a nucleoplasmic accumulation even in the absence of MG132, but without a buildup of the model substrate at the INM.

2.2.1 Buffers and reagents

- MG132 (MilliporeSigma)
- p97 inhibitor (CB-5083) (Selleckchem)

- Primary antibodies: anti-LBR (ab32535, Abcam); anti-nuclear pore complex proteins (mAb414) (ab24609, Abcam)
- Alexa Fluor secondary antibodies (Thermo Fisher Scientific)
- Fluoromount-G (Thermo Fisher Scientific)
- PBS
- Blocking solution: 5 mg/mL bovine serum albumin in PBS (MilliporeSigma)
- Hoechst 33342 (Thermo Fisher Scientific): 2.5 μg/mL in PBS
- Triton X-100 (MilliporeSigma)
- 16% Paraformaldehyde (Electron Microscopy Sciences)

2.2.2 Equipment

- Confocal microscope (Zeiss or similar).

2.2.3 Procedure

1. To assess if tagging with GFP S11 alters the localization of LBR1600*, place 12-mm round coverslips into a 12-well plate, seed 6×10^4 cells in each well and leave the cells to adhere on coverslips overnight. On the next day, treat cells with DMSO, 10 μ*M* MG132, 10 μ*M* p97 inhibitor, or MG132 plus p97 inhibitor for 4 h.
2. Fix the cells with 4% paraformaldehyde in PBS at room temperature for 15 min and wash with PBS 3 times. Note that the wash here means adding the PBS into the well, aspirating it, and repeating this. The washing requires no incubation time.
3. Permeabilize the cells with 0.2% Triton X-100 at room temperature for 5 min, and wash with PBS 3 times.
4. Create a humidified chamber by placing a wet filter paper in a 15 cm dish and put a piece of parafilm on top of the wet filter paper. Pipet 15 μL of BSA blocking solution for each coverslip on the parafilm and invert the coverslips bearing the fixed cells onto the drops of blocking solution. Incubate for 10 min. Note that the amount 15 μL is for a 12-mm round coverslip. If larger coverslips are used, the amount of reagent should be scaled up accordingly.
5. Pipet 15 μL of primary antibodies prepared in blocking solution onto the parafilm. Blot the edge of coverslips against kimwipes to remove excess solution and invert the coverslip onto the drop of primary antibodies. Incubate for 1 h at room temperature.
6. Dip the coverslips sequentially through three beakers filled with 150 mL PBS for a quick rinse and blot the edge of coverslips on kimwipes to remove buffer.

7. Pipet 15 μL of secondary antibodies prepared in blocking solution to the parafilm. Invert the washed coverslips onto the drop of Alexa Fluor-conjugated secondary antibody and incubate for 1 h in the dark or cover the dish with aluminum foil.
8. Repeat the washing described in step 6, and invert the coverslip onto a drop of Hoechst to stain DNA, incubate for 5 min in the dark.
9. Repeat the washing as in step 6.
10. Pipet 15 μL of Fluoromount-G on a microscope slide, and invert the coverslip from step 9 onto the drop of mounting media.
11. Absorb excess mounting media with filter paper and leave it to dry in the dark. After the slide is dry, seal the coverslip edges with clear nail polish.

2.2.3.1 Notes

1. Alexa Fluor and Hoechst dye are light sensitive and slides should therefore be stored in the dark.
2. Tagging the small GFP fragment S11 to LBR1600* does not alter its distribution. Note that the LBR1600*-S11 shows accumulation in nucleoplasm upon proteasome inhibition (Fig. 2B), which is similar to the untagged LBR1600* described in our previous study (Tsai et al., 2016).
3. An improved split-GFP system with lower background fluorescence was recently reported (Feng et al., 2017) and can likely be engineered analogously for live cell imaging applications. N-terminal tagging of LBR1600* with commonly used epitope tags is sufficient for microscopy using fixed samples (cf. Fig. 2B) and biochemical analysis (Sections 2.3 and 3). The use of a hemagglutinin (HA) tag is recommended, since the Flag tag slightly alters the LBR1600* localization, which might be attributable to charge repulsion between DNA and the negative net charge of the Flag tag.

2.3 Pulse-chase experiment for protein turnover analysis

This protocol relies on pulse-chase labeling using metabolic labeling with ^{35}S-Cys/Met and is adapted from previously described methods (Ernst, Mueller, Ploegh, & Schlieker, 2009). Together with SDS-PAGE and autoradiography, this method allows for the direct measurement of protein stability in a cellular setting. For laboratories not authorized for the use of radioactivity, cycloheximide-chase assays and immunoblotting are an alternative (cf. Hwang et al., 2017).

2.3.1 Buffers and reagents

- Starvation medium (100 mL Eagle's medium (D0422, Sigma), 1 mL 100 × penicillin/streptomycin (10,000 units/mL penicillin and 10,000 μg/mL streptomycin, 1 mL 100 × L-glutamine (200 m*M*), 10 mL dialyzed FBS)
- ^{35}S protein-labeling mix (NEG072007MC, Perkin Elmer)
- Recovery medium (3 mL normal cell culture medium with 30 μL of 100 × L-Cys and L-Met (50 m*M* L-Cys and 250 m*M* L-Met)
- Standard SDS-PAGE equipment
- Benzonase (MilliporeSigma)
- NET buffer (50 m*M* Tris–HCl, pH 7.5, 75 m*M* NaCl, 5 m*M* $MgCl_2$, and 0.5% Nonidet P-40)
- Dimethyl sulfoxide (DMSO)
- 5% (wt/vol) PPO (2,5-diphenyloxazole from MilliporeSigma) in DMSO
- Trichloroacetic acid (TCA)

2.3.2 Equipment

- Thermomixer (Eppendorf)
- Benchtop centrifuge (Eppendorf)
- SDS-PAGE equipment (Dan-Kar or similar)
- Power supply (Bio-Rad)
- Gel dryer (Bio-Rad)
- Stainless steel trays
- Large film cassettes (Thermo Fisher Scientific)
- Storm 860 Molecular Imager (GE Healthcare) or similar
- Imaging plates (Fujifilm)
- Carestream Kodak BioMax MR film (35 × 43 cm, MilliporeSigma)

2.3.3 Procedure

1. Seed HeLa cells in 10-cm plates at a density of 1.2×10^6 per plate the day before transfection.
2. On the next day, transfect the cells with plasmids encoding the gene of interest. Incubate overnight.
3. Trypsinize the cells, transfer them into 50-mL Falcon tubes, and spin at 800 × *g* for 3 min.
4. Resuspend cells in 10 mL starvation medium and centrifuge at 800 × *g* for 3 min.
5. Resuspend cells in 3 mL starvation medium and incubate at 37°C for 30 min. Gently swirl the tube every 5–10 min to keep the cells from settling.

6. Meanwhile, thaw ^{35}S protein-labeling mix in a fume hood, label 2-mL tubes for all samples and time points, add 0.5 mL ice-cold PBS to each tube, and prepare recovery medium.
7. Spin down cells at 800 × *g* for 3 min and resuspend cells with 300 μL starvation medium.
8. Add 30 μL of ^{35}S protein-labeling mix to the cells in the fume hood using filter tips, and incubate cells on a programmed Thermo-mixer (pulse shake (1 min off/4 s on) at 500 rpm) for 10 min.
9. Stop labeling reaction by addition of 3 mL of recovery medium to each tube.
10. Transfer 900 μL cell suspension to a 2-mL tube with 500 μL ice-cold PBS and spin down cells at 800 × *g* for 3 min at 4°C.
11. Wash cells with 1 mL cold PBS, spin down cells at 800 × *g* for 3 min at 4°C, aspirate PBS, and freeze the cell pellets at −80°C (time point 0 h).
12. Take aliquots after 0.5 and 1 h, repeat the PBS wash described in step 11 when harvesting each aliquot, and store them at −80°C.
13. After all time points are collected, add 100 μL of 1% SDS/PBS to cell pellets and vortex tubes vigorously five times with a quick pulse for 5 s each time; incubate samples at 50°C for 15 min.
14. Cool down samples to room temperature, briefly spin the tubes at 10,000 × *g* for 30 s to collect droplets on tube walls and add 4 μL of diluted benzonase (2 μL benzonase stock (500U) in 50 μL PBS) into each tube. Vortex tubes and incubate at room temperature for 20 min to eliminate DNA.
15. Inactivate benzonase by putting samples at 100°C for 2–3 min, briefly spin down the lysate, add 1 mL of NET buffer to tubes, vortex, and centrifuge at 16,000 × *g* at 4°C for 10 min.
16. Determine counts per minute (cpm) of the zero time point ($t = 0$) of each time series. *Note*: since later samples of each time series will be standardized relative to the zero time point ($t = 0$) of the corresponding series, this step 16 only needs to be performed on the $t = 0$ samples.
    a. Using a pencil, mark out a long piece of Whatman paper with enough 1.5 × 1.5 cm squares to accommodate as many time-zero time points as present. Only the time-zero time point of each set is measured. Spot 10 μL of each $t = 0$ lysate to the center of each square and leave it to air dry in the fume hood for 10 min.
    b. Immerse the paper in 5% TCA in a plastic container big enough to accommodate the whole piece of the Whatman paper for 10 min.
    c. Immerse the paper in 100% ethanol for 1 min.

   d. Immerse the paper in acetone for 1 min and let it dry in the fume hood for 10 min.
   e. Cut out each square with a scissor and place it onto the bottom of a scintillation vial.
   f. Add 5 mL scintillation solution and count the radioactivity (1–3 million cpm would be a typical range).
17. Standardize the volume of $t=0$ samples to the one with the lowest cpm by taking the necessary volume to give equal cpm for all samples. Continue to use the same volume for other time points of each series. Example: if there are two sets of samples A and B, each of them includes three time points (0, 0.5, and 1 h). Measure the cpm of $t=0$ samples in both A and B as described earlier. If $t=0$ of A has a reading of 2 million cpm and B has a reading of 2.5 million cpm, transfer 900 μL of A into a new 1.5-mL tube. The residual volume (~100 μL or more) will be disposed along with the old tube to avoid transfer of the pellet material into the immunoprecipitation step, which would increase the background. Once the volume of all $t=0$ samples is determined, we take the same volume for other time points in the same set. In this example case, we transfer 900 μL lysates from each time point in series A to a new tube and transfer 720 μL from each sample in the B series to a new tube, respectively. Adjust all volumes to 1 mL with NET buffer/0.1% SDS.
18. Preclear lysates of proteins that bind directly to protein A by incubating each tube with 30 μL protein-A beads (equilibrated with NET buffer/0.1% SDS). Rock at 4°C for 1 h and then sediment the beads by centrifugation at $2400 \times g$ for 1 min.
19. Transfer 950 μL supernatant to a new tube and start immunoprecipitation with specific antibody at the desired dilution and 30 μL protein-A beads.
20. After 3 h or overnight incubation, wash beads with 1.3 mL NET buffer/0.1% SDS 3 times.
21. Aspirate the buffer with a fine needle after last wash, add 65 μL of 1 × SDS sample buffer, and incubate at 65°C for 5 min. Spin for 1 min and load the supernatant onto an SDS-PAGE gel.

2.3.4 SDS-PAGE and autoradiography

The detailed protocol for gel development and fluorography was previously described (Ploegh, 2001).

1. Run an SDS-PAGE gel (32 cm × 25 cm × 1 mm, L × W × T) overnight at 70–90 V in 1 × Laemmli Resolving Gel Buffer.

2. Carefully transfer gel into a stainless steel tray containing DMSO and incubate for at least 30 min with gentle shaking.
3. Change to refresh DMSO and shake for another 30 min.
4. Change to 5% (w/v) PPO in DMSO and shake for 1 h.
5. Wash the gel 4 times with water for 3 min each time.
6. Transfer the gel carefully onto filter paper in a gel dryer and dry the gel at 80°C for 2 h.
7. Transfer the dried gel into a large film cassette and develop with a film at −80°C or an imaging plate at room temperature.

2.3.4.1 Notes

1. A typical pulse-chase result is provided in Fig. 2C and D. In the presence of p97QQ or MG132 inhibitor, LBR1600*-S11 was stabilized as expected. Notice that tagging S11 to LBR1600* and expressing GFP1–10 in the system does not interfere with the turnover of LBR1600* as evidenced by the half-life of LBR1600*-S11 (about 10–15 min), which is very similar to the naturally occurring disease allele (LBR1600*) determined previously using anti-LBR antibodies (Tsai et al., 2016).
2. The spilt-GFP system can also be monitored by FACS analysis (Fig. 2E), which can be exploited to seed cells in single cell format to generate clonal cell lines. Moreover, FACS separation can be applied as a readout in the context of genetic screens (cf. Timms et al., 2016).
3. The PPO step can be omitted if gels are not imaged using photographic films (e.g., by using imaging plates).

3. Fractionation of nuclei into NE and nucleoplasm

This protocol was developed to rapidly separate nucleoplasm from the NE membrane fraction to establish a biochemical readout for the generation of a dislocated or "soluble" LBR1600* degradation intermediate previously observed by light microscopy (Tsai et al., 2016) (cf. Fig. 2B, second panel). This procedure is loosely based on a method introduced in Emig, Schmalz, Shakibaei, and Buchner (1995) with several modifications to increase the speed of the fractionation procedure and to avoid the use of an ultracentrifuge. In brief, cells are homogenized by repeated passing through a needle and subjected to a low-speed centrifugation over a sucrose cushion to pellet the nuclei. Isolated nuclei are then subjected to heparin/nuclease treatment and centrifugation, which results in the release of nucleoplasm into the supernatant (Fig. 3A).

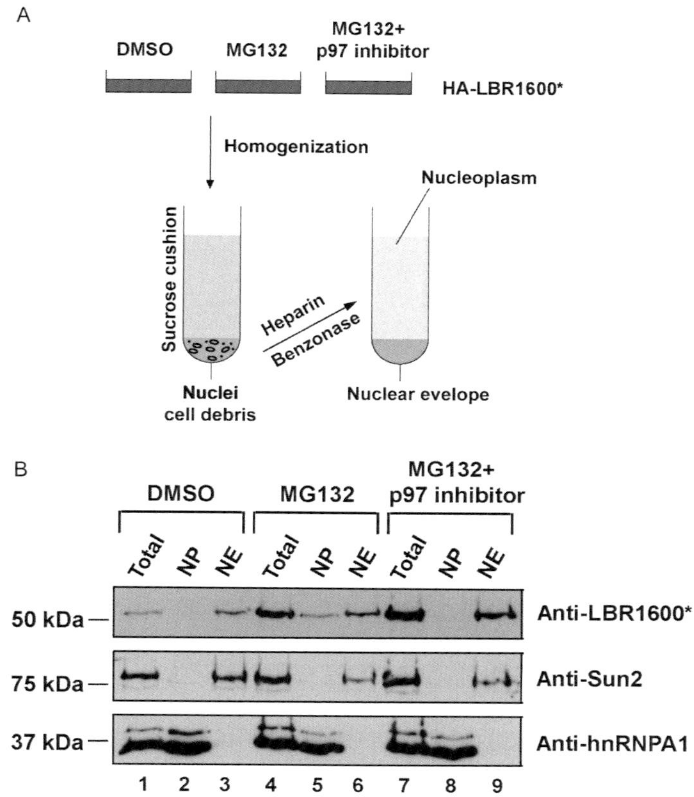

Fig. 3 Flow chart describing the isolation and the fractionation of nuclei into nuclear envelope and nucleoplasm. (A) A stable cell line expressing HA-LBR1600* was treated either with DMSO, 10 μ*M* MG132, or 10 μ*M* p97 inhibitor plus MG132 for 4 h, and homogenized by passing through a 25G needle. The homogenates were sedimented through a low-speed sucrose cushion, and the resulting pellets were resuspended in buffer including heparin and benzonase. After the final centrifugation, the supernatant contains nucleoplasm, and the resulting pellet is the nuclear envelope fraction. (B) Five percent of each fraction in each condition was subjected to SDS-PAGE and immunoblotted with anti-LBR, and the indicated control antibodies to assess the quality of the fractionation. We used anti-Sun2 as a marker for the nuclear envelope fraction, and anti-hnRNPA1 as the marker for nucleoplasm. Total: 5% of cells saved for loading control before fractionation, *NE*, nuclear envelope; *NP*, nucleoplasm.

3.1 Buffers and reagents

- PBS
- Buffer A (10 m*M* HEPES, pH 7.4, 250 m*M* sucrose, 2 m*M* $MgCl_2$) supplemented with 1 m*M* phenylmethanesulfonyl fluoride (PMSF).
- STM 0.9 (Sucrose-Tris-Magnesium) buffer: (50 m*M* Tris–HCl, pH 7.4, 0.9 *M* sucrose, 5 m*M* $MgCl_2$) supplemented with 1 m*M* PMSF.

- TP buffer: (10 m*M* Tris–HCl, pH 8.0, 10 m*M* Na_2HPO_4, 2 m*M* $MgCl_2$) supplemented with heparin (24 mg/12 mL buffer), benzonase nuclease (1 µL/0.5 mL buffer), 1 m*M* PMSF, and 1 × protease inhibitor cocktail without EDTA (Roche).
- hnRNP A1 antibody (ab5832, Abcam), Sun2 antibody (06–1038, MilliporeSigma)
- Benzonase (MilliporeSigma)
- 10 m*M* MG132 (1000 ×) (MilliporeSigma)
- 10 m*M* p97 inhibitor CB-5083 (1000 ×) (Selleckchem)

3.1.1 Notes

1. PMSF is not stable in aqueous solution; make a fresh 100 m*M* stock in ethanol, and dilute 100-fold into buffer A immediately prior to beginning the experiment.
2. Protease inhibitor cocktail can be dissolved in 1 mL water (50 ×) and diluted into TP buffer before starting the experiment.

3.2 Equipment

- Benchtop centrifuge (Sorvall Legend X1R, Thermo Fisher Scientific or similar)
- Microcentrifuge (Eppendorf or similar)
- Standard SDS-PAGE equipment (Bio-Rad or similar)

3.3 Procedure

1. Treat three 10-cm plates of HeLa cells (6×10^6 cells/plate) with 10 µL DMSO, 10 µ*M* MG132, or 10 µ*M* of both p97 inhibitor and MG132 in 10 mL growth medium. Incubate for 4 h in a cell culture incubator.
2. Trypsinize the cells, transfer cells into 15-mL tubes, and spin at 500 × *g* for 5 min at 4°C.
3. Resuspend the cells with 5 mL cold PBS, and set 250 µL cells from each sample aside as input loading controls. Lyse the input fractions with SDS-PAGE loading buffer, incubate at 65°C for 5 min, and store at −20°C.
4. Spin down the remaining cells at 500 × *g* for 5 min at 4°C.
5. Resuspend the three pellets in 5 mL of buffer A and spin at 500 × *g* for 5 min at 4°C.
6. Discard the supernatants and resuspend the pellets in 100 µL of buffer A. Incubate on ice for 10 min. Prepare three 15 mL tubes filled with 10 mL of chilled STM 0.9 buffer on ice.

7. Homogenize the cells via passing through a 1 mL syringe with a 25-gauge needle 10 times and gently layer each homogenate directly onto the top of the STM 0.9 buffer in the 15-mL tubes. Centrifuge the tubes at 1000 × *g* for 10 min at 4°C.
8. Remove 9 mL supernatant from each tube and repeat the centrifugation as in step 7. After this step, a solid white pellet should be observed at the bottom of the tube.
9. Aspirate the supernatants and wash the pellets with 0.5 mL of TP buffer without supplements. Spin at 1000 × *g* for 10 min at 4°C and remove the supernatants.
10. Resuspend the pellets in another 0.5 mL of TP buffer with all supplements (heparin, benzonase, and protease inhibitors), and rock at 4°C for 2.5 h.
11. Centrifuge at 15,000 × *g* for 45 min at 4°C. The supernatant is the nucleoplasmic fraction. Transfer the supernatants into fresh 1.5-mL tubes. Remove a 50 μL aliquot from each for SDS-PAGE analysis; these are then mixed with an equal volume of 2 × SDS-PAGE sample buffer, heat the samples at 65°C for 5 min, and store at −20°C.
12. Wash the pellet with 0.5 mL of TP buffer, spin at 15,000 × *g* for 10 min at 4°C, and lyse the final pellets with 50 μL SDS-PAGE sample buffer and heat at 65°C for 5 min. Pellets include membrane fractions enriched for NE and ER membranes. Note that additional steps (see Emig et al., 1995) would be required to achieve a more homogenous preparation of NEs if experiments require high purity of nuclear membranes; however, these are not necessary to monitor the membrane extraction step of LBR1600*.
13. Resolve 5% of each fraction by SDS-PAGE and immunoblot with antibodies against LBR and the markers.

3.3.1 Notes

The input cell number can be scaled up if immunoprecipitation will be performed after the fractionation. The buffer volumes in the procedure should be proportional to the input cell number to ensure that there is enough buffer to extract the nucleoplasm. In some cases, if there is not enough TP buffer or incubation time, the last membrane fraction will contain intact nuclei, evident by the viscous solution caused by DNA release when the final pellets were lysed. This has to be avoided since a failure to separate membranes from nucleoplasm can lead to invalid conclusions. It is therefore critical to use appropriate markers (cf. Fig. 3B) to control for the identity of the fractions. To assess the quality of the fractionation, hnRNPA1 and Sun2 are recommended as markers. hnRNPA1 is one of the most abundant nuclear proteins, localizes at the nucleoplasm, binds to mRNA, and is shuttled between the nucleus and cytoplasm (Purice & Taylor, 2018). Sun2 is an INM protein (Lee & Burke, 2018) and serves as a marker of the insoluble membrane fraction. A typical result is provided in Fig. 3B, where Sun2 was only present in the membrane fractions and hnRNPA1 was only present in nucleoplasm fractions, indicating the fractionation was successful. In accordance with our previous study and microscopy data (cf. Fig. 2B and (Tsai et al., 2016)), LBR1600* appears to be dislocated from the NE and degraded in the nucleoplasm. The LBR1600* turnover can be inhibited at the steps of membrane dislocation or degradation using p97 inhibitor or proteasome inhibitor MG132, respectively. In the presence of MG132, a portion of LBR1600* is observed in the soluble nucleoplasmic fraction, and it is barely detected in the DMSO treatment (cf. Fig 3B, lane 5 and lane 2); however, in the presence of p97 inhibitor, LBR1600* was stabilized and trapped in the membrane fraction as expected (cf. Fig. 3B, lane 5 and lane 8; Fig. 2B, last panel).

4. Concluding remarks

Dysfunctions of INM proteins give rise to many human diseases known as nuclear envelopathies (Dauer & Worman, 2009). While the involvement of protein quality control systems in pathologies is firmly established (Labbadia & Morimoto, 2015), much remains to be learned about the protein quality control mechanisms at the INM. The set of methods described here establishes both model substrates and readouts for measuring protein turnover in the NE. The split-GFP approach (Section 2) provides a widely applicable platform for large-scale experimental approaches to studying NE quality control; for example, it should allow the identification of novel INM turnover machinery through forward genetic screens (Timms et al., 2016). The biochemical fractionation described in Section 3, in conjunction with suitable inhibitors or dominant-negative alleles, will facilitate the identification of LBR1600*-associated proteins, and can be employed in proteomics-based approaches analogous to studies that successfully uncovered mammalian ERAD components (Christianson et al., 2011; Ernst et al., 2011).

In principle, our experimental platform can be successfully adapted for many other INM-resident proteins. However, it should be noted that some optimization might be required to achieve appropriate expression levels and that engineering of fusion constructs entails the testing of, for example, various linker sequences or lengths to achieve authentic localization to the INM.

Acknowledgments

We thank members of Schlieker laboratory for comments on the manuscript. This work was supported by NIH grants R01GM114401 and R01GM126835 to C.S.

References

Anderson, D. J., Le Moigne, R., Djakovic, S., Kumar, B., Rice, J., Wong, S., et al. (2015). Targeting the AAA ATPase p97 as an approach to treat cancer through disruption of protein homeostasis. *Cancer Cell*, *28*(5), 653–665. https://doi.org/10.1016/j.ccell.2015.10.002.

Boni, A., Politi, A. Z., Strnad, P., Xiang, W., Hossain, M. J., & Ellenberg, J. (2015). Live imaging and modeling of inner nuclear membrane targeting reveals its molecular requirements in mammalian cells. *The Journal of Cell Biology*, *209*(5), 705–720. https://doi.org/10.1083/jcb.201409133.

Cabantous, S., Terwilliger, T. C., & Waldo, G. S. (2005). Protein tagging and detection with engineered self-assembling fragments of green fluorescent protein. *Nature Biotechnology*, *23*(1), 102–107. https://doi.org/10.1038/nbt1044.

Christianson, J. C., Olzmann, J. A., Shaler, T. A., Sowa, M. E., Bennett, E. J., Richter, C. M., et al. (2011). Defining human ERAD networks through an integrative mapping strategy. *Nature Cell Biology*, *14*(1), 93–105. https://doi.org/10.1038/ncb2383.

Dauer, W. T., & Worman, H. J. (2009). The nuclear envelope as a signaling node in development and disease. *Developmental Cell*, *17*(5), 626–638. https://doi.org/10.1016/j.devcel.2009.10.016.

Davis, H. E., Rosinski, M., Morgan, J. R., & Yarmush, M. L. (2004). Charged polymers modulate retrovirus transduction via membrane charge neutralization and virus aggregation. *Biophysical Journal*, *86*(2), 1234–1242. https://doi.org/10.1016/S0006-3495(04)74197-1.

Deng, M., & Hochstrasser, M. (2006). Spatially regulated ubiquitin ligation by an ER/nuclear membrane ligase. *Nature*, *443*(7113), 827–831. https://doi.org/10.1038/nature05170.

Emig, S., Schmalz, D., Shakibaei, M., & Buchner, K. (1995). The nuclear pore complex protein p62 is one of several sialic acid-containing proteins of the nuclear envelope. *The Journal of Biological Chemistry*, *270*(23), 13787–13793.

Ernst, R., Claessen, J. H., Mueller, B., Sanyal, S., Spooner, E., van der Veen, A. G., et al. (2011). Enzymatic blockade of the ubiquitin-proteasome pathway. *PLoS Biology*, *8*(3), e1000605. https://doi.org/10.1371/journal.pbio.1000605.

Ernst, R., Mueller, B., Ploegh, H. L., & Schlieker, C. (2009). The otubain YOD1 is a deubiquitinating enzyme that associates with p97 to facilitate protein dislocation from the ER. *Molecular Cell*, *36*(1), 28–38. https://doi.org/10.1016/j.molcel.2009.09.016.

Feng, S., Sekine, S., Pessino, V., Li, H., Leonetti, M. D., & Huang, B. (2017). Improved split fluorescent proteins for endogenous protein labeling. *Nature Communications*, *8*(1), 370. https://doi.org/10.1038/s41467-017-00494-8.

Foresti, O., Rodriguez-Vaello, V., Funaya, C., & Carvalho, P. (2014). Quality control of inner nuclear membrane proteins by the Asi complex. *Science*, *346*(6210), 751–755. https://doi.org/10.1126/science.1255638.

Hwang, J., Walczak, C. P., Shaler, T. A., Olzmann, J. A., Zhang, L., Elias, J. E., et al. (2017). Characterization of protein complexes of the endoplasmic reticulum-associated degradation E3 ubiquitin ligase Hrd1. *The Journal of Biological Chemistry*, *292*(22), 9104–9116. https://doi.org/10.1074/jbc.M117.785055.

Kamiyama, D., Sekine, S., Barsi-Rhyne, B., Hu, J., Chen, B., Gilbert, L. A., et al. (2016). Versatile protein tagging in cells with split fluorescent protein. *Nature Communications*, *7*, 11046. https://doi.org/10.1038/ncomms11046.

Kasbekar, D. P. (2018). A cross-eyed geneticist's view I. Making sense of the lamin B receptor, a chimeric protein. *Journal of Biosciences*, *43*(2), 235–237.

Khmelinskii, A., Blaszczak, E., Pantazopoulou, M., Fischer, B., Omnus, D. J., Le Dez, G., et al. (2014). Protein quality control at the inner nuclear membrane. *Nature*, *516*(7531), 410–413. https://doi.org/10.1038/nature14096.

Labbadia, J., & Morimoto, R. I. (2015). The biology of proteostasis in aging and disease. *Annual Review of Biochemistry*, *84*, 435–464. https://doi.org/10.1146/annurev-biochem-060614-033955.

Lee, Y. L., & Burke, B. (2018). LINC complexes and nuclear positioning. *Seminars in Cell & Developmental Biology*, *82*, 67–76. https://doi.org/10.1016/j.semcdb.2017.11.008.

Li, X., Roberti, R., & Blobel, G. (2015). Structure of an integral membrane sterol reductase from *Methylomicrobium alcaliphilum*. *Nature*, *517*(7532), 104–107. https://doi.org/10.1038/nature13797.

Makatsori, D., Kourmouli, N., Polioudaki, H., Shultz, L. D., McLean, K., Theodoropoulos, P. A., et al. (2004). The inner nuclear membrane protein lamin B receptor forms distinct microdomains and links epigenetically marked chromatin to the nuclear envelope. *The Journal of Biological Chemistry*, *279*(24), 25567–25573. https://doi.org/10.1074/jbc.M313606200.

Mehrtash, A. B., & Hochstrasser, M. (2019). Ubiquitin-dependent protein degradation at the endoplasmic reticulum and nuclear envelope. *Seminars in Cell & Developmental Biology*. https://doi.org/10.1016/j.semcdb.2018.09.013, in press.

Nikolakaki, E., Mylonis, I., & Giannakouros, T. (2017). Lamin B receptor: Interplay between structure, function and localization. *Cell*, *6*(3). https://doi.org/10.3390/cells6030028.

Ohba, T., Schirmer, E. C., Nishimoto, T., & Gerace, L. (2004). Energy- and temperature-dependent transport of integral proteins to the inner nuclear membrane via the nuclear pore. *The Journal of Cell Biology*, *167*(6), 1051–1062. https://doi.org/10.1083/jcb.200409149.

Olins, A. L., Rhodes, G., Welch, D. B., Zwerger, M., & Olins, D. E. (2010). Lamin B receptor: Multi-tasking at the nuclear envelope. *Nucleus*, *1*(1), 53–70. https://doi.org/10.4161/nucl.1.1.10515.

Oosterwijk, J. C., Mansour, S., van Noort, G., Waterham, H. R., Hall, C. M., & Hennekam, R. C. (2003). Congenital abnormalities reported in Pelger-Huet homozygosity as compared to Greenberg/HEM dysplasia: Highly variable expression of allelic phenotypes. *Journal of Medical Genetics*, *40*(12), 937–941.

Ploegh, H. L. (2001). One-dimensional isoelectric focusing of proteins in slab gels. *Current Protocols in Protein Science*, *00*(1), 10.2.1–10.2.8. https://doi.org/10.1002/0471140864.ps1002s00.

Prakash, A., Sengupta, S., Aparna, K., & Kasbekar, D. P. (1999). The erg-3 (sterol delta14,15-reductase) gene of *Neurospora crassa*: Generation of null mutants by repeat-induced point mutation and complementation by proteins chimeric for human lamin B receptor sequences. *Microbiology*, *145*(Pt. 6), 1443–1451. https://doi.org/10.1099/13500872-145-6-1443.

Purice, M. D., & Taylor, J. P. (2018). Linking hnRNP function to ALS and FTD pathology. *Frontiers in Neuroscience*, *12*, 326. https://doi.org/10.3389/fnins.2018.00326.

Pyrpasopoulou, A., Meier, J., Maison, C., Simos, G., & Georgatos, S. D. (1996). The lamin B receptor (LBR) provides essential chromatin docking sites at the nuclear envelope. *The EMBO Journal*, *15*(24), 7108–7119.

Rose, A., & Schlieker, C. (2012). Alternative nuclear transport for cellular protein quality control. *Trends in Cell Biology*, *22*(10), 509–514. https://doi.org/10.1016/j.tcb.2012.07.003.

Shibata, Y., & Morimoto, R. I. (2014). How the nucleus copes with proteotoxic stress. *Current Biology*, *24*(10), R463–R474. https://doi.org/10.1016/j.cub.2014.03.033.

Silve, S., Dupuy, P. H., Ferrara, P., & Loison, G. (1998). Human lamin B receptor exhibits sterol C14-reductase activity in *Saccharomyces cerevisiae*. *Biochimica et Biophysica Acta*, *1392*(2–3), 233–244.

Timms, R. T., Menzies, S. A., Tchasovnikarova, I. A., Christensen, L. C., Williamson, J. C., Antrobus, R., et al. (2016). Genetic dissection of mammalian ERAD through comparative haploid and CRISPR forward genetic screens. *Nature Communications*, 7, 11786. https://doi.org/10.1038/ncomms11786.

Tsai, P. L., Zhao, C., Turner, E., & Schlieker, C. (2016). The lamin B receptor is essential for cholesterol synthesis and perturbed by disease-causing mutations. *eLife*, *5*, e16011. https://doi.org/10.7554/eLife.16011.

Turner, E. M., & Schlieker, C. (2016). Pelger-Huet anomaly and Greenberg skeletal dysplasia: LBR-associated diseases of cholesterol metabolism. *Rare Disease*, *4*(1), e1241363, https://doi.org/10.1080/21675511.2016.1241363.

Ungricht, R., Klann, M., Horvath, P., & Kutay, U. (2015). Diffusion and retention are major determinants of protein targeting to the inner nuclear membrane. *The Journal of Cell Biology*, *209*(5), 687–703. https://doi.org/10.1083/jcb.201409127.

Ungricht, R., & Kutay, U. (2015). Establishment of NE asymmetry-targeting of membrane proteins to the inner nuclear membrane. *Current Opinion in Cell Biology*, *34*, 135–141. https://doi.org/10.1016/j.ceb.2015.04.005.

Ungricht, R., & Kutay, U. (2017). Mechanisms and functions of nuclear envelope remodelling. *Nature Reviews. Molecular Cell Biology*, *18*(4), 229–245. https://doi.org/10.1038/nrm.2016.153.

Van Engelenburg, S. B., & Palmer, A. E. (2010). Imaging type-III secretion reveals dynamics and spatial segregation of *Salmonella* effectors. *Nature Methods*, 7(4), 325–330. https://doi.org/10.1038/nmeth.1437.

Webster, B. M., & Lusk, C. P. (2016). Border safety: Quality control at the nuclear envelope. *Trends in Cell Biology*, *26*(1), 29–39. https://doi.org/10.1016/j.tcb.2015.08.002.

Worman, H. J., Yuan, J., Blobel, G., & Georgatos, S. D. (1988). A lamin B receptor in the nuclear envelope. *Proceedings of the National Academy of Sciences of the United States of America*, *85*(22), 8531–8534.

Ye, Y., Meyer, H. H., & Rapoport, T. A. (2001). The AAA ATPase Cdc48/p97 and its partners transport proteins from the ER into the cytosol. *Nature*, *414*(6864), 652–656. https://doi.org/10.1038/414652a.

Ye, Q., & Worman, H. J. (1994). Primary structure analysis and lamin B and DNA binding of human LBR, an integral protein of the nuclear envelope inner membrane. *The Journal of Biological Chemistry*, *269*(15), 11306–11311.

CHAPTER FOUR

Assays for dissecting the in vitro enzymatic activity of yeast Ubc7

Bayan Mashahreh[a], Yuval Reiss[a], Reuven Wiener[b], Tommer Ravid[a,*]
[a]Department of Biological Chemistry, The Institute of Life Sciences, The Hebrew University of Jerusalem, Jerusalem, Israel
[b]Department of Biochemistry and Molecular Biology, Institute for Medical Research Israel-Canada (IMRIC), Hebrew University School of Medicine, Jerusalem, Israel
*Corresponding author: e-mail address: tommer.ravid@mail.huji.ac.il

Contents

Abstract

Ubiquitin (Ub)-mediated protein degradation is a key cellular defense mechanism that detects and eliminates defective proteins. A major intracellular site of protein quality control degradation is the endoplasmic reticulum (ER), hence the term ER-associated degradation, or endoplasmic reticulum-associated degradation (ERAD). Yeast ERAD is composed of three Ub-protein conjugation complexes, named according to their E3 Ub-protein ligase components, Hrd1, Doa10, and the Asi complex, which resides at the nuclear envelope (NE). These ER/NE membrane-associated RING-type E3 ligases recognize and ubiquitylate defective proteins in cooperation with the E2 conjugating enzyme Ubc7 and the obligatory Ubc7 cofactor Cue1. Interaction of Ubc7 with the RING domains of its cognate E3 Ub-protein ligases stimulates the formation of isopeptide (amide) Ub–Ub linkages. Each isopeptide bond is formed by transfer of an Ubc7-linked

Methods in Enzymology, Volume 619
ISSN 0076-6879
https://doi.org/10.1016/bs.mie.2018.12.035
© 2019 Elsevier Inc.
All rights reserved.

activated Ub to a lysine side chain of an acceptor Ub. Multiple Ub transfer reactions form a poly-Ub chain that targets the conjugated protein for degradation by the proteasome. To study the mechanism of Ub–Ub bond formation, this reaction is reconstituted in a cell-free system consisting of recombinant E1, Ub, Ubc7, its cofactor Cue1, and the RING domain of either Doa10 or Hrd1. Here we provide detailed protocols for the purification of the required recombinant proteins and for the reactions that produce an Ub–Ub bond, specifically, the formation of an Ubc7~Ub thiolester (Ub charging) and subsequent formation of the isopeptide Ub–Ub linkage (Ub transfer). These protocols also provide a useful guideline for similar in vitro ubiquitylation reactions intended to explore the mechanism of other Ub-conjugation systems.

1. Introduction

Endoplasmic reticulum-associated degradation (ERAD) is a key protein quality control mechanism, involving ubiquitin (Ub) conjugation at the outer leaflet of the ER membrane and the nuclear envelope (NE), followed by degradation by the 26S proteasome (Amm, Sommer, & Wolf, 2014; Mehrtash & Hochstrasser, 2018). Early studies of ERAD largely focused on the identification of components of the degradation machinery as well as their substrates. These initial studies led to the view that in addition to degrading defective ER/NE proteins, the ERAD system also eliminates aberrant proteins from other subcellular compartments, mainly nuclear and cytoplasmic (Hochstrasser & Varshavsky, 1990; Metzger, Maurer, Dancy, & Michaelis, 2008; Ravid, Kreft, & Hochstrasser, 2006). Despite the abundance and heterogeneity of misfolded proteins, the ERAD ubiquitylation machinery employs only a few Ub-ligation complexes. In fact, in the bakers' yeast *Saccharomyces cerevisiae* ERAD system, there are three such complexes, specified by their E3 ligase components, Hrd1, Doa10, and the Asi complex (Bordallo, Plemper, Finger, & Wolf, 1998; Carvalho, Goder, & Rapoport, 2006; Denic, Quan, & Weissman, 2006; Foresti, Rodriguez-Vaello, Funaya, & Carvalho, 2014; Hampton, Gardner, & Rine, 1996; Khmelinskii et al., 2014; Swanson, Locher, & Hochstrasser, 2001) (Fig. 1). Each of the three yeast ERAD E3 ligase complexes ubiquitylates a distinct set of substrates, with minor overlaps (Mehrtash & Hochstrasser, 2018; Vashist & Ng, 2004).

Generally, E3 Ub-protein ligases serve as the substrate recognition site of the Ub-ligation system. They bind the substrate and the relevant E2 conjugating enzyme, enabling efficient Ub transfer from E2 to substrate (Lorick et al., 1999; Seol et al., 1999; Skowyra et al., 1999). This reaction is enhanced by the E3 RING domains that allosterically stimulate the E2

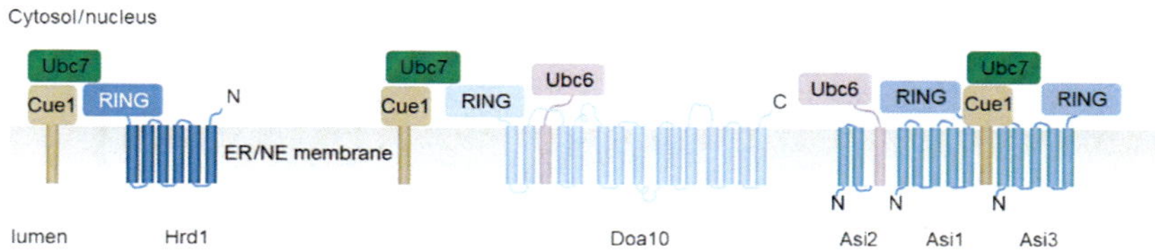

Fig. 1 *Basic components of the baker's yeast ERAD machinery*. The three yeast ERAD pathways are named according to their cognate E3 ligases, Hrd1, Doa10, and the Asi complex. All are multipass ER/NE-membrane proteins, and all, except for Asi2, contain a RING domain. These E3s cooperate with the E2 enzyme Ubc7 for Ub conjugation. Ubc7, which is primarily a cytosolic protein, interacts with Cue1 that recruits the enzyme to the ER/NE membrane and stimulates its enzymatic activity. Substrate ubiquitylation by the Doa10 and the Asi pathways additionally requires the activity of Ubc6, which transfers the initial Ub to an acceptor site on its substrate. This Ub is the primary acceptor site for poly-Ub chain buildup by Ubc7. ER, endoplasmic reticulum; NE, nuclear envelope.

Ub-transfer reaction (Metzger, Pruneda, Klevit, & Weissman, 2013). All three ERAD E3 ligases cooperate with the E2 enzyme Ubc7 during Ub conjugation (Chen, Johnson, Sommer, Jentsch, & Hochstrasser, 1993; Hampton & Bhakta, 1997; Khmelinskii et al., 2014). Protein degradation in the Doa10 and the Asi pathways also require the activity of Ubc6, a noncanonical C-terminally anchored membrane E2 conjugating enzyme (Khmelinskii et al., 2014; Lenk et al., 2002; Lester, Farquharson, Russell, & Houston, 2000). Ubc6 was shown to attach the initial Ub molecule to side chains of either Lys or Ser/Thr residues of Doa10 substrates (Weber et al., 2016). The initial Ub is thought to be the obligatory acceptor site for subsequent Ubc7-catalyzed polyubiquitylation.

All ERAD Ub-conjugation pathways target proteins for proteasomal degradation. Accordingly, Ubc7 primarily forms poly-Ub chains linked via the ε-amino group of Lys 48 (Bagola et al., 2013). Ubc7 activity requires an additional ER-resident cofactor, termed Cue1 (Biederer, Volkwein, & Sommer, 1997). By binding to Ubc7, this transmembrane adapter protein brings the otherwise cytosolic E2 and its cognate, membrane-anchored E3 ligases into close proximity, facilitating their interaction. In binding to Ubc7, Cue1 also stabilizes the E2 by preventing its self-ubiquitylation and subsequent proteasomal degradation (Ravid & Hochstrasser, 2007).

The Ubc7 enzymatic activity per se is composed of two successive reactions (Fig. 2): Initially, Ub binds covalently to Cys 89 at the active site of Ubc7 via trans-thiolation of the Ub from the E1 Ub-activating enzyme (Hershko, Heller, Elias, & Ciechanover, 1983). The formation of an Ubc7~Ub thioester bond is defined as E2 charging. The primary

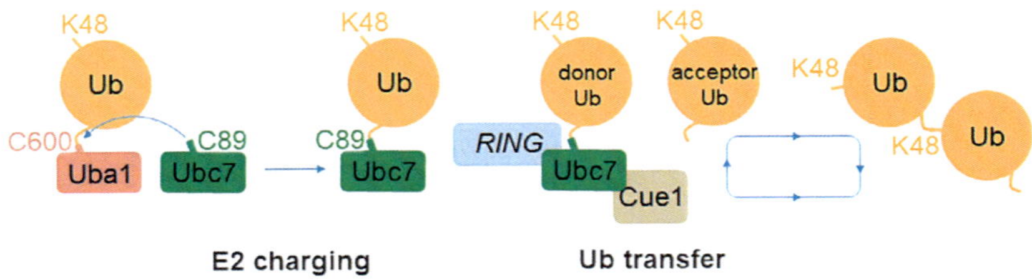

Fig. 2 *Elementary steps of Ubc7 enzymatic activity.* Ub-charging: The C-terminal carboxylate of Ub binds to the active-site Cys 89 side chain of Ubc7 via trans-thiolation from Cys 600 of the E1. Ub-transfer: The ε-amino group of Lys 48 of Ub attacks the thioester bond, leading to the formation of an isopeptide (amide) bond between Lys 48 of the acceptor Ub and Gly76 of the donor Ub. Chain elongation (poly-Ub chain formation) proceeds through multiple Ub transfer reactions.

thioester-linked Ub then serves as a donor in an Ub transfer reaction where an isopeptide bond is formed between the formerly thioester-linked carboxyl group of Gly76 of Ub and a lysine ε-amino group of either an acceptor substrate or an Ub molecule within a nascent Ub chain. Thus, chain elongation (poly-Ub chain formation) proceeds through multiple Ub transfer reactions.

An uncommon property of Ubc7 is its ability to efficiently form poly-Ub chains in the absence of an acceptor substrate, possibly attributed to the ability of Ubc7 to retain poly-Ub chains bound to its catalytic Cys, prior to *en bloc* conjugation to the degradation substrate (Li, Tu, Brunger, & Ye, 2007; Ravid & Hochstrasser, 2007). Furthermore, poly-Ub chain elongation by yeast Ubc7 can proceed even without an E3 RING ligase, albeit at a reduced rate: We have previously shown that the RING domains of Doa10 and Hrd1 enhance Ubc7-mediated Ub transfer by aligning the donor Ub on Ubc7 optimally for a nucleophilic attack by the ε-amino group of the acceptor molecule (Cohen, Wiener, Reiss, & Ravid, 2015). In contrast, Cue1 is absolutely required for Ubc7-mediated Ub transfer: The U7BR (Ubc7-binding region) of Cue1 stimulates both Ubc7 charging and Ub transfer (Metzger, Liang, et al., 2013), while the CUE domain is essential for the alignment of the E2∼Ub donor and the Ub acceptor in a catalytically productive manner (von Delbrück et al., 2016). Fig. 3 shows a proposed model for the minimal Doa10 complex required for efficient in vitro Ub-conjugation activity. This model can serve as a platform to explore putative interaction surfaces between Ubc7 and its cofactors, as previously described (Cohen et al., 2015).

The striking ability of Ubc7 to form free Lys 48-linked Ub chains, which relies on its physiological interactors—the RING domains of its cognate E3s

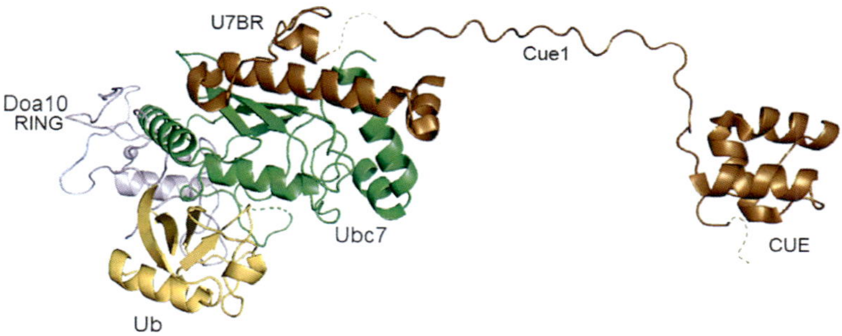

Fig. 3 *A structural model for the active Doa10 complex.* This model has been constructed in silico by docking the RING domain of gp78 (the mammalian ortholog of yeast Hrd1; PDB structure 2LXP) (Das et al., 2013) to yeast Ubc7 PDB structure 4JQU, using GalaxyTongDock_A (Ko, Park, Heo, & Seok, 2012). Ub PDB structure 1UBQ (Johnston, Larsen, Cook, Wilkinson, & Hill, 1997) was then docked to the obtained gp78/Ubc7 complex. Next, the NMR-derived structure of Doa10 (PDB structure 2M6M deposition to the PDB, unpublished) was superimposed on the gp78 RING domain, while the PDB structure 4JQU (Metzger, Liang, et al., 2013), containing Ubc7 and U7BR was superimposed on the docked Ubc7. Finally, the CUE domain (PDB NMR structure 2MYX) (von Delbrück et al., 2016) was arbitrarily added, based on the model proposed by von Delbrück et al. (2016). The actual position of the CUE domain relative to U7BR remains obscure. Broken line indicates structurally undetermined regions.

and the U7BR and CUE domains of Cue1 (Fig. 3)—provides an excellent experimental tool to explore molecular aspects of the E2 mechanism of action. Herein we describe in detail methods to produce the protein components required for in vitro Ubc7-mediated polyubiquitylation as well as protocols for the dissection of the reaction into its elementary steps.

2. Protein production and purification

This section describes procedures to produce and purify recombinant Ub, Uba1, Ubc7, Cue1, and the RING domains of Doa10 and Hrd1. These are the minimal components of Ubc7-mediated polyubiquitylation.

2.1 General considerations for recombinant protein production and purification

2.1.1 Bacteria used

- All proteins described herein are cloned in expression vectors under control of the T7 promoter and subsequently expressed in the bacterial strain BL21 Rosetta™ (DE3)pLysS.

2.1.2 Bacterial growth media and reagents

- Growth medium-LB (rich medium). Ingredients (w/v): 1% tryptone, 0.5% yeast extract, and 1% NaCl. For LB plates, 1.5% Bacto agar is added.
- Medium for *E. coli* expression of recombinant proteins-Terrific Broth (TB). Ingredients (w/v): 1.2% tryptone, 2.4% yeast extract, 0.94% dipotassium phosphate (K_2P0_4), 0.22% monopotassium phosphate (KH_2P0_4), 0.4% glycerol.
- Antibiotics:
  - Chloramphenicol (CHL) at final concentration of 17 μg/mL. Used to maintain the pACYC-based plasmid, carrying a T7 lysozyme gene derivative in the bacterial strain BL21 Rosetta™ (DE3) pLysS.
  - Ampicillin (AMP) at final concentration of 100 μg/mL. Used to maintain plasmids with AMP-resistance gene, carrying recombinant protein genes.
  - Kanamycin (KAN) at final concentration of 50 μg/mL. Used to maintain plasmids with KAN-resistance gene, carrying recombinant protein genes.
- Isopropyl β-D-1-thiogalactopyranoside (IPTG) at final concentration of 0.5 m*M*. Used to induce recombinant protein production.

2.1.3 Recombinant protein induction

1. Transform a plasmid, encoding the desired recombinant protein (Table 1), into bacterial strain BL21 Rosetta™ (DE3) pLysS. Seed on a selective LB plate containing the appropriate antibiotic.

Table 1 Plasmids used in this study

Plasmid	Relevant markers	Source
pTR120	pET-42b-Doa10 RING (1 – 112)	M. Hochstrasser
pTR525	pTYB2-Ubc7-2HA	R. Hampton
pTR526	pTYB2-Ubc7-2HA-Cue1$^{\Delta TM}$	R. Hampton
pTR635	pREST-6His-mUba1	A. Ciechanover
pTR839	pET14-Ub	A. Ciechanover
pTR1475	pET-42b-Hrd1 RING plus flanking regions (287–531)	Ravid Lab
pTR1543	pET14-Flag-Ub	Ravid Lab
pTR1544	pET14-Flag-Ub^{R48}	Ravid Lab

2. Pick a single antibiotic-resistant colony and grow a starter culture overnight (O/N) at 37°C.
3. The following morning dilute the culture 1:100 in 0.5–2 L of TB medium and grow to mid-late log phase: optical density $(OD)_{600}$ of ~0.8.
4. Cool down the culture by shaking for 1 h at 17°C.
5. Add IPTG to a final concentration of 0.5 m*M* and grow O/N at 17°C while shaking.
6. Harvest the cells by centrifugation at 6000 × *g* for 15 min at 4°C. Wash once more with ice-cold PBS and keep the cell pellet at −80°C until use.

2.1.4 Cell extraction

1. Lyse cells using a Microfluidizer high shear processor, according to the manufacturer instructions.
2. Clear the cell extract by centrifugation at 20,000 × *g* for 30 min at 4°C. Transfer the cleared supernatant to a clean vessel.
3. (optional) Further clear the extract by filtration, first through a 0.45-μ*M* filter and then through a 0.2-μ*M* filter.

2.1.5 Notes

1. Induce O/N protein expression only when cells are at mid-late log phase (OD_{600} ~ 0.8) at 16–18°C. Cultures exceeding log phase cell density, or growing at higher temperatures, will accumulate increased proportions of dead cells, resulting in reduced efficiency of protein expression.
2. For cell lysis, the microfluidizer can be replaced with a French-press high-pressure cell homogenizer or a sonicator equipped with an appropriate probe.
3. For complete lysis of bacterial cells using a microfluidizer: three passages are required. Cool the extract on ice for several minutes between cycles to prevent overheating.

2.2 Production and purification of Ub

Recombinant human Ub is purified from bacteria by precipitation in perchloric acid according to the method of Pickart and Raasi (2005). This method is based on the unique properties of the Ub molecule, which is remarkably resistant to denaturing conditions (Ciechanover, Hod, & Hershko, 1978). Thus, when treated with a strong acid, almost all bacterial proteins precipitate, while the majority of Ub remains soluble.

2.2.1 Equipment

- 500-mL beaker
- Magnetic stir bar
- Magnetic stirrer
- Dialysis tubing with a Molecular Weight Cut Off (MWCO) of 3–5 kDa
- Centrifugal filters with a MWCO of 3–5 kDa

2.2.2 Reagents and buffers

- Plasmid pET14-Ub encoding the human Ub ORF (pTR839) was a gift from A. Ciechanover (Technion, Israel). A Flag peptide tag-coding sequence was added upstream of and in frame with the sequence encoding Ub to produce plasmid pTR1543 by transfer PCR, as previously described (Erijman, Shifman, & Peleg, 2014).
- Cell lysis buffer: 50 m*M* Tris–HCl, pH 7.6, 1 m*M* phenylmethylsulfonyl fluoride (PMSF), freshly added from a 200 m*M* stock in EtOH, that was kept at −20°C.
- Dialysis buffer: 10 m*M* Tris–HCl, pH 7.6

2.2.3 Procedure

1. Transform the Ub-encoding plasmid into bacterial cells and spread on a selective LB plate containing AMP and CHL.
2. Grow a single colony from a fresh (O/N) plate in 30 mL of TB medium containing AMP and CHL at 37°C to OD_{600} ~0.6. Transfer the bacterial culture to 4°C and incubate O/N while shaking.
3. For subsequent cell culture growth and subsequent recombinant protein induction, follow steps 2–5 in Section 2.1.3.
4. Solubilize the cell pellet in 50 mL ice-cold lysis buffer (per 1 L culture), and lyse cells following Steps 1–2 under Section 2.1.4.
5. Transfer the cleared supernatant to an appropriate glass vessel, pre-chilled on ice, and add a magnetic stir bar.
6. Place the vessel containing the cell lysate in an ice bath and place it on top of a magnetic stirrer. Keep stirring throughout the entire purification process.
7. Add dropwise 0.5 mL of 70% perchloric acid to 50 mL of cleared extract while stirring. Upon addition of the perchloric acid, the solution will become increasingly cloudy, ultimately turning completely milky. Once the addition of perchloric acid is completed, continue stirring for a further 5 min.

8. Centrifuge the acid-treated lysate at 20,000 × *g* for 20 min at 4°C. The resulting supernatant contains Ub, while the pellet contains virtually the entire bacteria proteome.
9. (optional)Transfer the supernatant to a clean tube and centrifuge once more to remove residual insoluble material.
10. Transfer the cleared solution into a dialysis tube and dialyze O/N at 4°C against 2 L of dialysis buffer with at least one buffer change.
11. Concentrate the dialysate using a centrifugal filter unit to a concentration of 0.17 OD_{280} (~1 mg/mL for Ub). Dispense the concentrated protein into convenient aliquots (<50 μL) and store at −80°C.

2.2.4 Notes

1. A typical final Ub yield from a 1 L of TB culture is 50–70 mg.
2. For optimal induction of Ub in bacteria, use fresh cultures and keep cells in log phase.
3. The Ub concentration is determined using its extinction coefficient value at 280 nm. This value can be determined empirically or calculated based on the Ub amino acid sequence. An extinction coefficient value of 1490 mol/L per cm was obtained using ProtParam (Gasteiger et al., 2005). This gives a percent solution extinction coefficient value (Abs 0.1%) of 0.174 mg/mL per cm. Accordingly, an absorption value of a purified Ub solution of 0.174 OD_{280} is equivalent to a concentration of 1 mg/mL.
4. If desired, Ub can be further purified by high-pressure liquid chromatography using a Superdex 75 column. We found it unnecessary for in vitro ubiquitylation reactions, since Ub was the only protein that we detected by Coomassie Blue staining of the purified material.
5. For long-term storage, Ub can be kept as a powder. Dialyze the final Ub solution against water and then freeze-dry using a commercially available lyophilizer or similar devices (Ó'Fágáin & Colliton, 2017). Keep the Ub powder desiccated until use, upon which dissolve it in water at 10 mg/mL.

2.3 Purification of E1 Ub-activating enzyme

For E1 purification, we employ a plasmid encoding a 6His-tagged mouse E1 enzyme (mUba1), although the 6His tag is not used for purification. Instead, we use covalent affinity purification on an immobilized Ub-Sepharose resin (Haas, Warms, Hershko, & Rose, 1982) (generation of this resin is described in Section 2.3.1). This method takes advantage of the two-step Ub-activation reaction, specifically, the formation of an Ub~AMP adduct, and ATP breakdown followed by the formation of a thioester bond between Ub and the Cys active site of mUBA1 (Fig. 4A). The breakdown of ATP to AMP, coupled to pyrophosphate (PPi) hydrolysis, drives the reaction and the subsequent formation of the high energy E1 ~ Ub thioester. Accordingly, the purification procedure is a two-step process (Fig. 4B): (a) ATP-dependent formation of immobilized Ub ~ E1 is driven by the rapid hydrolysis of inorganic PPi, catalyzed by added pyrophosphatase. (b) E1 enzyme release from the Ub resin, achieved by forcing the reverse reaction in the presence of excess of the E1 ~ Ub-activation reaction products, AMP and PPi.

In vitro ubiquitylation reaction indicates that both the metal (Ni^{2+}–NTA) and the covalent Ub affinity procedures yield an active E1 enzyme (Fig. 4C, right panel). Still, a striking difference in terms of purity is observed by Coomassie Blue staining (Fig. 4C, left panel). While E1 purification by

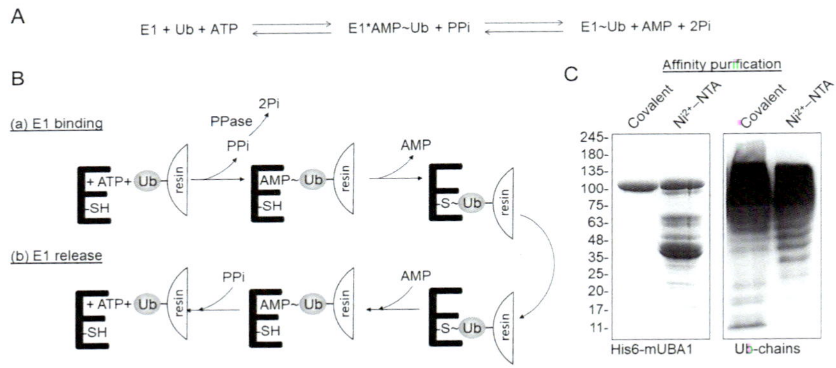

Fig. 4 Purification of mouse UBA1. (A) Enzymatic steps of Ub activation by E1. The first step of Ub activation is the formation of an adenylate adduct and the release of pyrophosphate (PPi). Next, Ub is transferred to a Cys residue in the E1 catalytic domain, forming a labile thioester bond with the release of AMP. (B) Scheme for the purification of mUBA1 by affinity binding to an Ub resin. Bacterial cell lysate is incubated with immobilized Ub in the presence of ATP and pyrophosphatase, resulting in mUBA1 binding to the resin and the release of two phosphate molecules. Next, the Ub is transferred to the catalytic Cys residue of mUBA1. Release of the E1 from the resin takes place following incubation with AMP and PPi that reverse the Ub-activation reaction, resulting in the formation of ATP. (C) Comparing mUBA1 protein purified by affinity chromatography using Ub resin "covalent" or by binding to a Ni^{2+}–NTA column. *Left:* Coomassie staining of mUBA1, separated on 5% SDS–PAGE, after parallel purification by the two indicated methods. *Right:* Measurement of purified mUBA1 activity. The poly-ubiquitylation reaction mix contains mUBA1, Ubc7/Cue1, Flag-Ub, and ATP. The components were incubated for 30 min at 30°C, followed by separation on 15% SDS–PAGE and immunoblotting with an anti-Flag monoclonal antibody (Sigma, catalog number F3165).

metal binding yields multiple protein bands, the covalent Ub affinity procedure yield a protomer with MW of ~110kDa, which correspond to mUBA1. Essentially, in the covalent Ub affinity procedure, every E1 molecule is a fully active enzyme. This is because both binding to and release from the Ub resin requires the E1 catalytic activity and hence an intact enzyme.

2.3.1 Preparation of Ub-affinity resin

2.3.1.1 Equipment

- 1.6 × 13 cm glass column
- Sintered glass filtering funnel
- 500-mL Erlenmeyer with suitable filter rubber adapter

2.3.1.2 Buffers and reagents

- 100 mg of dried Ub, self-made as described earlier.
- *N*-hydroxysuccinimide (NHS)-activated Sepharose (NAS)
- 1 m*M* HCl
- Coupling buffer: 0.1 *M* sodium bicarbonate pH 8.0, 0.5 *M* NaCl
- Blocking solution: 0.1 *M* Tris–HCl pH 8.0, 0.5 *M* NaCl
- Wash buffer: 0.1 *M* sodium acetate pH 4.0, 0.5 *M* NaCl
- Column storage buffer: 0.1 *M* Tris–HCl pH 7.2, 0.02% sodium azide

2.3.1.3 Procedure

1. Dissolve 100 mg of freeze-dried purified Ub in 20 mL coupling buffer. Keep a small aliquot to determine coupling efficiency later.
2. Resuspend 5 g of dried NAS beads in 20 mL of ice-cold 1 m*M* HCl. The beads will swell approximately twofold to ~10 mL.
3. Pour the slurry onto a sintered glass funnel and drain by applying vacuum (do not dry).
4. Wash the drained NHS beads rapidly with 200 mL of ice 1 m*M* HCl. Drain well.
5. Transfer the washed NHS resin "cake" into the dissolved Ub solution in a 50-mL plastic tube. Rotate the slurry gently at RT for 2 h.
6. Spin the slurry at 1500 × *g* for 1 min at RT and remove the solution containing uncoupled material. Keep an aliquot to determine coupling efficiency.
7. Determine coupling efficiency by measuring Ub concentration in the initial and postcoupling samples.

8. To block nonreacting NHS groups, add blocking solution and rotate the resin at RT for 1 h.
9. Transfer the resin into a glass column and then wash at 4°C with three alternating cycles of 20 column volumes (200 mL) each of wash buffer, followed by 20 column volumes of blocking solution. Keep the column in storage buffer at 4°C until use.

2.3.1.4 Notes

1. Purified Ub can also be obtained from a commercial vendor. If a commercial source is used the Ub solution should be dissolved in water at 10 mg/mL.
2. When kept as a powder, Ub is preserved for years. Anecdote: The covalent affinity-purified E1 presented in Fig. 4C was prepared using an Ub-Sepharose resin generated from freeze-dried Ub that had been prepared in A. Hershko's lab almost four decades ago.
3. The efficiency of Ub coupling to the NAS resin is typically >90%. It is tested by determination of Ub concentration as described in Section 2.2.4 before and after the coupling reaction, where aliquots are removed as described under Section 2.3.1.3 Steps 1 and 4, respectively.

2.3.2 Covalent-affinity purification of E1 enzyme

2.3.2.1 Equipment

Centrifugal filter units with MWCO of 50–75 kDa

2.3.2.2 Reagents and buffers

- For E1 purification, we use a pRSET parent vector expressing 6His epitope, followed by the mUba1 gene (6His-mUba1; pTR635 (Mori et al., 1997))
- Lysis buffer: 50 m*M* Tris–HCl pH 7.2, 1 m*M* DTT, 1 m*M* PMSF, protease inhibitor cocktail
- Equilibration buffer: 50 m*M* Tris–HCl pH 7.2, 2 m*M* ATP, 5 m*M* $MgCl_2$, 0.2 m*M* DTT
- Loading buffer: 50 m*M* Tris–HCl pH 7.2, 6 m*M* ATP, 12 m*M* MgCl, 6 units/mL inorganic pyrophosphatase
- Wash buffer 1: Tris–HCl pH 7.2, 1 *M* KCl
- Wash buffer 2: Tris–HCl pH 7.2
- Elution buffer: 50 m*M* Tris–HCl buffer pH 7.2, 2 m*M* AMP, 0.4 m*M* NaPPi
- Enzyme storage buffer: 20 m*M* Tris–HCl pH 7.2, 1 m*M* DTT

2.3.2.3 Procedure

1. Transform the mUba1-expressing plasmid into bacterial cells and select on LB plates containing CHL and AMP.
2. For recombinant protein production, follow Steps 1–5 under Section 2.1.3.
3. Produce a clear cell extract in 10 mL of E1 lysis buffer, as indicated in Section 2.1.4.
4. Equilibrate the Ub column using three column volumes (30 mL) of column equilibration buffer.
5. Dilute the E1 extract fivefold with column equilibration buffer. Apply the diluted lysate to the equilibrated column at a flow rate of ~0.5 mL/min. Gravitational flow usually suffices.
6. Wash with 3 column volumes (30 mL) of equilibration buffer.
7. Wash with 3 column volumes (30 mL) of wash buffer 1 (with KCl), followed by additional wash with three column volumes of wash buffer 2 (without KCl).
8. Elute the E1 by applying 5 column volumes of elution buffer (50 mL) and collect the eluate into a designated tube.
9. To change the solvent into a storage buffer, concentrate the E1 eluate to 0.5 mL using a centrifugal filter unit with MWCO of 50–75 kDa.
10. Dilute to 5 mL (10-fold dilution) by adding 4.5 mL storage buffer.
11. Reconcentrate to ~0.5 mL.
12. Redilute to 5 mL with storage buffer.
13. Dispense into suitable aliquots (50–100 μL) in 1.5-mL Eppendorf tubes and store at −80°C.

2.3.2.4 Notes

1. Ub resin maintenance: The Ub-Sepharose column, when washed properly after each purification experiment and kept at 4°C in a buffer containing 0.02% (W:V) azide, is stable and effective for years.
2. E1 is a very stable enzyme that retains full activity after multiple freeze-thaw cycles.

2.4 Purification of Ubc7 and Ubc7: Cue1 heterodimer

Purification of Ubc7 alone, or together with Cue1 is done by intein-mediated purification, employing the intein-mediated purification with an affinity chitin-binding tag (IMPACT) system from New England Biolabs, as previously described (Bazirgan, Garza, & Hampton, 2006). Only Ubc7 bears a Chitin-binding domain, but the copurification of Ubc7 and

Cue1 is enabled due to the high-affinity interaction between Ubc7 and Cue1 (Biederer et al., 1997), which is resistant to the high salt and reducing agent concentrations employed in this procedure.

2.4.1 Equipment

- Glass column (2.5 x 20 cm)
- Dialysis tubing with a MWCO of 12–14 kDa

2.4.2 Reagents and buffers

- The yeast Ubc7 coding sequence is extended at the C-terminus by two HA epitopes. The protein is expressed either alone (pTR525) or together with Cue1$^{\Delta TM}$ (Lacking amino acids 2–22 of its transmembrane domain) (pTR526) from the bi-cistronic plasmid pTYB2 (New England Biolabs). Both plasmids are a generous gift from Prof. R. Hampton, University of California, San Diego.
- Intein lysis buffer (ILB): 50 m*M* Tris–HCl pH 8.0, 0.5 *M* NaCl, 1 m*M* EDTA, 0.1% Triton X-100
- Enzyme storage buffer: 25 m*M* HEPES pH 7.4, 0.7 m*M* sodium phosphate, 137 m*M* NaCl, 5 m*M* KCl, 1 m*M* DTT, 10% glycerol

2.4.3 Procedure

1. Transform plasmids for Ubc7 or Ubc7/Cue1 expression into bacterial expression cells and select on LB plates containing AMP and CHL.
2. Follow Steps 1–5 under Section 2.1.3.
3. Resuspend the PBS-washed cell pellet in 25 mL of ILB with protease inhibitor cocktail.
4. Extract cells as indicated in Steps 1–3 under Section 2.1.4.
5. Add 20 mL of chitin slurry in 20% EtOH to a tube. Centrifuge the tube for 1 min at 1500 × *g* and carefully remove the ethanol.
6. Wash the chitin resin with 40 mL ILB. Centrifuge for 1 min at 1500 × *g* and carefully remove the ILB. Repeat twice.
7. Incubate the cell lysate with 10 mL of the washed chitin resin while rotating for 90 min at 4°C.
8. Pellet the resin by centrifugation at 1500 x *g* for 3 min. Resuspend the beads in 40 mL ILB and transfer into a glass column. Allow the beads to settle and then wash the column with 500 mL of ILB.
9. Resuspend the resin in 5 mL of ILB containing 50 m*M* DTT and transfer to an appropriate 50-mL tube. Incubate at 4°C O/N while rotating. At this step cleavage occurs and the Ubc7-2HA (or the

Ubc7-2HA/Cue1$^{\Delta TM}$ complex) is released from the chitin resin. Pellet the resin by centrifugation at 1500 x *g* for 3 min.

10. Transfer the supernatant into a dialysis tube with an MWCO of 12–14 kDa. Dialyze against 2 L of storage buffer O/N at 4°C with one buffer exchange.
11. To remove insoluble material, centrifuge the dialyzed protein solution at 75,000 x *g* for 1 h at 4°C.
12. Concentrate the purified protein to ~0.5 mL (2–3 mg/mL) using a centrifugal filter with an MWCO of 10–15 kDa. Dispense into suitable aliquots (<50 μL) to minimize repeated freezing–thawing cycles and store at −80°C until use.

2.5 Purification of Doa10 and Hrd1 RING domains

Purification of GST-Hrd1 and GST-Doa10 RING domains (Hrd1R and Doa10R) (aa 287–531 and 1–112, respectively), employs affinity purification on immobilized glutathione.

2.5.1 Materials

- Dialysis tubing with an MWCO of 12–14 KDa
- Disposable polypropylene 10 mL column

2.5.2 Reagents and buffers

- Plasmid pTR120 encodes GST-tagged Doa10R (aa 1–112). A generous gift from M. Hochstrasser (Yale University)
- Plasmid pTR1475 encodes GST-tagged Hrd1 RING domain (aa 287–531)
- Lysis buffer: 10 m*M* Tris–HCl pH 8.0, 150 m*M* NaCl, 0.2% Triton, 1 m*M* DTT
- Wash buffer: 50 m*M* Tris–HCl pH 7.5, NaCl 150 m*M*, 4 m*M* DTT
- Storage buffer: 50 m*M* Tris buffer pH 8, 150 m*M* NaCl, 1 m*M* DTT, 10% glycerol

2.5.3 Procedure

1. Transform plasmids for RING Doa10 or RING Hrd1 expression into bacterial expression cells and select on LB plates containing KAN and CHL.
2. Grow and harvest cells and induce protein expression as in Steps 1–5 under Section 2.1.3.

3. Resuspend pelleted cells in ice-cold lysis buffer containing protease inhibitor cocktail. Extract cells and produce a cleared lysate as describe in steps 1–3 under Section 2.1.4.
4. Wash 2 mL of slurry Glutathione affinity resin in 20% EtOH (Novagen) 3 × with wash buffer.
5. Add the cleared lysate to the washed resin mix and then incubate at 4°C while rotating for 90 min.
6. Spin the resin at 1000 x *g* for 1 min, remove the supernatant, then resuspend resin in 5 mL of lysis buffer and transfer into a column. Drain the column (use gravitational flow to avoid drying of the resin).
7. Wash the column with 50 volumes (100 mL) of lysis buffer, followed by an additional wash with five volumes (10 mL) of wash buffer.
8. Elute bound proteins with 50 m*M* reduced glutathione in wash buffer.
9. Dialyze the glutathione eluate with dialysis tubing with a MWCO of 12–14 KDa at 4°C against 1 L of storage buffer. Aliquot samples and keep them frozen at −80°C until use.

2.5.4 Notes

1. When adding reduced glutathione to the wash buffer, check the pH to make sure it is not too acidic. If needed, neutralize with 1 N NaOH.
2. The final dialysis step in necessary to remove excess glutathione present in the elution buffer.
3. Dispense the final enzyme preparation into small aliquots (<50 μL) to minimize repeated freezing thawing cycles.

2.6 Synthesis of fluorescent Ub (Ub^{FL})

Fluorescein-labeled Ub is used in E2-charging assays. The advantage of using Ub^{FL} is that the amount of Ubc7 ~ Ub^{FL} conjugates can be easily determined using a Typhoon FLA laser scanner or similar device

2.6.1 Equipment

- Cation exchange chromatography set-up

2.6.2 Buffers and reagents

- Plasmid pET14-Ub (pTR839)
- fluorescein-5-maleimide (FL) (Thermo Scientific)
- Labeling solution: 150 m*M* NaCl, 20 m*M* Tris (pH 7.5), and 1 m*M* tris(2-carboxyethyl) phosphine (TCEP)
- β-Mercaptoethanol (BME)

2.6.3 Procedure

1. Replace the codon corresponding to Lys 48 in plasmid pET14-Ub of Ub with a Cys codon by standard site-directed mutagenesis.
2. Label 50–100 μ*M* (~200 μL) of Ub^{C48} with fourfold excess of FL by incubation in labeling buffer for 2 h at RT in the dark.
3. Stop the reaction by adding a 10-fold excess of BME. Further incubate for 5 min in the dark.
4. Purify the labeled protein using a PD miniTrap G25 gravity column (GE Healthcare) preequilibrated with labeling buffer, following the manufacturer protocol.
5. Determine FL coupling efficiency and Ub^{FL} concentration by measuring the optical density at 280 and 495 nm and calculating the 495/280 ratio, as described in the manufacturer protocol (Thermo Scientific).
6. Dispense the labeled Ub into small aliquots (<50 μL) and stored at −80° C until use.

3. Measurement of Ubc7 enzymatic activity

3.1 Measurement of Ubc7 poly-ubiquitylation activity

Efficient poly-Ub chain formation requires that the Ub-charged E2 molecule transfers its bound Ub to an ε-amino acceptor group (Hershko et al., 1983). In the unique case of substrate-independent polyubiquitylation, carried out by Ubc7 in vitro, free poly-Ub chains are initiated by formation of di-Ub adduct. In this reaction, the acceptor Ub provides an ε-amino group of Lys 48 side chain that acts as an acceptor for an incoming donor Ub molecule provided by E2~Ub.

Below is a comprehensive protocol for the in vitro analysis of Ubc7-mediated ubiquitylation. The procedure has been adapted from a previously published procedure (Bazirgan et al., 2006).

3.1.1 Reagents and buffers

- E1: 50 n*M* 6His-mUBA1, final concentration
- E2: 1 μ*M* Ubc7 with or without 1 μ*M* Cue1, final concentration
- E3: 1 μ*M* GST-Doa10 RING or GST-Hrd1 RING, final concentration
- 4 μ*M* Ub or FLAG-Ub, final concentration
- Reaction buffer: 50 m*M* Tris–HCl pH 7.6, 5 m*M* $MgCl_2$, 4 m*M* ATP

3.1.2 Procedure

1. Mix together the reaction buffer, Ubc7/Cue1, Ub, with or without E3 RING domains in a final volume of 30 μL.
2. Warm the reaction mix for 10 min at 30°C.

3. Initiate the ubiquitylation reaction by adding 50 nM E1 enzyme.
4. Keep reaction mix at 30°C for the desired time points (usually 10–120 min).
5. Stop the reactions by adding 4 × sample buffer with DTT to each aliquot. Incubate at 95°C for 5 min.
6. Resolve Ub conjugates by separating proteins by 5–15% SDS–PAGE.
7. Transfer to PVDF 0.45 μM membrane (Hybond™Amersham).
8. Reveal Ub conjugates by immunoblotting with anti-Ub or anti-Flag antibodies.

3.1.3 Notes

1. We detect mono- and poly-Ub by immunoblot analysis with primary rabbit anti-Ub (Dako, Glostrup, Denmark), an Agilent company (Santa Clara, USA), or with anti-Flag antibodies (Sigma).
2. Fig. 5B shows a typical in vitro ubiquitylation reaction with and without Doa10 RING domain. Time 0 refers to the time of E1 addition after 10 min of preincubation of all the components without the E1.

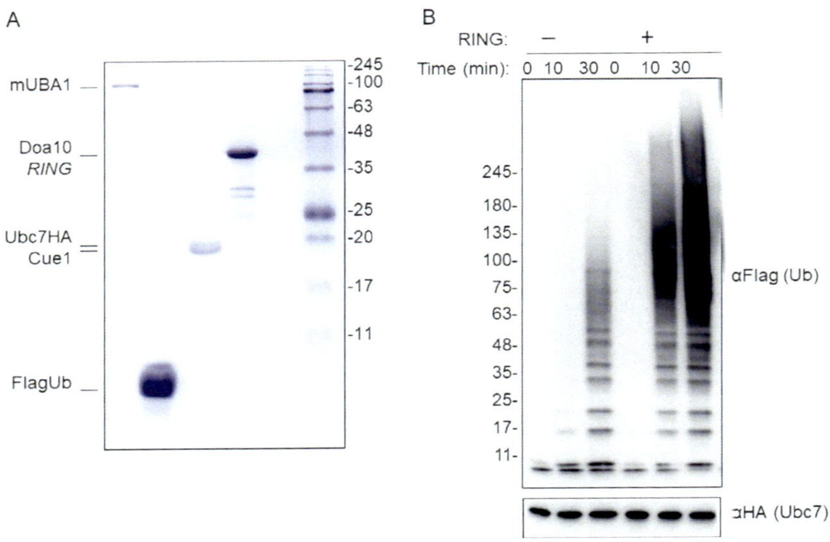

Fig. 5 In vitro formation of poly-ubiquitin chain. Ubc7-mediated poly-Ub chain formation with or without the Doa10 RING domain is determined by in vitro ubiquitylation reaction. (A) Purified components used for the in vitro reaction. (B) The different components were incubated at 30°C for 10 min without the E1. The reaction is initiated by adding the E1 (Time 0). Aliquots were taken at the indicated time points and proteins were separated on 5–15% SDS–PAGE, followed by immunoblotting using anti-Flag Abs.

3.2 Measurement of Ubc7 ∼ Ub thioester formation (Ubc7 charging)

The Ubc7 charging reaction utilizes either an Ub derivative that is labeled with fluorescein at position 48 (Ub^{FL}), facilitating direct measurement of E2 ∼ Ub, or Flag-tagged Ub^{R48} (pTR1544) requiring immunoblot analysis for detection. The Ub mutant in which Lys 48 has been replaced with Arg prevents it from acting as an efficient Ub acceptor (as is also true for Ub^{FL}). Hence, for most reactions employing Ub^{R48}, only one cycle of E2-thioester bond formation occurs. Below we describe in detail a method to produce biochemically Ubc7 ∼ Ub species

3.2.1 Reagents and buffers

- E1: 50 n*M* 6His-mUBA1, final concentration
- E2: 1 μ*M* Ubc7/Cue1, final concentration
- 20 μ*M* Ub^{FL} or $FlagUb^{R48}$, final concentration
- Reaction buffer: 30 m*M* HEPES pH 7.6, 10 m*M* $MgCl_2$, 100 m*M* NaCl, 200 μ*M* ATP

3.2.2 Procedure

1. Incubate E1, Ubc7(Cue1), and Ub^{FL} or $FlagUb^{R48}$ in 50 μL reaction buffer at 30°C.
2. Keep reaction mix at 30°C for the desired time points (usually 10–120 min).
3. Stop the reactions by adding 4 × sample buffer without DTT to each aliquot. Resolve Ubc7 ∼ Ub conjugates by separating by 15% non-reducing SDS–PAGE at 4°C with precooled running buffer.
4. Reveal Ub conjugates by either using a Typhoon laser scanner (GE Healthcare) for Ub^{FL} detection or by transfer to a PVDF 0.45 μ*M* membrane (Hybond™Amersham) and immunoblotting with anti-Flag antibodies for $FlagUb^{R48}$ detection.

3.2.3 Notes

1. This reaction is enhanced in the presence of Cue1 via allosteric activation through its U7BR (Metzger, Liang, et al., 2013).
2. Use reaction mix incubated for the longest time point as a negative control for thioester bond formation. Add 4 × sample buffer without DTT and incubate at 95°C for 5 min.
3. For $FlagUb^{R48}$, only the unreduced sample shows a major band of ∼33 kDa corresponding to Ub at the active-site Cys of Ubc7 (Fig. 6). The ∼50 kDa low abundance band, shown in Fig. 6, may represent an Ubc7 ∼ Ub^{R48}–Ub^{R48} species with an Ub–Ub linkage different than Lys 48.

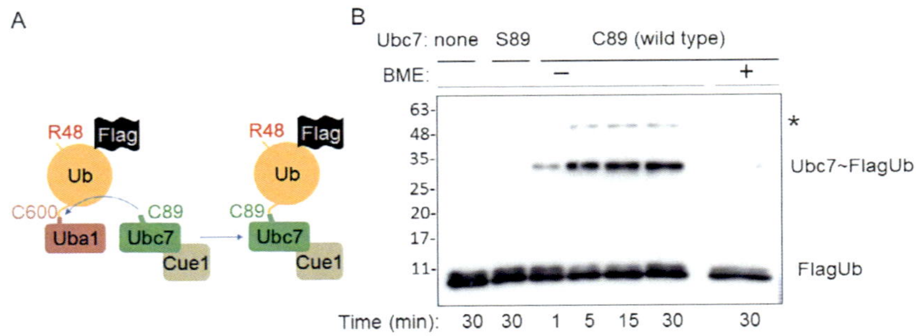

Fig. 6 Determination of Ubc7 charge reaction. (A) Ubiquitin trans-thiolation is determined using $FlagUb^{R48}$ that cannot function as an acceptor for poly-Ub chain formation. The thioester bond between Ub and the active-site Cys of mUBA1 is subjected to nucleophilic attack by the thiolate of Cys89 on Ubc7 active site, to form a covalent Ubc7 ~ $FlagUb^{R48}$ adduct. (B) Samples are taken after 1, 5, 15, and 30 min. An Ubc7 ~ $FlagUb^{R48}$ adduct is determined by immunoblotting with anti-Flag Abs, after separation on 15% SDS–PAGE under reducing and nonreducing conditions. Reactions with no Ubc7 or with active-site mutant (S89) serve as negative controls. BME, *β-mercaptoethanol*; *, Unidentified band.

3.3 Measurement of a single ubiquitin transfer reaction

Detailed mechanistic studies on Ub chain formation require tools that enable the detection of a single Ub transfer reaction. To this end, one has to overcome the ability of the wild-type Ub to serve as both donor and acceptor. This can be done by generating Ub mutants that can only serve as donor or acceptor. For instance, Ub that lacks the last two Gly residues at its C-terminus ($Ub^{\Delta GG}$) can only serve as acceptor; and Ub that has a mutation at Lys 48 (Ub^{R48}) can only serve as donor Ub. Previously it has been suggested that Ube2g2, the mammalian orthologue of Ubc7, functions as a homodimer during poly-Ub chain elongation (Liu et al., 2014). The emerging model suggests that two Ube2g2 molecules bind to each other in a front-to-back orientation with each Ube2g2 protomer carrying an Ub molecule covalently at the active site, one serves as a donor and the other as an acceptor.

To test if yeast Ubc7 acts as a dimer in a similar fashion, we set up a reaction using as an acceptor $Ub^{\Delta GG}$, which cannot bind covalently to the E2. Below is a protocol for detecting di-Ub produced by a single Ub transfer reaction, employing a preformed donor Ubc7 ~ Ub and uncharged $Ub^{\Delta GG}$ as an acceptor. We find that, in contrast to Ube2g2, transfer of donor Ub to an acceptor Ub by Ubc7 do not require that both molecules form a thioester bond. This is evident by the ability of both Ub and $Ub^{\Delta GG}$ to act as equivalent Ub acceptors (Fig. 7B).

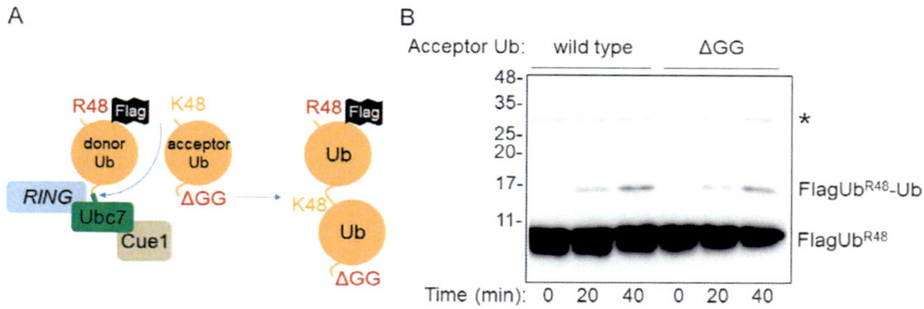

Fig. 7 Determining a single Ub transfer reaction. (A) The formation of Ub–Ub adduct is measured using $FlagUb^{R48}$ that functions as a donor, and a mutant Ub lacking Gly75 and Gly76 (ΔGG) as an acceptor. The Ub~Ubc7 thioester bond is subjected to nucleophilic attack by the ε-amino group on Lys 48 of the acceptor Ub, to form a covalent $FlagUb\text{-}Ub^{\Delta GG}$ adduct. (B) Samples are taken at time 0 and after 20 and 40 min. The newly formed di-Ub adduct is determined after separation on 15% SDS–PAGE by immunoblotting with anti-Flag Abs. The reaction with wild-type Ub is shown for comparison purposes. *, Unidentified band.

3.3.1 Reagents and buffers

- E1: 50 n*M* 6His-mUBA1, final concentration
- 2 μ*M* Ubc7, final concentration
- 20 μ*M* Ub, final concentration
- 20 μ*M* $Ub^{\Delta GG}$, final concentration
- 2 μ*M* Ubc7/Cue1 complex, final concentration
- 20 μ*M* $FlagUb^{R48}$, final concentration
- 5000 mU/mL Apyrase (New England Biolabs)
- Reaction buffer: 40 m*M* HEPES pH 7.6, 100 m*M* NaCl, 10 m*M* $MgCl_2$, 200 μ*M* ATP

3.3.2 Procedure

1. Formation of Ubc7:Cue1-Ub donor complex: Incubate Ubc7/Cue1 complex with $FlagUb^{R48}$ and E1, in reaction buffer at total volume of 60 μL for 30 min at 30°C.
2. Add Apyrase and incubate at 30°C for further 5 min to deplete ATP.
3. As an acceptor, add untagged Ub or $Ub^{\Delta GG}$ in reaction buffer at total volume of 30 μL.
4. Incubate the "donor" and "acceptor" mixture at 30°C for the desired time periods.
5. Stop the single cycle transfer reaction by addition of 4 × SDS–PAGE loading buffer containing 200 mg/mL DTT.

6. Heat at 95°C for 5 min, spin down and resolve samples in a 15% SDS–PAGE acrylamide gel.
7. Transfer to PVDF 0.45 μ*M* membrane (Hybond™Amersham).
8. Subject the membrane to immunoblot analysis with anti-Flag. The di-Ub adduct is detected as a distinct ~20 kDa band.

3.3.3 Notes

1. The incubation time required for formation of visible di-Ub should be determined empirically as it depends on both the amount of Ubc7 in the charging reaction and the amount of the respective complexes used for the transfer reaction.
2. To test the contribution of E3 ligases to Ub transfer, perform in parallel identical reaction where one contains, either purified Doa10R or Hrd1R GST fusion proteins (at a final concentration of 2.5 μ*M*).

4. Final remarks

The Ub system was discovered nearly 40 years ago through reconstitution of the enzymatic reaction in a cell-free system (Ciechanover et al., 1978), and the principles of the Ub conjugation system were resolved by using purified E1, E2, and E3 enzymes (Hershko et al., 1983). Since then, genetics and cell biology approaches have dominated Ub-system research, achieving the identification of multiple system components, their physiological function and mode of regulation. Nevertheless, the investigation of the mode of action of Ub enzymes still largely depends on our ability to isolate enzymes and test their activity under controlled conditions in vitro. In this chapter, we have described methods to purify proteins required to investigate the intrinsic activity of Ubc7, a key component of the ERAD Ub-conjugation system. These methods have been used by us to decipher the relative contribution of Ubc6 and Ubc7 to polyubiquitylation by the Doa10 pathway (Weber et al., 2016), and to assess the distinct mode of activation of Ubc7 by its cognate E3 ligases, Hrd1 and doa10 (Cohen et al., 2015).

At present, several protein components of the Ub system are available from commercial resources. However, relying on these resources for extensive in vitro research may be problematic due to their limited repertoire and expense. Alternatively, the methods described herein can yield a large variety of intact and mutant proteins. Notably, the same principles for E1 binding to the resin and its release (Section 2.3) can be implemented for purifying E1 components of other Ub-like conjugation systems, such as Neddylation,

SUMOylation, or UFMylation. In addition, the protocol for Ub purification, which is based on the unique properties of the molecule that grant it resistance to various stress conditions, is simple and relatively fast; thus it can be used for purification of multiple Ub variants and/or other Ub-like proteins.

Acknowledgment

We wish to thank Dr. A. Shiber and Dr. I Cohen for their immense contribution to setting up the protein purification and the in vitro assays. The authors thank A. Ciechanover, R. Hampton, and M. Hochstrasser for plasmids. This work is kindly supported by the Israel Science Foundation (grants 786/08, 119/14).

References

Amm, I., Sommer, T., & Wolf, D. H. (2014). Protein quality control and elimination of protein waste: The role of the ubiquitin-proteasome system. *Biochimica et Biophysica Acta, 1843*(1), 182–196. https://doi.org/10.1016/j.bbamcr.2013.06.031.

Bagola, K., von Delbruck, M., Dittmar, G., Scheffner, M., Ziv, I., Glickman, M. H., et al. (2013). Ubiquitin binding by a CUE domain regulates ubiquitin chain formation by ERAD E3 ligases. *Molecular Cell, 50*(4), 528–539. https://doi.org/10.1016/j.molcel.2013.04.005.

Bazirgan, O. A., Garza, R. M., & Hampton, R. Y. (2006). Determinants of RING-E2 fidelity for Hrd1p, a membrane-anchored ubiquitin ligase. *The Journal of Biological Chemistry, 281*(51), 38989–39001.

Biederer, T., Volkwein, C., & Sommer, T. (1997). Role of Cue1p in ubiquitination and degradation at the ER surface. *Science, 278*, 1806–1809.

Bordallo, J., Plemper, R. K., Finger, A., & Wolf, D. H. (1998). Der3p/Hrd1p is required for endoplasmic reticulum-associated degradation of misfolded lumenal and integral membrane proteins. *Molecular Biology of the Cell, 9*(1), 209–222.

Carvalho, P., Goder, V., & Rapoport, T. A. (2006). Distinct ubiquitin-ligase complexes define convergent pathways for the degradation of ER proteins. *Cell, 126*(2), 361–373.

Chen, P., Johnson, P., Sommer, T., Jentsch, S., & Hochstrasser, M. (1993). Multiple ubiquitin-conjugating enzymes participate in the in vivo degradation of the yeast MAT alpha 2 repressor. *Cell, 74*(2), 357–369.

Ciechanover, A., Hod, Y., & Hershko, A. (1978). A heat-stable polypeptide component of an ATP-dependent proteolytic system from reticulocytes. *Biochemical and Biophysical Research Communications, 81*(4), 1100–1105.

Cohen, I., Wiener, R., Reiss, Y., & Ravid, T. (2015). Distinct activation of an E2 ubiquitin-conjugating enzyme by its cognate E3 ligases. *Proceedings of the National Academy of Sciences of the United States of America, 112*(7), E625–E632. https://doi.org/10.1073/pnas.1415621112.

Das, R., Liang, Y. H., Mariano, J., Li, J., Huang, T., King, A., et al. (2013). Allosteric regulation of E2:E3 interactions promote a processive ubiquitination machine. *The EMBO Journal, 32*(18), 2504–2516. https://doi.org/10.1038/emboj.2013.174.

Denic, V., Quan, E. M., & Weissman, J. S. (2006). A luminal surveillance complex that selects misfolded glycoproteins for ER-associated degradation. *Cell, 126*(2), 349–359.

Erijman, A., Shifman, J. M., & Peleg, Y. (2014). A single-tube assembly of DNA using the transfer-PCR (TPCR) platform. *Methods in Molecular Biology, 1116*, 89–101. https://doi.org/10.1007/978-1-62703-764-8_7.

Foresti, O., Rodriguez-Vaello, V., Funaya, C., & Carvalho, P. (2014). Quality control of inner nuclear membrane proteins by the Asi complex. *Science*, *346*(6210), 751–755. https://doi.org/10.1126/science.1255638.

Gasteiger, E., Hoogland, C., Gattiker, A., Duvaud, S. e., Wilkins, M. R., Appel, R. D., et al. (2005). Protein identification and analysis tools on the ExPASy server. In J. M. Walker (Ed.), *The proteomics protocols handbook* (pp. 571–607). Totowa, NJ: Humana Press.

Haas, A. L., Warms, J. V., Hershko, A., & Rose, I. A. (1982). Ubiquitin-activating enzyme. Mechanism and role in protein-ubiquitin conjugation. *The Journal of Biological Chemistry*, *257*(5), 2543–2548.

Hampton, R. Y., & Bhakta, H. (1997). Ubiquitin-mediated regulation of 3-hydroxy-3-methylglutaryl-CoA reductase. *Proceedings of the National Academy of Sciences of the United States of America*, *94*(24), 12944–12948.

Hampton, R. Y., Gardner, R. G., & Rine, J. (1996). Role of 26S proteasome and HRD genes in the degradation of 3-hydroxy-3-methylglutaryl-CoA reductase, an integral endoplasmic reticulum membrane protein. *Molecular Biology of the Cell*, 7(12), 2029–2044.

Hershko, A., Heller, H., Elias, S., & Ciechanover, A. (1983). Components of ubiquitin-protein ligase system. Resolution, affinity purification, and role in protein breakdown. *The Journal of Biological Chemistry*, *258*(13), 8206–8214.

Hochstrasser, M., & Varshavsky, A. (1990). In vivo degradation of a transcriptional regulator: The yeast a2 repressor. *Cell*, *61*(4), 697–708.

Johnston, S. C., Larsen, C. N., Cook, W. J., Wilkinson, K. D., & Hill, C. P. (1997). Crystal structure of a deubiquitinating enzyme (human UCH-L3) at 1.8 a resolution. *The EMBO Journal*, *16*(13), 3787–3796.

Khmelinskii, A., Blaszczak, E., Pantazopoulou, M., Fischer, B., Omnus, D. J., Le Dez, G., et al. (2014). Protein quality control at the inner nuclear membrane. *Nature*, *516*(7531), 410–413. https://doi.org/10.1038/nature14096.

Ko, J., Park, H., Heo, L., & Seok, C. (2012). GalaxyWEB server for protein structure prediction and refinement. *Nucleic Acids Research*, *40*(Web Server issue), W294–W297. https://doi.org/10.1093/nar/gks493.

Lenk, U., Yu, H., Walter, J., Gelman, M. S., Hartmann, E., Kopito, R. R., et al. (2002). A role for mammalian Ubc6 homologues in ER-associated protein degradation. *Journal of Cell Science*, *115*, 3007–3014. Pt 14.

Lester, D., Farquharson, C., Russell, G., & Houston, B. (2000). Identification of a family of noncanonical ubiquitin-conjugating enzymes structurally related to yeast UBC6. *Biochemical and Biophysical Research Communications*, *269*(2), 474–480. https://doi.org/10.1006/bbrc.2000.2302.

Li, W., Tu, D., Brunger, A. T., & Ye, Y. (2007). A ubiquitin ligase transfers preformed polyubiquitin chains from a conjugating enzyme to a substrate. *Nature*, *446*(7133), 333–337.

Liu, W., Shang, Y., Zeng, Y., Liu, C., Li, Y., Zhai, L., et al. (2014). Dimeric Ube2g2 simultaneously engages donor and acceptor ubiquitins to form Lys48-linked ubiquitin chains. *The EMBO Journal*, *33*(1), 46–61. https://doi.org/10.1002/embj.201385315.

Lorick, K. L., Jensen, J. P., Fang, S., Ong, A. M., Hatakeyama, S., & Weissman, A. M. (1999). RING fingers mediate ubiquitin-conjugating enzyme (E2)-dependent ubiquitination. *Proceedings of the National Academy of Sciences of the United States of America*, *96*(20), 11364–11369.

Mehrtash, A. B., & Hochstrasser, M. (2018). Ubiquitin-dependent protein degradation at the endoplasmic reticulum and nuclear envelope. *Seminars in Cell & Developmental Biology*. https://doi.org/10.1016/j.semcdb.2018.09.013.

Metzger, M. B., Liang, Y. H., Das, R., Mariano, J., Li, S., Li, J., et al. (2013). A structurally unique E2-binding domain activates ubiquitination by the ERAD E2, Ubc7p, through multiple mechanisms. *Molecular Cell*, *50*(4), 516–527. https://doi.org/10.1016/j.molcel.2013.04.004.

Metzger, M. B., Maurer, M. J., Dancy, B. M., & Michaelis, S. (2008). Degradation of a cytosolic protein requires endoplasmic reticulum-associated degradation machinery. *The Journal of Biological Chemistry*, *283*(47), 32302–32316. doi:M806424200. (pii). https://doi.org/10.1074/jbc.M806424200.

Metzger, M. B., Pruneda, J. N., Klevit, R. E., & Weissman, A. M. (2013). RING-type E3 ligases: Master manipulators of E2 ubiquitin-conjugating enzymes and ubiquitination. *Biochimica et Biophysica Acta—Molecular Cell Research*, *1843*, 47–60. https://doi.org/10.1016/j.bbamcr.2013.05.026.

Mori, S., Tanaka, K., Kanaki, H., Nakao, M., Anan, T., Yokote, K., et al. (1997). Identification of an ubiquitin-ligation system for the epidermal-growth-factor receptor—herbimycin a induces in vitro ubiquitination in rabbit-reticulocyte lysate. *European Journal of Biochemistry*, *247*(3), 1190–1196.

Ó'Fágáin, C., & Colliton, K. (2017). Storage and Lyophilization of pure proteins. In D. Walls & S. T. Loughran (Eds.), *Protein chromatography: Methods and protocols* (pp. 159–190). New York, NY: Springer New York.

Pickart, C. M., & Raasi, S. (2005). Controlled synthesis of polyubiquitin chains. In *Vol. 399. Methods in enzymology* (pp. 21–36). Academic Press.

Ravid, T., & Hochstrasser, M. (2007). Autoregulation of an E2 enzyme by ubiquitin-chain assembly on its catalytic residue. *Nature Cell Biology*, *9*(4), 422–427.

Ravid, T., Kreft, S. G., & Hochstrasser, M. (2006). Membrane and soluble substrates of the Doa10 ubiquitin ligase are degraded by distinct pathways. *The EMBO Journal*, *25*(3), 533–543 (pii). https://doi.org/10.1038/sj.emboj.7600946.

Seol, J. H., Feldman, R. M., Zachariae, W., Shevchenko, A., Correll, C. C., Lyapina, S., et al. (1999). Cdc53/cullin and the essential Hrt1 RING-H2 subunit of SCF define a ubiquitin ligase module that activates the E2 enzyme Cdc34. *Genes & Development*, *13*(12), 1614–1626.

Skowyra, D., Koepp, D. M., Kamura, T., Conrad, M. N., Conaway, R. C., Conaway, J. W., et al. (1999). Reconstitution of G1 cyclin ubiquitination with complexes containing SCFGrr1 and Rbx1. *Science*, *284*(5414), 662–665.

Swanson, R., Locher, M., & Hochstrasser, M. (2001). A conserved ubiquitin ligase of the nuclear envelope/endoplasmic reticulum that functions in both ER-associated and Matalpha2 repressor degradation. *Genes & Development*, *15*(20), 2660–2674.

Vashist, S., & Ng, D. T. (2004). Misfolded proteins are sorted by a sequential checkpoint mechanism of ER quality control. *The Journal of Cell Biology*, *165*(1), 41–52.

von Delbrück, M., Kniss, A., Rogov, V. V., Pluska, L., Bagola, K., Löhr, F., et al. (2016). The CUE domain of Cue1 aligns growing ubiquitin chains with Ubc7 for rapid elongation. *Molecular Cell*, *62*(6), 918–928. https://doi.org/10.1016/j.molcel.2016.04.031.

Weber, A., Cohen, I., Popp, O., Dittmar, G., Reiss, Y., Sommer, T., et al. (2016). Sequential poly-ubiquitylation by specialized conjugating enzymes expands the versatility of a quality control ubiquitin ligase. *Molecular Cell*, *63*(5), 827–839. https://doi.org/10.1016/j.molcel.2016.07.020.

CHAPTER FIVE

Methods for genetic analysis of mammalian ER-associated degradation

Dara E. Leto, Ron R. Kopito*
Department of Biology, Stanford University, Stanford, CA, United States
*Corresponding author: e-mail address: kopito@stanford.edu

Contents

Abstract

Identification and degradation of misfolded proteins by the ubiquitin-proteasome system (UPS) is crucial for maintaining proteostasis, but only a handful of UPS components have been linked to the recognition of specific substrates. Studies in *Saccharomyces cerevisiae* using systematic perturbation of nonessential genes have uncovered UPS components that recognize and ubiquitylate model substrates of the UPS; however, similar analyses in metazoans have been limited. In this chapter, we describe methods for using CRISPR/Cas9 technology combined with genome-wide high complexity single guide (sgRNA) libraries and a transcriptional shutoff strategy for phenotypic selection based on kinetic measurements of protein turnover to identify the genes required to degrade model clients of the mammalian ER-associated degradation system. We also discuss considerations for screen design, execution, and interpretation.

Methods in Enzymology, Volume 619
ISSN 0076-6879
https://doi.org/10.1016/bs.mie.2019.01.006
© 2019 Elsevier Inc.
All rights reserved.

1. Introduction

Newly synthesized secretory and membrane proteins undergo conformational maturation within the endoplasmic reticulum (ER), where they are subject to strict quality control before being trafficked to distal compartments of the secretory system. Proteins that fail to correctly fold or assemble are identified, ubiquitylated, removed from the ER, and destroyed by the cytosolic ubiquitin-proteasome system (UPS) in a process referred to as ER-associated degradation (ERAD) (Olzmann, Kopito, & Christianson, 2013). Deployment of misfolded, aggregation-prone proteins to other secretory system compartments or the extracellular space can have catastrophic effects on cellular and organismal health (Guerriero & Brodsky, 2012). Determining how the ERAD system distinguishes terminally misfolded proteins within the highly heterogeneous population of nascent proteins that traverse the ER is therefore essential for understanding the molecular basis of many human diseases.

Genetic analyses performed in yeast were instrumental in uncovering the core machinery and basic mechanisms of ERAD. A series of landmark genetic screens analyzing the degradation of short-lived proteins, CPY*, Hmg2p, and Matα2, in randomly mutagenized cells, identified the membrane-embedded E3 ligases, Hrd1 and Doa10, that are required for essentially all ERAD in yeast (Hampton, Gardner, & Rine, 1996; Knop, Finger, Braun, Hellmuth, & Wolf, 1996; Swanson, Locher, & Hochstrasser, 2001). Although highly successful, these studies, which relied on direct assessment of substrate turnover by immunoblotting (Knop et al., 1996) or indirect assessment of protein turnover by growth-based selection (Hampton, Gardner, et al., 1996; Swanson et al., 2001), were limited in throughput and sensitivity. An optical reporter composed of an Hmg2p variant fused to GFP was developed to enable direct, high-throughput measurements of ERAD function in real-time in intact cells (Hampton, Koning, Wright, & Rine, 1996). Using this strategy, mutants with increased cell fluorescence due to impaired degradation can be identified by fluorometry, live-cell microscopy, or flow cytometry. Fusion of fluorescent proteins to a wide variety of short-lived proteins has subsequently been used to uncover genes required to degrade diverse misfolded proteins via ERAD (Buschhorn, Kostova, Medicherla, & Wolf, 2004; Cronin, Khoury, Ferry, & Hampton, 2000; Medicherla, Kostova, Schaefer, & Wolf, 2004; Neal et al., 2018) and other UPS-dependent systems in yeast (Heck, Cheung, & Hampton, 2010).

The mammalian ERAD system is far more complex than yeast ERAD, likely reflecting the expanded size and diversity of the secretory proteome that it regulates. At least a dozen ERAD E3 ligases and a large cohort of accessory factors have been identified, and while presumably recognizing unique subsets of misfolded proteins, most of these E3 ligases are poorly characterized and have few known substrates (Ruggiano, Foresti, & Carvalho, 2014). Early attempts by us and others to use genetic screens to identify and functionally dissect the molecular players in mammalian ERAD relied on steady-state measurements of fluorescent ERAD reporter levels for phenotypic selections and RNAi/shRNA libraries to perturb gene function (Burr et al., 2011; Christianson et al., 2011; Stagg et al., 2009; van de Weijer et al., 2014). While successful at identifying genes that are strongly required for degradation of ERAD reporters, these screens used focused gene perturbation libraries that probed only a subset of the genome, suffered from limited sensitivity, and displayed high rates of false positives and false negatives. Recent advances in genome editing technology, notably the development of efficient and precise CRISPR/Cas9-based gene knockouts, high complexity genome-wide single guide RNA (sgRNA) libraries, and pooled screening methodologies, now permit unbiased whole-genome quantitative phenotype analysis of complex pathways in mammalian cells. Application of these technologies to study mammalian ERAD has enabled highly sensitive detection of genes affecting protein turnover, empowering the discovery of novel ERAD components and uncovering surprising diversity in the mammalian ERAD pathways that recognize and handle distinct misfolded proteins (Ju et al., 2018; Leto et al., 2019; Stefanovic-Barrett et al., 2018; Timms et al., 2016).

Here, we describe protocols that utilize cell lines with acute, controllable expression of a misfolded protein and a phenotypic selection based on kinetic measurements of fluorescent reporter turnover in intact cells to identify genes that regulate protein degradation by mammalian ERAD. Several recent publications summarize available gene perturbation reagents and provide detailed protocols for introducing gene targeting libraries and preparing next-generation sequencing libraries (Deans et al., 2016; Kampmann, 2018; Kampmann, Bassik, & Weissman, 2014; Morgens, Deans, Li, & Bassik, 2016; Morgens et al., 2017; Shang, Wang, Fan, & Wang, 2017). We focus on methods for generating and validating reporter cell lines and for performing fluorescence-activated cell sorting (FACS)-based phenotypic selections. While this chapter focuses on the study of ERAD, the methods and reagents described are widely applicable for genetic analysis of UPS-dependent degradation in mammalian cells.

2. Materials

2.1 Reagents

1. TetON cell line. If not available, a TetON line can be generated using reagents and protocols from Takara Bio, Inc.
2. Lentiviral plasmid encoding the ERAD substrate under the control of a tetracycline (tet)-inducible promoter and a selection marker under the control of a constitutive promoter (see Note 1). To enable rapid transcriptional shutoff, we express the ERAD substrate from an mRNA without a downstream poly(A) tail, which destabilizes the mRNA.
3. Lentiviral plasmid encoding Cas9 WT or a Cas9 variant (Addgene).
4. Third generation lentiviral plasmids: pVSVG, pMDL, pRSV (Addgene).
5. sgRNA library.
6. HEK293T cells and K562 cells (ATCC).
7. DMEM cell culture medium—with and without phenol red (Corning).
8. RPMI cell culture medium—with and without phenol red (Corning).
9. Fetal bovine serum (FBS) (Sigma-Aldrich).
10. 100 × L-glutamine (200 m*M*; Sigma-Aldrich).
11. 1 × sterile Phosphate-buffered saline (PBS) (Corning).
12. Trypsin–EDTA (GIBCO).
13. Cryovials (Corning).
14. TransIT-LT1 transfection reagent (Mirus Bio).
15. Polybrene (Sigma-Aldrich).
16. Sterile 0.45 μ*M* Millex Durapore PVDF syringe filters (Sigma-Aldrich).
17. 5-mL sterile Luer lock syringes.
18. Selection antibiotic(s).
19. Doxycycline (Sigma-Aldrich).
20. Tubes: 5-mL FACS tubes (Corning), sterile 5-mL FACS tubes with cell strainer snap cap (Corning), 1.5-mL microcentrifuge tubes (Fisher Scientific), sterile 14- and 50-mL conical tubes (Corning).
21. Sterile tissue culture-treated plates: 10 cm, 6-well, and 12-well (Corning).
22. Sterile tissue culture-treated flasks: T25 and T225 (Corning).
23. UPS inhibitors: MG132 (Sigma-Aldrich), NMS-873 (Sigma-Aldrich), TAK-243 (MedChem Express); resuspend inhibitors in DMSO (Fisher Scientific).
24. Emetine (Sigma-Aldrich).
25. β-Mercaptoethanol (Sigma-Aldrich).
26. BCA Protein Assay Kit (Pierce).

2.2 Equipment

1. Sonicator (Fisher Scientific).
2. Flow cytometer equipped with 405, 488, and 532 lasers.
3. Cell sorter equipped with 405, 488, and 532 lasers.
4. Cryo-freezing containers.
5. Centrifuge with a swinging bucket rotor and plate adapters.

2.3 Solutions

1. 1% SDS lysis buffer: 1% SDS, 50 m*M* Tris–HCl pH 7.5, 150 m*M* NaCl, 2.5 m*M* EDTA, 100 m*M* PMSF, cOmplete EDTA-free protease inhibitor cocktail at the manufacturer's recommended concentration (Sigma-Aldrich; Catalog #11873580001).
2. 1% TX-100 solubilization buffer: 1% TX-100, 50 m*M* Tris–HCl pH 7.5, 150 m*M* NaCl, 5 m*M* EDTA, 20 n*M* *N*-ethylmaleimide, 1 m*M* PMSF, cOmplete EDTA-free protease inhibitor cocktail.
3. 1% SDS solubilization buffer: 1% SDS, 50 m*M* Tris–HCl pH 7.5, 150 m*M* NaCl, 5 m*M* EDTA, 20 n*M* *N*-ethylmaleimide, 1 m*M* PMSF, cOmplete EDTA-free protease inhibitor cocktail.
4. Complete tissue culture medium: DMEM supplemented with 10% FBS or RPMI supplemented with 10% FBS and 2 m*M* L-glutamine.
5. Cell freezing medium: 10% DMSO, 90% FBS.
6. 5 × Laemmli buffer: 250 m*M* Tris–HCl, pH 7.0, 10% SDS, 20% glycerol, 0.05% bromphenol blue.

2.4 Software

1. FlowJo (FlowJo, LLC) or equivalent flow cytometry analysis package.
2. ImageStudio (LI-COR Biosciences) or equivalent image analysis software.
3. Bowtie version 1.2.2 (Langmead, Trapnell, Pop, & Salzberg, 2009).
4. casTLE version 1.0 (Morgens et al., 2016) or equivalent screen analysis statistical package.

3. Methods

3.1 Considerations for screen design

3.1.1 Cell type

The cell type used for screening should be selected based on biological and technical considerations. CRISPR/Cas9-mediated gene knockout screens are dependent on the nonhomologous end-joining (NHEJ) DNA repair pathway and must be conducted in a cell line in which this pathway is active. The availability of a comprehensive sgRNA library to target genes in a given species must also be evaluated when selecting a cell line for screening. Genome-wide genetic screens require that a large population of cells (typically $>10^8$) be efficiently virally transduced with an sgRNA library and passaged for several generations. The use of FACS to enrich cell populations introduces additional considerations, including the tendency of cells to form clumps, their viability when incubated at high densities on ice for several hours, and their propensity to shear during cell sorting. Suspension cell lines such as K562 (Bassik et al., 2013), Jurkat (Birsoy et al., 2015), and Ramos (Morgens et al., 2017) are commonly used for genetic screens because they can be cultured at high densities, do not require trypsinization for passaging, and grow stably over many generations. However, adherent cell lines, including HeLa (Savidis et al., 2016), HEK293T (Ju et al., 2018), HepG2 (Xia et al., 2016), and mouse embryonic stem cells (Ruiz et al., 2016), have also been used successfully for pooled forward genetic screening.

3.1.2 Dynamic range

To identify genes that affect protein turnover, it is essential that the ERAD reporter displays a sufficient dynamic range to allow for efficient separation of cell populations with different degradation phenotypes. A strongly over-expressed ERAD reporter can display a decreased dynamic range due to saturation of the degradation machinery or aggregation. Constitutive over-expression of exogenous proteins may constitute a selective pressure to drive cell adaptation (Neal et al., 2018), a problem that is likely exacerbated by the use of genomically unstable cancer cells for genetic screening.

Reporter dynamic range is also influenced by the protein's intrinsic half-life, with longer-lived proteins displaying smaller changes in steady-state levels than shorter-lived proteins when degradation is impaired. Many ERAD substrates have relatively long half-lives of >30 min and display small changes in steady-state levels, even upon complete inhibition of their degradation. This is a commonly encountered situation that has limited the resolution of previous genetic analyses of mammalian ERAD.

To overcome this limitation, we recently described a screening strategy that uses the tetracycline (tet) ON system to transiently express fluorescent ERAD reporters (Leto et al., 2019). The reporter is expressed from a tet-inducible promoter and without a downstream poly(A) tail, allowing for rapid induction and subsequent elimination of reporter mRNA by adding and then removing the tetracycline analog, doxycycline (dox). Destabilizing reporter mRNA by excluding a poly(A) tail provides several benefits, including enabling rapid transcript elimination and an overall decrease in reporter expression levels (thus minimizing artifacts due to saturation of degradation machinery or aggregation). We have found that for ERAD reporters, especially for those with half-lives >30 min, evaluating fluorescence levels after shutting off new synthesis greatly expands reporter dynamic range and improves the recovery of modest modulators of reporter degradation in pooled genetic screens (see Fig. 1). However, it should be noted that mutations in RNA decay pathways will also increase fluorescence by stabilizing reporter mRNA; thus, these screen artifacts should be removed from downstream analysis and all screen hits should be carefully validated by testing their effect on reporter protein turnover using a translation shutoff (or pulse-chase) assay.

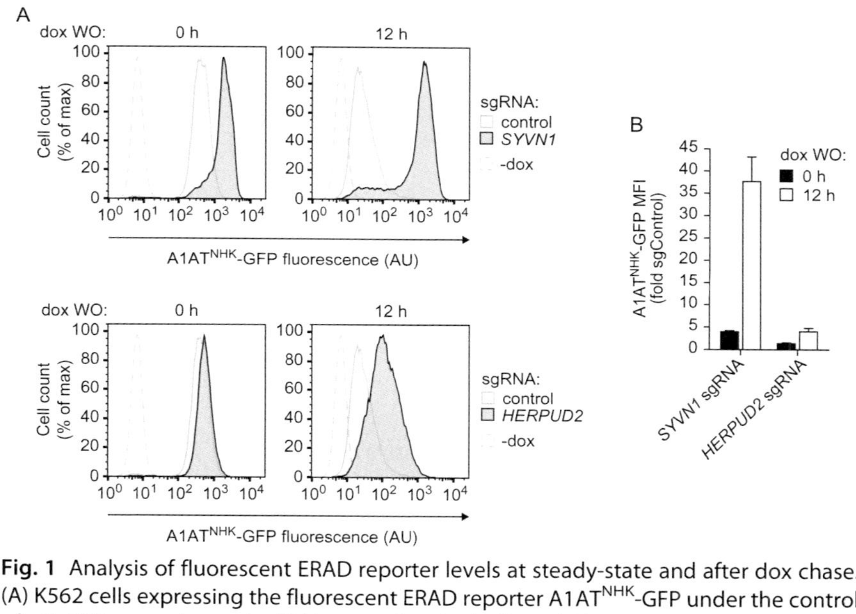

Fig. 1 Analysis of fluorescent ERAD reporter levels at steady-state and after dox chase. (A) K562 cells expressing the fluorescent ERAD reporter A1AT^{NHK}-GFP under the control of a TetON promoter and sgRNAs targeting the indicated ERAD components were subjected to dox washout as described in Section 3.2.2.4. GFP fluorescence was measured at the indicated times by flow cytometry. (B) Quantification of (A). Data are the mean of three independent experiments ± SEM. *Reprinted from Leto, D. E., Morgens, D. W., Zhang, L., Walczak, C. P., Elias, J. E., Bassik, M. C., et al. (2019). Genome-wide CRISPR analysis identifies substrate-specific conjugation modules in ER-associated degradation.* Molecular Cell, 73, *1–13 with permission from Elsevier.*

3.1.3 Cas9 variant and sgRNA libraries

Before undertaking a genetic screen, researchers should consider which screening approach to use. Variants of Cas9 now enable CRISPR-based loss-of-function and gain-of-function screens. A variety of sgRNA libraries that vary in sgRNA and target gene number, nontargeting sgRNA number and type, selection markers, and targeted species have been and continue to be developed at a rapid pace, and the user is advised to carefully evaluate the available options. Although genome-wide sgRNA libraries that contain many unique sgRNAs per gene provide more discovery power and robust datasets, screens using these libraries are labor and resource intensive; thus, the library used for screening must be carefully selected in consideration of the goals and technical limitations of the study. Several recent reviews discuss in detail the advantages and disadvantages of available Cas9 variants and sgRNA libraries (Kampmann, 2018; Shang et al., 2017).

3.2 Inducible ERAD reporter cell line generation

3.2.1 Lentiviral preparation and transduction

1. Plate 750,000 low passage number HEK293T cells in 2 mL of complete DMEM medium in one well of a sterile tissue culture-treated six-well plate. Incubate the cells overnight in a humidified incubator at 37°C and 5% CO_2.
2. The next day, set up a DNA transfection mix by adding 4.5 µL of TransIT-LT1 transfection reagent to 150 µL of serum-free DMEM. Mix and incubate at room temperature for 5 min.
3. To the TransIT-LT1/DMEM mix, add 0.75 µg of tet-inducible ERAD reporter expression plasmid and 0.75 µg of third generation lentiviral packaging mix (1:1:1 mix of pMDL, pRSV, and pVSVG). Mix and incubate the DNA transfection mix at room temperature for 20 min.
4. Add the DNA transfection mix dropwise to the HEK293T cells. Return the cells to the incubator.
5. 24 h after transfection, add 2 mL of complete DMEM medium to the transfected cells, then return the cells to the incubator.
6. 72 h after transfection, collect the cell medium containing the viral particles (or lentiviral supernatant) and pass it through a sterile 0.45-µ*M* syringe filter.
7. Infect TetOn cells with ERAD reporter lentivirus:
   a. To virally transduce suspension cells, collect cells by centrifuging in a swinging bucket rotor at 800 × *g* at 4°C for 5 min. Remove the

medium and resuspend the cell pellet in lentiviral supernatant supplemented with 8 μg/mL polybrene (see Notes 2 and 3). Plate the resuspended cells into one well of a sterile 12-well plate. Centrifuge the plate in a swinging bucket rotor fitted with a plate adapter at $1000 \times g$ at 33°C for 2 h. Remove the viral supernatant and resuspend the cells in fresh growth medium. Grow the cells in a humidified incubator at 37°C and 5% CO_2.

b. To transduce adherent cells, split the cells the day before so that they are ~50% confluent on the day of transduction. Infect the cells by adding the viral supernatant supplemented with 8 μg/mL polybrene (see Notes 2 and 3) and grow in a humidified incubator at 37°C and 5% CO_2. 24 h later, remove the viral supernatant and replace with fresh complete medium.

8. 72 h after viral transduction, begin selecting cells by culturing in complete medium supplemented with the appropriate antibiotic (see Note 4). Replace the medium every 3–4 days and split the cells as needed to prevent them from becoming confluent.

9. Continue passaging the cells until the population is healthy and growing normally and a control (untransduced) population is dead.

10. It is optimal to select stable cell lines that exhibit a narrow distribution of fluorescence intensity in the presence of dox. A homogenous polyclonal population can be enriched by FACS, or alternatively, a monoclonal population can be isolated by single cell sorting or limiting dilution using established protocols.

11. After isolating a dox-inducible ERAD reporter cell line, a Cas9 variant can be introduced using the above protocol for lentiviral packaging and cell transduction. The cells expressing Cas9 should be enriched by drug selection or FACS before proceeding with genetic screening.

3.2.2 Cell line validation

Before proceeding with a genetic screen, the dox-inducible ERAD reporter cell line should be evaluated to ensure that the reporter is not expressed in the absence of inducer and, in the presence of inducer, is synthesized correctly, is not aggregated, is rapidly degraded, and is dependent on the UPS for turnover. These initial validation steps are also used to determine the dynamic range of the reporter at steady-state and optimize the expression parameters, including concentration of inducer and length of induction.

3.2.2.1 Optimization of reporter induction

1. Stimulate healthy, exponentially proliferating TetOn ERAD reporter cells with dox by plating the cells in medium containing varying concentration of dox (from 0 to 1 μg/mL). Also include a control sample of cells that do not express the ERAD reporter.
   a. For adherent lines, seed the cells in 12-well tissue culture-treated sterile plates. Plate cells 24 h before adding dox so that they will be ~30%–50% confluent on the day of treatment. To initiate reporter expression, replace the medium with fresh complete medium containing dox.
   b. For suspension lines, collect the cells by centrifuging in a swinging bucket rotor at 800 × *g* at room temperature for 5 min. Remove the medium and resuspend the cells to a final density of ~250,000 cells/mL in complete medium containing dox.
2. Incubate the cells in a humidified incubator at 37°C and 5% CO_2. 16–24 h after treating with dox, collect the cells and transfer into 5-mL FACS tubes. Pellet the cells by centrifuging in a swinging bucket rotor at 800 × *g* at 4°C for 5 min. Resuspend the cells in 1 mL cold PBS, then place on ice. Protect the cells from light by covering the tubes with foil.
3. Analyze at least 20,000 cells per condition by flow cytometry.
4. Use FlowJo or comparable flow cytometry analysis software to calculate a median fluorescence intensity for each sample.
   a. Confirm that the reporter is only expressed in the presence of dox by comparing the fluorescence of TetOn ERAD reporter cells cultured in medium lacking dox with control cells that do not contain the reporter.
   b. To determine the optimal concentration of dox to use for subsequent experiments, select the lowest dose that results in a tight fluorescence distribution.
5. To determine the optimal length of time for reporter induction, repeat the above procedure using a standard concentration of dox and varying the hours of treatment. For subsequent experiments, choose the time point at which the population has reached steady-state levels and displays a tight, homogeneous distribution by flow cytometry analysis.

3.2.2.2 Analysis of reporter solubility by sequential detergent extraction of protein lysates

1. Treat TetOn ERAD reporter cells with dox using the optimized conditions (see Section 3.2.2.1).
2. Collect 1×10^6 cells in a 1.5-mL microcentrifuge tube. Pellet the cells by centrifuging at 800 × *g* at 4°C for 2 min, then aspirate the medium

using a vacuum aspirator. Add 1 mL of ice-cold PBS, centrifuge at 800 × *g* at 4°C for 2 min, then aspirate the PBS using a vacuum aspirator.

3. Resuspend the cell pellet in 100 μL of 1% TX-100 solubilization buffer. Rotate at 4°C for 20 min, then centrifuge at 20,000 × *g* at 4°C for 20 min.
4. Collect the supernatant and move into a new 1.5-mL microcentrifuge tube. For long-term storage, flash-freeze the supernatant by submerging in a liquid nitrogen bath and store at −80°C.
5. Wash the insoluble pellet from step 3 three times with 200 μL of 1% TX-100 solubilization buffer.
6. Sonicate the washed pellet in 100 μL of 1% SDS solubilization buffer using a tip sonicator on setting 2.5 for 20 s. Centrifuge at 20,000 × *g* at 18°C for 20 min. Collect the supernatant and move into a new 1.5-mL microcentrifuge tube. For long-term storage, flash-freeze the supernatant by submerging in a liquid nitrogen bath and store at −80°C.
7. Thaw samples on ice. To each sample, add 5 × Laemmli buffer and β-mercaptoethanol to a final concentration of 2.5%. Incubate at 65°C for 15 min.
8. Separate equal amounts of total cell lysates by SDS-PAGE and visualize the ERAD reporter using a standard immunoblotting protocol.

3.2.2.3 Analysis of reporter half-life and UPS dependence by translation shutoff

3.2.2.3.1 Suspension cell lines

1. Introduce 10 mL of the TetOn reporter cells at 500,000 cells/mL into a sterile tissue culture-treated T25 flask. Induce reporter expression using the optimized conditions (see Section 3.2.2.1).
2. Immediately before halting translation by adding emetine, collect a zero time point sample by collecting 1 mL of culture in a cold 1.5-mL microcentrifuge tube. To the remaining culture, add emetine to a final concentration of 20 μ*M* and place back in the incubator. Pellet the cells collected at time zero by centrifuging at 800 × *g* at 4°C for 2 min, then aspirate the medium using a vacuum aspirator. Add 1 mL of ice-cold PBS, centrifuge at 800 × *g* at 4°C for 2 min, then aspirate the PBS using a vacuum aspirator.
   a. For flow cytometry analysis, resuspend the cell pellet in 250 μL of ice-cold PBS and incubate on ice in the dark.
   b. For immunoblot analysis, flash-freeze the cell pellet by submerging in a liquid nitrogen bath, then store the cell pellet at −80°C.
3. To the rest of the culture, add emetine to a final concentration of 20 μ*M* (see Note 5). Incubate the culture in a humidified incubator at 37°C and 5% CO_2. For use of UPS inhibitors, see Note 6.

4. Repeat step 2 at additional time points (see Note 7).
5. To determine reporter half-life by flow cytometry, analyze and quantify reporter levels in each sample as described in Section 3.2.2.1. Quantified data are typically represented as percentage of the signal at the zero time point.
6. To determine reporter half-life by immunoblot analysis:
   a. Thaw cell pellets on ice, then resuspend each cell pellet in 150 μL of 1% SDS lysis buffer.
   b. Sonicate samples using a tip sonicator on setting 2.5 for 20 s or until samples are no longer viscous.
   c. Centrifuge the lysates at 20,000 × *g* at room temperature for 15 min. Transfer the supernatant into a new microcentrifuge tube and discard the pelleted material.
   d. Measure the protein concentration of each sample using a BCA Protein Assay Kit according to manufacturer's instructions. Normalize the samples by protein concentration by diluting with 1% SDS lysis buffer.
   e. To each sample, add 5 × Laemmli buffer and β-mercaptoethanol to a final concentration of 2.5%. Incubate the samples at 65°C for 15 min.
   f. Separate equal amounts of total cell lysates using a standard SDS-PAGE protocol. Visualize the ERAD reporter using a standard immunoblotting protocol (see Notes 8–9).
   g. Quantify the reporter using an appropriate immunoblot analysis package (see Note 10). Quantified data are typically represented as percentage of the signal at the zero time point.

3.2.2.3.2 Adherent cell lines

1. 1 day before dox treatment, seed the TetOn reporter cells in tissue culture-treated 12-well plates so that they will be ~30%–50% confluent on the day of treatment. Incubate overnight in a humidified incubator at 37°C and 5% CO_2.
2. The next day, initiate reporter expression by replacing the medium with fresh medium containing dox. Use the optimized induction conditions as described in Section 3.2.2.1.
3. At the desired time points, add emetine to a final concentration of 20 μ*M* to one well of cells (see Note 7). Between time points, incubate the cells in a humidified incubator at 37°C and 5% CO_2.
4. To stop the emetine chase, aspirate the medium from the cells using a vacuum aspirator. Wash cells once with 1 mL of ice-cold PBS.

5. To analyze reporter turnover by flow cytometry:
   a. Collect the cells by adding 100 μL of trypsin–EDTA to each well. Once the cells have detached from the plate, add 1 mL of complete medium to each well and transfer the resuspended cells into a cold 1.5-mL microcentrifuge tube.
   b. Pellet the cells by centrifuging at 800 × *g* at 4°C for 2 min, then aspirate the medium using a vacuum aspirator. Add 1 mL of ice-cold PBS, centrifuge at 800 × *g* at 4°C for 2 min, then aspirate the PBS using a vacuum aspirator.
   c. Resuspend each cell pellet in 250 μL of ice-cold PBS.
   d. Filter the cell using a 5-mL tube capped with a 35-μ*M* nylon mesh filter. Incubate the cells on ice in the dark until all time points have been collected.
   e. Analyze the cells and quantify the fluorescence intensity as described in Sections 3.2.2.1 and 3.2.2.3.1, step 5.
6. To analyze reporter turnover by immunoblotting:
   a. Add 150 μL of 1% SDS lysis buffer to each well. Use the wide end of a 1-mL pipette tip to scrape the cells into the buffer.
   b. Collect the lysates from step 6a in a 1.5-mL microcentrifuge tube.
   c. Process and analyze the samples as described in Section 3.2.2.3.1, steps 6c–h.

3.2.2.4 Optimization of reporter turnover readout by transcription shutoff

This is essentially a variation of the protocols provided in Sections 3.2.2.3.1 and 3.2.2.3.2. Here, new reporter synthesis is halted by specifically shutting off reporter transcription, rather than using a broad translation inhibitor such as emetine to block all new protein synthesis in the cell.

The rate of transcription shutoff is dependent on the stability of the reporter transcript, the concentration of transcriptional inducer (in our case, dox) added to cells, and the intrinsic half-life of the fluorescent reporter. When synthesized from a destabilized, poly(A)-less transcript, ERAD reporter fluorescence levels typically remain constant for approximately 3 h after removing dox and then begin to decline at a rate proportional to the half-life of the reporter (Fig. 2). If fluorescence levels do not begin to decline within 2–4 h after removing dox, decrease the amount of dox added to the cells and retest the transcription shutoff kinetics.

1. Plate TetOn ERAD reporter cells and treat the cells with dox as described in Section 3.2.2.3.1 (for suspension cells) or 3.2.2.3.2 (for adherent cells).
2. Immediately before halting transcription by removing dox, collect a zero time point sample for flow cytometry analysis as described in

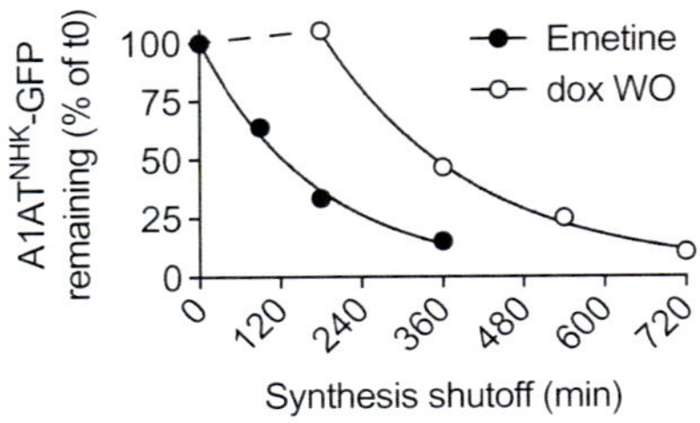

Fig. 2 Fluorescent ERAD reporter decay after shutting off reporter transcription by removing dox (dox WO) or halting reporter translation with emetine. K562 cells expressing the fluorescent ERAD reporter A1AT^{NHK}-GFP under the control of a TetON promoter were treated with dox for 16h to induce reporter expression. The mean fluorescence intensity (MFI) of the reporter cells was measured at the indicated times after dox washout (dox WO) or emetine addition. Data are representative of two independent experiments. *Reprinted from Leto, D. E., Morgens, D. W., Zhang, L., Walczak, C. P., Elias, J. E., Bassik, M. C., et al. (2019). Genome-wide CRISPR analysis identifies substrate-specific conjugation modules in ER-associated degradation.* Molecular Cell, 73, *1–13 with permission from Elsevier.*

Section 3.2.2.3.1 (for suspension cells) or 3.2.2.3.2 (for adherent cells). Resuspend cells in complete medium lacking phenol red, then incubate the cells on ice and protected from light.

3. Shut off reporter transcription in the remaining cells by removing dox:
   a. For adherent cells, aspirate the medium using a vacuum aspirator and wash the cells twice with 2mL of medium lacking supplements, then add 2mL of complete medium lacking dox and replace the cells in the humidified incubator.
   b. For suspension cells, collect the cells in a sterile 14-mL conical tube. Pellet the cells by centrifuging in a swinging bucket rotor at 800 × *g* at room temperature for 5min. Aspirate the medium using a vacuum aspirator, add 10mL of medium without supplements, and centrifuge in a swinging bucket rotor at 800 × *g* at room temperature for 5min. Repeat the wash once, then aspirate medium using a vacuum aspirator. Resuspend the cells in 10mL of complete medium lacking dox, then plate the cells into a sterile tissue culture-treated T25 flask and place back in the incubator.
4. Collect a sample of cells every 2h for 12h. Incubate cells on ice in the dark. (See Note 11).
5. Analyze and quantify reporter levels in each sample by flow cytometry as described in Sections 3.2.2.1 and 3.2.2.3.1, step 5. If cell fluorescence does not decline within 4h of dox removal, this procedure should be repeated using a lower concentration of dox.

3.2.3 Conducting a screen

Genetic screening in a pooled format using CRISPR/Cas9 consists of (1) generation of a cell population expressing an sgRNA library; (2) performing the functional screening; and (3) sample processing and data analysis (Fig. 3). Because detailed protocols for steps 1 and 3 have been published elsewhere (Deans et al., 2016; Kampmann et al., 2014; Morgens et al., 2016, 2017), here we focus on the functional screening method that we use to identify genes involved in protein degradation.

To identify genetic modulators of ERAD reporter degradation, genome-edited cells harboring sgRNAs are fractionated into low and high fluorescence populations by FACS after a period of transcription shutoff. Cells that express sgRNAs targeting genes that promote protein degradation are enriched in the high fluorescence population; cells that express sgRNAs targeting genes that slow protein degradation are enriched in low fluorescence population; and cells that express sgRNAs that do not affect protein turnover will be equally represented in both populations. While sorting cells based on the fluorescence of an unstable reporter protein alone will effectively identify genes that affect protein degradation, genes that alter protein synthesis will also be detected. To remove genes that affect protein synthesis, a second, long-lived fluorescence protein can be introduced, and cells can be sorted based on the ratio of the fluorescence signals (see Note 1).

The protocol below is optimized for genetic screening in K562 cells but can be adapted for screening in other cell lines.

1. Infect TetOn reporter cells expressing Cas9, or a Cas9 variant, with the sgRNA library and select for library-infected cells as previously described (Kampmann et al., 2014; Morgens et al., 2017).
2. Grow cells in complete RPMI medium in the absence of a selection reagent for 72 h in a humidified incubator at 37°C and 5% CO_2. Collect cells by centrifuging in a swinging bucket rotor at $800 \times g$ at room temperature for 5 min, then resuspend in cell freezing medium to a final density of 5×10^6–1×10^7 cells/mL (see Note 12). Aliquot cells into cryovials, then place the vials in a cryo-freezing container and store overnight in a −80°C freezer.
3. 8–24 h later, move the cryovials containing the cells into a liquid nitrogen freezer for long-term storage.
4. To initiate the screen, thaw enough aliquots of the library-infected cells to ensure 1000-fold coverage of the library elements (see Note 13). Resuspend the cells in complete RPMI medium and expand the culture

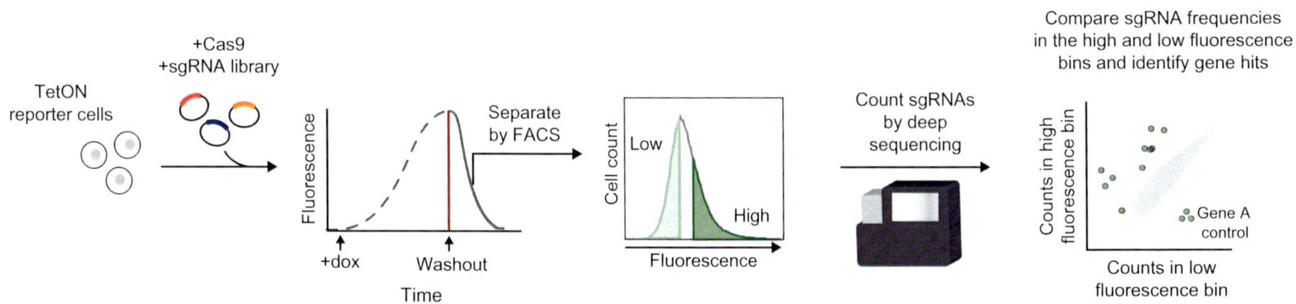

Fig. 3 Genetic screening strategy. TetON reporter cells are infected with a Cas9 variant and an sgRNA library. Fluorescent ERAD reporter expression is induced by treating with dox, then reporter synthesis is shut off by removing dox. When ~20%–30% of the reporter remains, cells are sorted into low and high fluorescence bins by fluorescence activated cell sorting (FACS). sgRNA frequences in each cell population are determined by deep sequencing. sgRNA frequencies are compared to identify statistically significant gene hits that alter ERAD reporter degradation rate when deleted.

for 48 h or until the cell number has increased to at least 4000-fold coverage of the library elements and the cells are doubling at a normal rate.

5. Collect the cells in sterile 50-mL conical tubes and pellet by centrifuging in a swinging bucket rotor at 800 × *g* at room temperature for 5 min. Resuspend the cells at a final density of 350,000 cells/mL in complete RPMI medium supplemented with dox and plate into sterile tissue culture-treated T225 flasks.
6. Grow the cells in a humidified incubator at 37°C and 5% CO_2 for the experimentally determined optimal induction time (see Section 3.2.2.1).
7. Collect the cells in sterile 50-mL conical tubes and pellet by centrifuging in a swinging bucket rotor at 800 × *g* at room temperature for 5 min. Aspirate the medium using a vacuum aspirator, replace with RPMI medium without supplements, and centrifuge in a swinging bucket rotor at 800 × *g* at room temperature for 5 min. Repeat the wash once, then aspirate the medium. Resuspend the cells in complete RPMI medium lacking dox and plate into sterile tissue culture-treated T225 flasks.
8. Incubate cells in a humidified incubator at 37°C and 5% CO_2 until cell fluorescence levels decrease to approximately 20%–30% of the level at the zero time point (see Note 14).
9. Collect the cells in sterile 50-mL conical tubes and pellet by centrifuging in a swinging bucket rotor at 800 × *g* at room temperature for 5 min. Aspirate the medium using a vacuum aspirator, then resuspend the cells in RPMI medium without phenol red and supplemented with 0.5% FBS.
10. Pass the cells through a 5-mL tube capped with a 35-μ*M* nylon mesh filter, then place the tubes on ice to halt reporter degradation. Cover the tubes with foil to protect the cells from light.
11. Separate the cells into the highest ~30% and lowest ~30% fluorescence by FACS (see Note 15). Sort the cells into chilled 15-mL conical tubes containing 5 mL of medium supplemented with 20% FBS. Collect enough cells to maintain at least 500-fold representation of the library elements in each bin.
12. At the end of the sort, analyze on the cell sorter a small sample of each cell population to check the sort purity.
13. Collect the cells by centrifuging in a swinging bucket rotor at 3000 × *g* at 4°C for 15 min. Remove the medium using a 14-mL pipette. Add 5 mL of PBS to each 14-mL conical tube and pellet the cells by centrifuging in a swinging bucket rotor at 3000 × *g* at 4°C for 15 min. Remove the PBS using a 14-mL pipette.

14. Flash-freeze the cell pellets by submerging in a liquid nitrogen bath and store at −80°C.
15. Extract the genomic DNA from the cells, prepare the sequencing libraries, and perform deep sequencing as previously described (Deans et al., 2016).
16. Align sgRNAs to the reference genome sequence using Bowtie (Langmead et al., 2009) and compare sgRNA frequencies in the high and low fluorescence populations using a statistics package such as casTLE (Morgens et al., 2016) (see Note 16).
17. Following the completion of a genetic screen, the results of the screen should be validated by repeating the genetic screen at least once and by individually resting positive and negative screen hits (see Note 17).

3.3 Outlook

Despite considerable progress in elucidating the biochemical mechanisms by which substrates are conjugated to polyUb chains and destroyed by 26S proteasomes, how structurally distinct variants of conformationally compromised quality control substrates are recognized and committed to destruction by the UPS remains poorly understood. Here, we describe a functional genomic approach that employs a phenotypic selection based on protein turnover rate and CRISPR/Cas9-mediated gene perturbations to map the genetic pathways by which proteins are degraded by the UPS. Our recent application of genome-wide gene knockout analyses to study mammalian ERAD precisely identified unique subsets of known ERAD components required for the destruction of topologically diverse misfolded proteins and uncovered novel players in ERAD (Leto et al., 2019), demonstrating that CRISPR/Cas9-based genetic screening is a powerful new approach for deconvolving substrate selective degradation pathways. To date, genetic analyses of protein degradation performed in mammalian cells have relied on single gene knockouts. The application of new high-performance gene perturbation tools to study protein degradation, such as CRISPRi to inhibit gene expression (rather than resulting in irreversible gene deletion), CRISPRa to activate gene expression, double knockout sgRNA libraries for pairwise gene perturbations, and gene editors for mutagenesis studies, provides exciting opportunities to further deconvolve the complex cellular pathways that regulate the proteome.

4. Notes

1. In addition to degradation rate, the fluorescence levels of ERAD reporters are influenced by biosynthetic processes. Genes affecting chromatin state, transcription, gene splicing, and translation are common artifacts in genetic screens, likely due to pleiotropic effects of their perturbation on the reporter and on machinery required for the folding or degradation of the reporter. To control for general effects on gene expression, a constitutively expressed compatible stable fluorescent protein can be expressed and, during cell sorting, used to bin the cells based on the ratiometric fluorescence signal. Many available sgRNA libraries contain a fluorescent marker protein. Alternatively, a fluorogenic marker protein expressed from a constitutive promoter can be included on the plasmid encoding the ERAD reporter.
2. Early passage number, exponentially proliferating cells, should be used for viral transduction. Cells should be screened for mycoplasma contamination frequently, but at a minimum before lentiviral infection and after selecting stable clonal lines.
3. The lentiviral transduction should be optimized for each cell type by adjusting the amount of virus added to a constant number of cells. To minimize propagation of multiply-infected cells, use a concentration of virus that results in a target infection of ≤40%. As a starting point, we typically infect 500,000 K562 cells with 0.5 mL of fresh viral supernatant. If the virus will not be used immediately, it can be stored at −80C and titered after thawing.

   3–5 days after transduction, measure the percentage of cells infected by treating cells with 1 μg/mL of dox for 16–24 h, then analyzing fluorescence by flow cytometry or microscopy.
4. Before beginning selection, determine the lowest concentration of antibiotic required to kill nontransduced cells within 7–14 days by conducting a kill curve analysis.
5. We prefer to use the translation inhibitor emetine because the lyophilized chemical and stock solutions prepared in water are stable for long periods of time when stored at −20°C. Alternatively, cycloheximide can be added to cells to a final concentration of 50 μg/mL to shut off translation. Cycloheximide powder should be stored in the dark and

with dessicant at −20°C. To prepare a cycloheximide stock solution, resuspend the lyophilized chemical in DMSO or EtOH and store in the dark at −20°C for up to 3 months.

6. To determine whether degradation of a particular substrate is dependent on the UPS, a UPS inhibitor (e.g., 10 μ*M* MG132 to block proteasome activity; 5 μ*M* NMS-873 to inhibit VCP (Chou et al., 2014); or 1 μ*M* TAK-243 to inhibit E1 activity (Barghout et al., 2019)) together with emetine can be added to cells. To minimize cytotoxicity and secondary, nonspecific effects of inhibitors, we try to minimize the duration of drug treatments to 4 h or less.

7. Because protein half-lives vary greatly, the appropriate emetine chase length must be empirically tested for each ERAD reporter. As a starting point, collecting cells at 0, 30, 60, and 120 min after emetine addition is suitable for many substrates.

8. In addition to immunoblotting for the ERAD reporter, researchers should immunoblot the same membrane using an antibody against a loading control (e.g., GAPDH or tubulin). For reporter quantification, the reporter signal should be normalized to the loading control signal in each gel lane.

9. We prefer to use fluorescence detection, rather than chemiluminescence detection, to quantitate protein levels by immunoblotting because fluorophore signal is more linearly correlated with protein concentration and has a wider dynamic range of detection (Eaton et al., 2014). In either case, researchers should be careful to ensure that the protein signal is within the linear range of detection by including serial dilutions of the of cell lysates when immunoblotting (Taylor & Posch, 2014).

10. Fluorescence or chemiluminescence signals can be quantified using standard image analysis packages such as ImageStudio or ImageJ.

11. In our experience, reporter turnover is efficiently halted and cells remain viable when incubated on ice for extended periods of time (8–12 h) in medium containing FBS. Nevertheless, conditions should be tested and optimized for each cell line. Cells can be analyzed by flow cytometry in medium containing FBS and lacking phenol red. Alternatively, cells can be pelleted and resuspended in PBS right before flow cytometry analysis.

12. The optimal cell density for freezing should be determined empirically for each cell line. Ensure that cells are healthy and exponentially proliferating before freezing.

13. To avoid population bottlenecks and increased noise in the screen data, the library elements should be represented at 1000-fold at all times. For example, for a library containing 100,000 sgRNAs, at least 1×10^8 cells should be maintained at each stage of screening.
14. Cells are typically incubated without dox for 6–12 h, depending on the half-life of the fluorescent reporter protein.
15. Sorting parameters should be optimized for each cell line to maximize sort speed while minimizing cell lysis. We sort K562 cells using an 85 μm nozzle at ~10–12,000 events/s using the four-way purity precision mode on a BD Aria cell sorter.
16. Many open-source statistical tools for analyzing screen results have been described, including casTLE (Morgens et al., 2016), Mann–Whitney (Bassik et al., 2013), MAGeCK (Chou et al., 2014), STARS (Doench et al., 2016), RSA (Konig et al., 2007), and RIGER (Luo et al., 2008). We refer readers to the primary publications when selecting an analysis package.
17. If the genetic screening results are robust, we typically see a very high rate of hit validation (often 100%). We recommend selecting a mix of strong, intermediate, and negative gene hits for individual retest using a translation shutoff assay. We typically retest 5–20 hits per screen (dictated by the concordance between screen replicates and the biological interest of the screen results) using at least two unique sgRNAs per gene. The efficacy of the sgRNAs used for validation should be confirmed by western blot. If an antibody is not available, gene editing frequency can be evaluated using an assay such as qPCR or Tracking of Indels by Decomposition (TIDE) (Brinkman, Chen, Amendola, & van Steensel, 2014).

Acknowledgments

The authors thank former and current members of the Kopito lab and David Morgens (Stanford University, Stanford, CA), Michael Bassik (Stanford University, Stanford, CA), and Cathy Crumpton (Stanford University, Stanford, CA) for essential help developing protocols. This work was supported by Grants from the National Institutes of Health and the Alpha-1 Foundation to D.E.L. and a Grant from the National Institutes of Health to R.R.K.

References

Barghout, S. H., Patel, P. S., Wang, X., Xu, G. W., Kavanagh, S., Halgas, O., et al. (2019). Preclinical evaluation of the selective small-molecule UBA1 inhibitor, TAK-243, in acute myeloid leukemia. *Leukemia*, *33*, 37–91. https://doi.org/10.1038/s41375-018-0167-0.

Bassik, M. C., Kampmann, M., Lebbink, R. J., Wang, S., Hein, M. Y., Poser, I., et al. (2013). A systematic mammalian genetic interaction map reveals pathways underlying ricin susceptibility. *Cell*, *152*(4), 909–922. https://doi.org/10.1016/j.cell.2013.01.030.

Birsoy, K., Wang, T., Chen, W. W., Freinkman, E., Abu-Remaileh, M., & Sabatini, D. M. (2015). An essential role of the mitochondrial electron transport chain in cell proliferation is to enable aspartate synthesis. *Cell*, *162*(3), 540–551. https://doi org/10.1016/j.cell.2015.07.016.

Brinkman, E. K., Chen, T., Amendola, M., & van Steensel, B. (2014). Easy quantitative assessment of genome editing by sequence trace decomposition. *Nucleic Acids Research*, *42*(22), e168. https://doi.org/10.1093/nar/gku936.

Burr, M. L., Cano, F., Svobodova, S., Boyle, L. H., Boname, J. M., & Lehner, P. J. (2011). HRD1 and UBE2J1 target misfolded MHC class I heavy chains for endoplasmic reticulum-associated degradation. *Proceedings of the National Academy of Sciences of the United States of America*, *108*(5), 2034–2039. https://doi.org/10.1073/pnas.1016229108.

Buschhorn, B. A., Kostova, Z., Medicherla, B., & Wolf, D. H. (2004). A genome-wide screen identifies Yos9p as essential for ER-associated degradation of glycoproteins. *FEBS Letters*, *577*(3), 422–426. https://doi.org/10.1016/j.febslet.2004.10.039.

Chou, T. F., Bulfer, S. L., Weihl, C. C., Li, K., Lis, L. G., Walters, M. A., et al. (2014). Specific inhibition of p97/VCP ATPase and kinetic analysis demonstrate interaction between D1 and D2 ATPase domains. *Journal of Molecular Biology*, *426*(15), 2886–2899. https://doi.org/10.1016/j.jmb.2014.05.022.

Christianson, J. C., Olzmann, J. A., Shaler, T. A., Sowa, M. E., Bennett, E. J., Richter, C. M., et al. (2011). Defining human ERAD networks through an integrative mapping strategy. *Nature Cell Biology*, *14*(1), 93–105. https://doi.org/10.1038/ncb2383.

Cronin, S. R., Khoury, A., Ferry, D. K., & Hampton, R. Y. (2000). Regulation of HMG-CoA reductase degradation requires the P-type ATPase Cod1p/Spf1p. *The Journal of Cell Biology*, *148*(5), 915–924.

Deans, R. M., Morgens, D. W., Okesli, A., Pillay, S., Horlbeck, M. A., Kampmann, M., et al. (2016). Parallel shRNA and CRISPR-Cas9 screens enable antiviral drug target identification. *Nature Chemical Biology*, *12*(5), 361–366. https://doi.org/10.1038/nchembio.2050.

Doench, J. G., Fusi, N., Sullender, M., Hegde, M., Vaimberg, E. W., Donovan, K. F., et al. (2016). Optimized sgRNA design to maximize activity and minimize off-target effects of CRISPR-Cas9. *Nature Biotechnology*, *34*(2), 184–191. https://doi.org/10.1038/nbt.3437.

Eaton, S. L., Hurtado, M. L., Oldknow, K. J., Graham, L. C., Marchant, T. W., Gillingwater, T. H., et al. (2014). A guide to modern quantitative fluorescent western blotting with troubleshooting strategies. *Journal of Visualized Experiments*, (93), e52099. https://doi.org/10.3791/52099.

Guerriero, C. J., & Brodsky, J. L. (2012). The delicate balance between secreted protein folding and endoplasmic reticulum-associated degradation in human physiology. *Physiological Reviews*, *92*(2), 537–576. https://doi.org/10.1152/physrev.00027.2011.

Hampton, R. Y., Gardner, R. G., & Rine, J. (1996). Role of 26S proteasome and HRD genes in the degradation of 3-hydroxy-3-methylglutaryl-CoA reductase, an integral endoplasmic reticulum membrane protein. *Molecular Biology of the Cell*, 7(12), 2029–2044.

Hampton, R. Y., Koning, A., Wright, R., & Rine, J. (1996). In vivo examination of membrane protein localization and degradation with green fluorescent protein. *Proceedings of the National Academy of Sciences of the United States of America*, *93*(2), 828–833.

Heck, J. W., Cheung, S. K., & Hampton, R. Y. (2010). Cytoplasmic protein quality control degradation mediated by parallel actions of the E3 ubiquitin ligases Ubr1 and San1. *Proceedings of the National Academy of Sciences of the United States of America* *107*(3), 1106–1111. https://doi.org/10.1073/pnas.0910591107.

Ju, L. G., Lin, X., Yan, D., Li, Q. L., Wu, M., & Li, L. Y. (2018). Characterization of WDR20: A new regulator of the ERAD machinery. *Biochimica et Biophysica Acta. Molecular Cell Research*, *1865*(7), 970–980. https://doi.org/10.1016/j.bbamcr.2018.04.006.

Kampmann, M. (2018). CRISPRi and CRISPRa screens in mammalian cells for precision biology and medicine. *ACS Chemical Biology*, *13*(2), 406–416. https://doi.org/10.1021/acschembio.7b00657.

Kampmann, M., Bassik, M. C., & Weissman, J. S. (2014). Functional genomics platform for pooled screening and generation of mammalian genetic interaction maps. *Nature Protocols*, *9*(8), 1825–1847. https://doi.org/10.1038/nprot.2014.103.

Knop, M., Finger, A., Braun, T., Hellmuth, K., & Wolf, D. H. (1996). Der1, a novel protein specifically required for endoplasmic reticulum degradation in yeast. *The EMBO Journal*, *15*(4), 753–763.

Konig, R., Chiang, C. Y., Tu, B. P., Yan, S. F., DeJesus, P. D., Romero, A., et al. (2007). A probability-based approach for the analysis of large-scale RNAi screens. *Nature Methods*, *4*(10), 847–849. https://doi.org/10.1038/nmeth1089.

Langmead, B., Trapnell, C., Pop, M., & Salzberg, S. L. (2009). Ultrafast and memory-efficient alignment of short DNA sequences to the human genome. *Genome Biology*, *10*(3), R25. https://doi.org/10.1186/gb-2009-10-3-r25.

Leto, D. E., Morgens, D. W., Zhang, L., Walczak, C. P., Elias, J. E., Bassik, M. C., et al. (2019). Genome-wide CRISPR analysis identifies substrate-specific conjugation modules in ER-associated degradation. *Molecular Cell*, *73*, 377–389.

Luo, B., Cheung, H. W., Subramanian, A., Sharifnia, T., Okamoto, M., Yang, X., et al. (2008). Highly parallel identification of essential genes in cancer cells. *Proceedings of the National Academy of Sciences of the United States of America*, *105*(51), 20380–20385. https://doi.org/10.1073/pnas.0810485105.

Medicherla, B., Kostova, Z., Schaefer, A., & Wolf, D. H. (2004). A genomic screen identifies Dsk2p and Rad23p as essential components of ER-associated degradation. *EMBO Reports*, *5*(7), 692–697.

Morgens, D. W., Deans, R. M., Li, A., & Bassik, M. C. (2016). Systematic comparison of CRISPR/Cas9 and RNAi screens for essential genes. *Nature Biotechnology*, *34*(6), 634–636. https://doi.org/10.1038/nbt.3567.

Morgens, D. W., Wainberg, M., Boyle, E. A., Ursu, O., Araya, C. L., Tsui, C. K., et al. (2017). Genome-scale measurement of off-target activity using Cas9 toxicity in high-throughput screens. *Nature Communications*, *8*, 15178. https://doi.org/10.1038/ncomms15178.

Neal, S., Jaeger, P. A., Duttke, S. H., Benner, C., Glass, C. K., Ideker, T., et al. (2018). The Dfm1 Derlin is required for ERAD retrotranslocation of integral membrane proteins. *Molecular Cell*, *69*(5), 915. https://doi.org/10.1016/j.molcel.2018.02.014.

Olzmann, J. A., Kopito, R. R., & Christianson, J. C. (2013). The mammalian endoplasmic reticulum-associated degradation system. *Cold Spring Harbor Perspectives in Biology*, *5*(9). https://doi.org/10.1101/cshperspect.a013185.

Ruggiano, A., Foresti, O., & Carvalho, P. (2014). Quality control: ER-associated degradation: Protein quality control and beyond. *The Journal of Cell Biology*, *204*(6), 869–879. https://doi.org/10.1083/jcb.201312042.

Ruiz, S., Mayor-Ruiz, C., Lafarga, V., Murga, M., Vega-Sendino, M., Ortega, S., et al. (2016). A genome-wide CRISPR screen identifies CDC25A as a determinant of sensitivity to ATR inhibitors. *Molecular Cell*, *62*(2), 307–313. https://doi.org/10.1016/j.molcel.2016.03.006.

Savidis, G., McDougall, W. M., Meraner, P., Perreira, J. M., Portmann, J. M., Trincucci, G., et al. (2016). Identification of zika virus and dengue virus dependency factors using functional genomics. *Cell Reports*, *16*(1), 232–246. https://doi.org/10.1016/j.celrep.2016.06.028.

Shang, W., Wang, F., Fan, G., & Wang, H. (2017). Key elements for designing and performing a CRISPR/Cas9-based genetic screen. *Journal of Genetics and Genomics*, *44*(9), 439–449. https://doi.org/10.1016/j.jgg.2017.09.005.

Stagg, H. R., Thomas, M., van den Boomen, D., Wiertz, E. J., Drabkin, H. A., Gemmill, R. M., et al. (2009). The TRC8 E3 ligase ubiquitinates MHC class I molecules before dislocation from the ER. *The Journal of Cell Biology*, *186*(5), 685–692. https://doi.org/10.1083/jcb.200906110.
Stefanovic-Barrett, S., Dickson, A. S., Burr, S. P., Williamson, J. C., Lobb, I. T., van den Boomen, D. J., et al. (2018). MARCH6 and TRC8 facilitate the quality control of cytosolic and tail-anchored proteins. *EMBO Reports*, *19*(5). https://doi.org/10.15252/embr.201745603.
Swanson, R., Locher, M., & Hochstrasser, M. (2001). A conserved ubiquitin ligase of the nuclear envelope/endoplasmic reticulum that functions in both ER-associated and Matalpha2 repressor degradation. *Genes & Development*, *15*(20), 2660–2674.
Taylor, S. C., & Posch, A. (2014). The design of a quantitative western blot experiment. *BioMed Research International*, *2014*, 361590. https://doi.org/10.1155/2014/361590.
Timms, R. T., Menzies, S. A., Tchasovnikarova, I. A., Christensen, L. C., Williamson, J. C., Antrobus, R., et al. (2016). Genetic dissection of mammalian ERAD through comparative haploid and CRISPR forward genetic screens. *Nature Communications*, *7*, 11786. https://doi.org/10.1038/ncomms11786.
van de Weijer, M. L., Bassik, M. C., Luteijn, R. D., Voorburg, C. M., Lohuis, M. A. M., Kremmer, E., et al. (2014). A high-coverage shRNA screen identifies TMEM129 as an E3 ligase involved in ER-associated protein degradation. *Nature Communications*, *5*, 3832. https://doi.org/10.1038/ncomms4832.
Xia, P., Zhang, X., Xie, Y., Guan, M., Villeneuve, D. L., & Yu, H. (2016). Functional toxicogenomic assessment of triclosan in human HepG2 cells using genome-wide CRISPR-Cas9 screening. *Environmental Science & Technology*, *50*(19), 10682–10692. https://doi.org/10.1021/acs.est.6b02328.

CHAPTER SIX

A dual system to manipulate protein levels for DNA replication- and cell cycle-related studies

Néstor García-Rodríguez[a,b,*], **Helle D. Ulrich**[c,*]
[a]Centro Andaluz de Biología Molecular y Medicina Regenerativa (CABIMER), Universidad de Sevilla-CSIC-Universidad Pablo de Olavide, Sevilla, Spain
[b]Departamento de Genética, Universidad de Sevilla, Sevilla, Spain
[c]Institute of Molecular Biology (IMB), Mainz, Germany
*Corresponding authors: e-mail address: nestor.garcia@cabimer.es; h.ulrich@imb-mainz.de

Contents

Abstract

Investigation of cell cycle-regulated processes often necessitates the rapid manipulation of individual protein levels in synchronized populations over the course of a cell cycle. In the budding yeast, the two major orthogonal approaches by which this is accomplished are conditional gene expression via inducible or repressible promoters and regulated metabolic destabilization of a protein of interest via ubiquitin-mediated degradation signals. Here, we describe an application of these principles to the investigation of DNA damage signaling during replication. Using a combination of conditional gene expression via a tetracycline-repressible promoter and inducible protein degradation via an auxin-regulated degron system, we have analyzed the cross talk

Methods in Enzymology, Volume 619
ISSN 0076-6879
https://doi.org/10.1016/bs.mie.2018.12.038

© 2019 Elsevier Inc.
All rights reserved.

between factors controlling the bypass of lesions during replication and the activation of the DNA damage checkpoint. Here, we describe the basic principles underlying our experimental system and provide a detailed protocol to analyze the cross talk between damage bypass and checkpoint signaling. Finally, we discuss the advantages and disadvantages of using this approach and point out possible applications to other regulatory events and other organisms.

1. Introduction

The ease of genetic manipulations in the budding yeast *Saccharomyces cerevisiae* has facilitated complex mechanistic studies addressing the functions and interactions of cellular factors in vivo. In addition to simple gene deletions, more subtle alterations to endogenous proteins have proven extremely valuable, such as the introduction of point mutations, the appendage of peptide tags that modulate the protein's properties, or the manipulation of gene regulatory sequences to alter their expression levels (Botstein & Fink, 2011; Duina, Miller, & Keeney, 2014).

Such investigations using yeast genetics have yielded and continue to provide fundamental insights into a variety of cellular processes, particularly the cell cycle and DNA replication (Duina et al., 2014). In this context, analysis is complicated by several inherent problems. On the one hand, many important cell cycle regulators and replication proteins are essential, which precludes the use of gene deletions. On the other hand, in order to be able to gather interpretable data in the context of such studies, the use of synchronized cell populations is essential, since it reduces heterogeneity in the cell cycle and allows the study of gene function at a particular time or cell cycle stage. This stage-specific analysis, however, often necessitates a rapid manipulation of protein levels within a single cell cycle.

In budding yeast, researchers use two major strategies to manipulate protein abundance: conditional gene expression via inducible or repressible promoters and regulated metabolic destabilization of a protein of interest via inducible, ubiquitin-mediated degradation signals (Faden, Mielke, Lange, & Dissmeyer, 2014). The combined use of these systems allows one to regulate more than one protein at a time, which is particularly useful for investigating the cross talk between different cellular pathways.

In this chapter, we describe an application of these principles to the investigation of DNA damage signaling during replication. Using a combination of conditional gene expression via a tetracycline-regulable promoter and inducible protein degradation via an auxin-regulated degron system, we have analyzed the cross talk between factors controlling the bypass of lesions during replication and the activation of the DNA damage checkpoint (Garcia-Rodriguez, Morawska, Wong, Daigaku, & Ulrich, 2018). In the following sections, we describe the basic principles underlying our experimental system and provide a detailed protocol to analyze the cross talk between damage bypass and checkpoint signaling. Finally, we point out limitations of using this approach, briefly discuss possible applications to other regulatory events, and mention options to expand the system to other organisms.

1.1 Principles of the auxin-inducible degron system

Functional studies usually rely on the deletion of the relevant gene to investigate the physiological roles of its protein product. However, this strategy is not applicable to genes that are essential for the organism's viability. To study such genes, conditional mutants, which retain or lack their function depending on the experimental conditions, are particularly useful. Traditional temperature-sensitive (*ts*) conditional mutants, produced by random mutagenesis of the gene of interest or through classical genetic screening, have been used for decades to study the function of essential genes. Yet, the difficulty of generating *ts* alleles and their potential "leakiness," i.e., the retention of residual activity at nonpermissive temperature, have limited their use. To address these problems, methods for producing conditional mutants by using inducible degradation signals, or degrons, have been developed. The first such system was described by Varshavsky and coworkers, who developed a portable temperature-inducible degron that can be used to create a *ts* mutant of any protein of interest (Dohmen & Varshavsky, 2005; Dohmen, Wu, & Varshavsky, 1994). This method relies on the N-end rule pathway, a ubiquitin-mediated degradation system that is based on a protein's amino (N)-terminus as a recognition signal for protein ubiquitylation and subsequent delivery to the proteasome (Varshavsky, 1996). In the context of the N-end rule degron system, upon a temperature shift, a degron module linked to the N-terminus of the target protein undergoes a conformational change, which eventually exposes a destabilizing N-terminal residue, thus inducing ubiquitylation by Ubr1, the ubiquitin protein ligase (E3) that governs the N-end rule. The system is very efficient and reversible, but it is limited to proteins that remain functional with the required N-terminal modification. Furthermore, it inevitably involves the application of heat stress, which might induce undesirable physiological changes in the cell, making this approach nonoptimal for replication- or cell cycle-related studies. For example, expression of two G1 cyclins required for the G1–S transition, *CLN1* and *CLN2*, was found to be repressed upon heat-shock treatment (Rowley, Johnston, Butler, Werner-Washburne, & Singer, 1993).

More recently, an auxin-inducible degron (AID) system has emerged as a powerful alternative to control protein levels in nonplant eukaryotes (Nishimura, Fukagawa, Takisawa, Kakimoto, & Kanemaki, 2009). This technique makes use of the auxin-dependent degradation pathway from plants that lead to the proteasomal degradation of the Aux/IAA family of transcriptional repressors (Dharmasiri, Dharmasiri, & Estelle, 2005; Kepinski & Leyser, 2005). In its physiological context, binding of the plant hormone auxin to the F-box protein TIR1 promotes the interaction of the E3 complex SCF^{TIR1} with the IAA proteins, which are then polyubiquitylated and degraded via the proteasome. The SCF complex, consisting of Skp1, Cullin, and an F-box protein that provides specificity, is highly conserved among eukaryotes (Zimmerman, Schulman, & Zheng, 2010). Thus, expressing the F-box protein TIR1 in nonplant cells is usually sufficient to reconstitute the AID system in yeast or vertebrates (Fig. 1A). This system allows rapid and reversible depletion of a protein of interest and is more flexible than the N-end rule pathway since the degron sequence, derived from the IAA17 protein, can be fused to the N- or C-terminus of the protein of interest. Moreover, this method avoids physiological problems associated with the stress response induced by a temperature shift, and it minimizes possible side effects of the treatment because auxin is biologically silent in nonplant eukaryotes (Nishimura et al., 2009). The AID system has been further developed by shortening the AID tag (AID* or miniAID) and appending a series of epitope tags to allow detection by fluorescence microscopy or commercially available antibodies (Kubota, Nishimura, Kanemaki, & Donaldson, 2013; Morawska & Ulrich, 2013). Of note, in the original approach TIR1 expression was driven by either a galactose-inducible *GAL1–10* promoter or the constitutive *ADH1* promoter (Nishimura et al., 2009). Overexpression of TIR1 can be problematic in studies of essential genes since it can lead to an auxin-independent depletion that may result in target proteins falling below the levels required for survival (Mendoza-Ochoa et al., 2019). In order to circumvent this problem, additional regulatory features have been introduced to control TIR1 expression more tightly, e.g., a β-estradiol-based expression system (Mendoza-Ochoa et al., 2019).

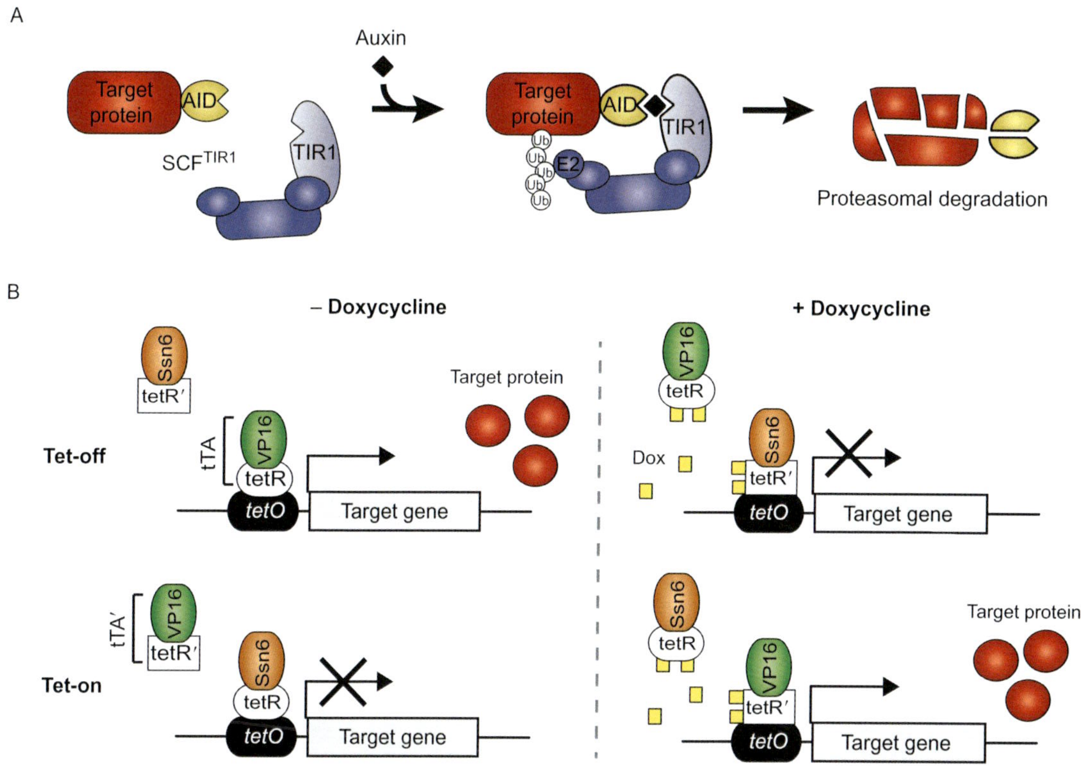

Fig. 1 Strategies to accomplish rapid manipulation of individual protein levels. (A) The auxin-inducible degron (AID) system. An exogenous auxin-responsive F-box protein, TIR1, associates with the host's endogenous SCF complex. In the presence of auxin, SCF^{TIR1} recognizes the AID-tagged target protein as a substrate, inducing its ubiquitylation and proteasomal degradation. (B) Tetracycline-regulable promoter systems in *S. cerevisiae*. In the Tet-off system, a fusion of tetR to a transcriptional activator domain (from VP16) to create the tTA activator induces expression of the target gene in the absence of doxycycline. Addition of the antibiotic causes release of tTA and binding of the repressive tetR′-Ssn6 construct. In the Tet-on system, the reverse arrangement of protein domains causes repression in the absence and induction in the presence of doxycycline.

1.2 Principles of Tet-regulated gene expression

An orthogonal strategy to characterize gene function by modulating protein abundance relies on the use of conditional expression systems that control individual gene activities specifically and in a temporally defined manner. Two types of regulable promoters are commonly used in *S. cerevisiae* (Maya, Quintero, de la, Munoz-Centeno, & Chavez, 2008). The first type is a native budding yeast promoter, including the widely used *GAL* promoters, activated by galactose and repressed by glucose (Johnston & Davis, 1984); the *CUP1* promoter, induced by addition of copper(II) ions (Karin et al., 1984); the *MET* promoters, responding to the concentration of methionine in the growth medium (Cherest, Nguyen, & Surdin-Kerjan, 1985; Sangsoda, Cherest, & Surdin-Kerjan, 1985); or the *PHO5* promoter, repressed by inorganic phosphate (Rogers, Lemire, & Bostian, 1982). Some of these promoters mediate a strong and rapid induction of protein synthesis and can be titrated by varying the concentration of the stimulus. However, they are subject to some limitations, including a basal activity of the promoter in the absence of the inducer, excessive expression levels that might cause nonphysiological responses, or unwanted metabolic alterations as a consequence of the required changes in the growth medium.

The second class of promoters consists of transcriptional elements derived from other organisms, such as bacteria or viruses. Among these, the so-called tetracycline-regulated promoters (Tet promoters) are the most popular. This system, first developed as a so-called Tet-off system in mammalian cells, makes use of a tetracycline transactivator (tTA) protein, created by the fusion of the *Escherichia coli* tetracycline repressor *tetR* to the activation domain of the VP16 protein from *Herpes simplex* virus. The resulting tTA protein binds to the tetracycline operator (*tetO*) sequences in its default state, thereby acting as a transcriptional activator in the absence of the antibiotic tetracycline or a commonly used analog, doxycycline. Upon addition of the antibiotic, gene expression is switched off by release of the tTA protein in its tetracycline-bound form (Gossen & Bujard, 1992). Subsequently, a Tet-on system was engineered by mutating the *tetR* moiety (*tetR′*), thus generating a modified tTA (tTA′) that only binds the operator in the presence of the antibiotic effector molecule (Gossen et al., 1995). The tetracycline-regulated promoter system has been successfully adapted to *S. cerevisiae* (Gari, Piedrafita, Aldea, & Herrero, 1997), and further optimization has reduced leakiness by expansion to a dual system in which a tTA activator is combined with a transcriptional repressor, Ssn6 (Fig. 1B). Here, a tight repression is achieved by combining tTA with tetR′-Ssn6, and a corresponding Tet-on system results from coexpression of tTA′ and tetR-Ssn6 (Belli, Gari, Piedrafita, et al., 1998). Hence, the Tet-promoter system provides many advantages because it allows rapid and tightly regulable expression with almost undetectable basal levels, requires no change in growth medium for induction or repression, avoids toxicity for cells, and has no detectable effect on global gene expression (Wishart, Hayes, Wardleworth, Zhang, & Oliver, 2005).

2. Application: Analysis of cross talk between DNA damage bypass and checkpoint signaling in the cell cycle

When cells are replicating their genome, the presence of DNA lesions poses a particularly serious problem since replication forks might stall or arrest. In this situation, cells employ DNA damage bypass pathways (also termed DNA damage tolerance or postreplicative repair) to ensure the complete duplication of the genome (Friedberg, 2005). The ubiquitin ligase Rad18 plays a central role in the regulation of DNA damage bypass by initiating the ubiquitylation of the replication clamp PCNA, which governs the choice of pathway used for the processing of DNA lesions (Hoege, Pfander, Moldovan, Pyrowolakis, & Jentsch, 2002; Moldovan, Pfander, & Jentsch, 2007; Ulrich, 2009). Importantly, DNA damage bypass can be delayed until after completion of replication without significant loss of viability (Daigaku, Davies, & Ulrich, 2010; Karras & Jentsch, 2010). Under these conditions, regions of single-stranded (ss) DNA, so-called daughter-strand gaps, accumulate in the newly replicated DNA. They can be filled postreplicatively once the pathway is reactivated. Insight into this phenomenon has strongly relied on the use of a regulable allele of Rad18, controlled by a Tet-off promoter system (Daigaku et al., 2010).

DNA checkpoints play a critical role in the maintenance of genomic integrity, not only by inducing a cell cycle arrest when genome integrity is compromised but also by mediating the repair of the damage (Nyberg, Michelson, Putnam, & Weinert, 2002). The signaling pathway elicited by replication stress is divided into two branches: the DNA damage checkpoint and the DNA replication checkpoint. In *S. cerevisiae*, both pathways are initiated by the sensor kinase Mec1 and converge on the effector kinase Rad53, which causes an overall amplification of the checkpoint response (Pardo, Crabbe, & Pasero, 2017).

Consistent with their protective role under conditions of replication stress, both PCNA ubiquitylation and checkpoint signaling are activated by a common signal: ssDNA covered by the ssDNA-binding replication protein A (RPA) complex at stalled replication intermediates or daughter-strand gaps behind replication forks (Davies, Huttner, Daigaku, Chen, & Ulrich, 2008; Garcia-Rodriguez et al., 2018; Zou & Elledge, 2003). Although the two reactions proceed independently of each other, several aspects of the checkpoint response have been shown to influence DNA damage bypass. In order to dissect the cross talk between the two pathways, we have designed an experimental setup that has allowed us to independently control the activities of DNA damage bypass and checkpoint signaling along the course of a cell cycle in a synchronized population (Garcia-Rodriguez et al., 2018). This setup relies on the combination of two previously characterized systems: a regulable DNA damage bypass system, in which *RAD18* expression is controlled by a Tet-repressible promoter, and the use of an auxin-inducible degradation allele of the checkpoint effector kinase Rad53 ($rad53^{AID*}$) (Daigaku et al., 2010; Morawska & Ulrich, 2013). As a simple genetic readout, recovery of cell viability upon reactivation of DNA damage bypass was measured under conditions where Rad53 was depleted at defined stages along the cell cycle (Garcia-Rodriguez et al., 2018).

3. Experimental procedures and materials

We describe here the experimental setup to reversibly and independently modulate the levels of two proteins, Rad18 and Rad53, by means of combined transcriptional regulation and auxin-mediated degron technology in the course of a cell cycle. The procedure follows the assay setup described by us before in the context of checkpoint activation and DNA damage bypass (Garcia-Rodriguez et al., 2018). As a read-out, we determine the recovery of cell viability after UV irradiation upon release into the cell cycle. Cell viability is monitored under conditions where *RAD18* expression is reactivated and Rad53 is depleted at different times along a single cell cycle. However, it is important to note that this setup can easily be modified for the analysis of other factors of interest or adapted to different experimental conditions relevant to a cell cycle-specific analysis or other processes that react to changes in the environmental conditions in a time-dependent manner.

3.1 Construction of yeast strains

To introduce Tet-regulated promoters and/or AID*-degron alleles, one will need *S. cerevisiae* strains that allow plasmid integration at the chromosomal *LEU2* and *URA3* loci, such as DF5 (*MATa, his3-Δ200, leu2-3,112, lys2-801, trp1-1, ura3-52*) (Finley, Ozkaynak, & Varshavsky, 1987) or W303 (*MATa, ade2-1, ura3-1, his3-11,15, trp1-1, leu2-3,112, can1-100, RAD5+*) (Thomas & Rothstein, 1989). The **a** mating type allows cell cycle synchronization by treatment with α-factor pheromone. We describe here a procedure that is based on DNA transformation methods, but alternative strategies, e.g., involving genetic crosses, are equally suitable. The plasmids described in this section are freely available from either Addgene (AID* system), the Yeast National BioResource Project (pNHK53), or Euroscarf (Tet system).

Substitution of the endogenous promoter of the gene of interest for the Tet promoter can be achieved by using a previously described promoter substitution cassette containing *kanMX4* as a selectable marker (Belli, Gari, Aldea, & Herrero, 1998) (Fig. 2A). Briefly, the pCM225 plasmid, encoding the *kanMX4* module followed by the tTA activator (VP16-*tetR*) under the control of the cytomegalovirus (CMV) promoter and the *tet*O_7 operator, serves as a template for PCR amplification using oligonucleotides with about 50bp homology to the target yeast DNA followed by sequences of the promoter substitution cassette. These oligonucleotides should be designed such that the forward oligonucleotide exhibits homology to a region about 100–200bp upstream of the target open reading frame (ORF) start codon, while the reverse oligonucleotide includes the target ORF start codon. The amplified PCR product is then transformed into yeast cells using the lithium acetate method for high-efficiency yeast transformation (Gietz, St Jean, Woods, & Schiestl, 1992). Homologous recombination of the amplified linear DNA into the chromosomal locus results in substitution of the promoter for the ORF with the *tet*O_7 promoter. Transformants are selected in YPD medium containing 200μg/mL geneticin (G418), and proper integration is confirmed by colony PCR, using one oligonucleotide that primes inside the cassette and one that is situated in the yeast genome adjacent to the integration site (Bergkessel & Guthrie, 2013). To achieve a more tightly regulated *tetO*-driven expression, the plasmid pCM244, containing the Ssn6 repressor coding sequence fused in frame to *tetR'* under control of the CMV promoter, is integrated at the chromosomal *LEU2* locus. The pCM244 plasmid is first linearized by digestion at the single *Eco*RV restriction site inside the plasmid *LEU2* gene to direct recombination to the endogenous *leu2* locus; cells are plated on medium lacking leucine to select for transformants. If an antibody against the target gene product is available, proper regulation should be verified by preparing total cell extracts from cultures grown in the presence or absence of 2μg/mL doxycycline, followed by Western blot analysis as

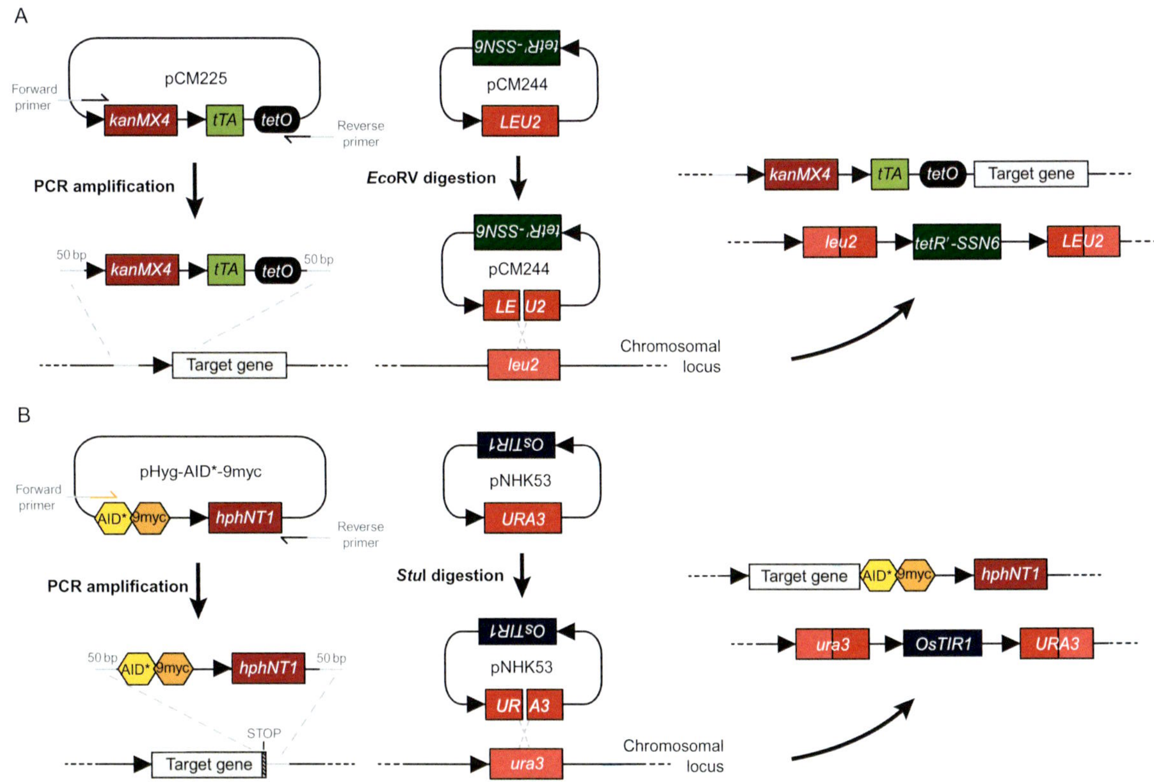

Fig. 2 Construction of yeast strains subject to dual regulation via the Tet-promoter and AID*-degron. Each strategy requires two genetic manipulations: the replacement of a short genomic sequence via homologous recombination with a PCR product generated from a plasmid template, and the integration of a linearized vector into a selectable marker gene. Coding regions are indicated by *boxes*, promoters by *arrowheads*. The resulting genotype is depicted on the right. (A) Construction of a Tet-repressible allele of a gene of interest, using plasmids pCM225 (encoding the tTA protein and containing the *tetO* regulatory sequence) and pCM244 (encoding tetR′-Ssn6). (B) Construction of an AID*-tagged allele of a gene of interest, using plasmids pHyg-AID*-9myc (containing the AID*-9myc sequence) and pNHK53 (encoding the TIR1 protein).

described below (Section 3.3). In the absence of a suitable antibody, phenotypic analysis under repressive conditions is advisable.

To create an auxin-inducible degron allele of the (second) gene of interest, a minimal AID tag (AID*) is introduced at the C-terminus of the target gene using a plasmid from the previously described set of AID*-tagging vectors (Morawska & Ulrich, 2013) (Fig. 2B). Briefly, the AID*-tagging cassette is amplified by PCR using forward and reverse oligonucleotides with about 50 bp homology to the region upstream and downstream of the STOP codon of the target gene, respectively, and followed by sequences of the AID*-tagging cassette. The AID* cassette includes an in-frame sequence encoding an epitope, such as GFP, myc, HA, or FLAG, for which monoclonal antibodies are commercially available. The amplified PCR product is then transformed into the strain already bearing the Tet-regulated allele as described earlier, and transformants are selected in YPD medium containing the antibiotic appropriate for the AID*-tagging cassette used (300 μg/mL hygromycin B for *hphNT1* or 100 μg/mL nourseothricin for *natNT2*). Note that AID* modules bearing the *kanMX4* cassette cannot be used for transformation into a host strain that already harbors the *Tet-RAD18* allele since that allele is already marked with *kanMX4*. If the use of such a cassette is desired, it can be separately transformed into an isogenic wild-type strain of the opposite mating type (α) and combined with the *Tet-RAD18* allele by mating and tetrad dissection. Proper integration is again verified by colony PCR, using appropriate oligonucleotides inside and outside the tagging cassette (Bergkessel & Guthrie, 2013). Expression of the tagged fusion protein should be verified by Western blotting of total cell extracts, using an antibody specific for the epitope tag used in the AID*-cassette, as described in Section 3.3.

Following verification of the desired AID* fusion allele, the DNA sequence coding for the F-box protein TIR1 from *Oriza sativa* (osTIR1) is integrated into the chromosomal *URA3* locus by integrative transformation of the plasmid pNHK53 (encoding *osTIR1-9myc* under control of the yeast *ADH1* promoter). Prior to transformation, the plasmid is linearized by digestion at the unique *Stu*I restriction site inside the plasmid-borne *URA3* gene. Transformants capable of growing on medium lacking uracil are selected. As alternatives to the osTIR1 F-box protein, vectors encoding alternative F-box proteins such as TIR1 from *Arabidopsis thaliana* (Nishimura et al., 2009) or AFB2 (Havens et al., 2012) can be used. Note that strains carrying a $RAD53^{AID*}$ allele should be constructed in cells lacking the ribonucleotide reductase inhibitor Sml1 to maintain viability upon auxin-induced degradation of Rad53. Efficient auxin-induced degradation of the target protein should at this point be verified by preparing total cell extracts in the absence or presence of auxin, followed by Western blotting as described in Section 3.3.

The final genotype of the strain used in the assays described below is: *MATa, his3-Δ200, leu2-3,112, lys2-801, trp1-1, ura3-52, CMVp-tTA-tetO$_7$-RAD18::kanMX, CMVp-tetR'-SSN6::LEU2, URA3::ADH1-osTIR1^{9myc}, sml1Δ::hisG-URA3-hisG, RAD53$^{AID*-9myc}$::hphNT1* (Garcia-Rodriguez et al., 2018). Note that in this particular case *ADH1-osTIR1^{9myc}* and *sml1Δ* were combined by mating and tetrad dissection.

3.2 Cell cycle assays and analysis of cell survival

3.2.1 Materials and equipment

- YPD liquid medium: 1% (w/v) Bacto-yeast extract, 2% (w/v) Bacto-peptone, 2% (w/v) glucose in water, sterilized by autoclaving
- YPD plates: YPD medium as above, containing 2% (w/v) Bacto-agar
- Water-agar plates: 2% (w/v) Bacto-agar in water
- Round (90 mm) and square (120 mm × 120 mm) Petri dishes
- Sterile distilled water
- Alpha-factor (Eurogentec): stock solution at 1 mg/mL in water (working concentration: 4–10 μg/mL)
- Doxycycline hydrochloride (SIGMA, #D3447): stock solution at 2 mg/mL in water, stored at −20°C (working concentration: 2 μg/mL)
- Auxin (indole-3-acetic acid sodium salt): stock solution at 250 m*M* in YPD medium, stored at −20°C (working concentration: 250 μ*M*)
- Culture tubes and flasks
- 1.5 mL microcentrifuge and 50-mL conical tubes
- Tabletop microcentrifuge and centrifuge for spinning 50-mL conical tubes
- Incubator (for plates) and shaking incubators (for flasks and tubes) at 30°C
- DNA cross-linker (Stratagene Stratalinker UV2400 or similar model)
- 96-Well plates
- Yeast strains (as described earlier)

3.2.2 Scheme A: Assessing the effects of checkpoint signaling on damage bypass during S phase

In this procedure, recovery of viability after UV damage is assessed upon reactivation of damage bypass by induction of Rad18 under conditions where checkpoint signaling is abolished by depletion of Rad53 before entry into S phase (Fig. 3A). Cell cycle profiles are analyzed in parallel to verify successful release for alpha-factor arrest and to determine the cell cycle stage at which the proteins of interest are degraded and/or reactivated.

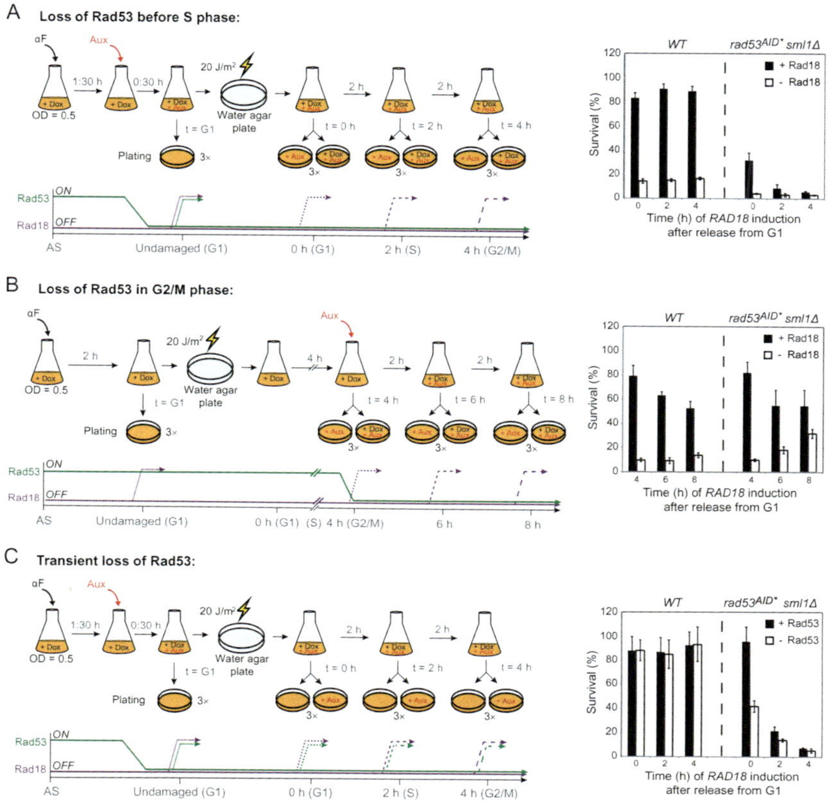

Fig. 3 Analysis of the cross talk between DNA damage bypass and checkpoint signaling in the course of a cell cycle. In these experiments, the activity of DNA damage bypass is manipulated by means of the *RAD18* gene under control of the *Tet-off* promoter, while checkpoint signaling is governed by modulating Rad53 protein levels via an AID*-degron. Before the beginning of the experiment, *RAD18* expression is repressed by the presence of doxycycline in the medium, while Rad53 is stably produced. The levels of Rad53 and Rad18 are then separately manipulated during synchronization of the culture in the G1 phase and/or after release into the cell cycle, and viability is scored by plating samples onto appropriate media (*αF*, alpha-factor; *Aux*, *+A*, auxin; *+D*, doxycycline). (A) Assessing the effects of checkpoint signaling on damage bypass during S phase. Degradation of Rad53 is induced during synchronization of cultures in G1, and viability in the absence of Rad53 is scored upon reactivation of *RAD18* expression. (B) Assessing the effects of checkpoint signaling on damage bypass after completion of S phase. Degradation of Rad53 is only induced once the cultures have reached G2/M phase, and recovery of viability in the absence of the checkpoint kinase is scored upon reactivation of *RAD18* expression in G2-arrested cells over extended times. (C) Assessing the effects of checkpoint reactivation during S phase on damage bypass. Degradation of Rad53 is induced during synchronization of cultures in G1, but the Rad53 kinase is reactivated in parallel to *RAD18* induction at defined stages of the cell cycle to assess the reversibility of the effects of its depletion. *This figure was adapted with modifications from Garcia-Rodriguez, N., Morawska, M., Wong, R. P., Daigaku, Y., & Ulrich, H. D. (2018). Spatial separation between replisome- and template-induced replication stress signaling.* The EMBO Journal, 37, *e98369*.

Day 1:

1. Inoculate a single colony into 2–3 mL YPD in a culture tube. Grow to saturation at 30°C overnight.

Day 2:

2. Prepare two square plates containing 2% agar in water per strain. Prepare three YPD plates, three YPD plates containing auxin (1 m*M*), and three YPD plates containing auxin + doxycycline (2 μg/mL), per strain. This number of plates (nine) is used for the first time point. For each additional time point, prepare three additional YPD + auxin plates and three YPD + auxin + doxycycline plates.
3. 14 h before the start of the time course experiment, prepare two cultures per strain: inoculate cells from overnight saturated cultures at dilutions of 1:5000 and 1:10,000 into 100-mL flasks containing 25 mL YPD supplemented with doxycycline in order to shut off *RAD18* transcription. The dilution factor may have to be readjusted according to the growth rate of the specific strain. Place the cultures at 30°C in a shaking incubator.

Day 3:

4. After 14 h of incubation, measure the OD_{600} of each culture. For each strain, select the culture with an OD_{600} ~ 0.5 (discard the other one) and add alpha-factor to a final concentration of 4–10 μg/mL to synchronize cells in G1 phase. Incubate at 30°C for 1.5 h. To verify successful synchronization, yeast cells should be visualized under a light microscope to ensure that most cells (>90%) have arrested with cell projections (shmoo-shaped).
5. At 1.5 h after alpha-factor addition, auxin is added to the cultures to a final concentration of 250 μ*M* in order to induce $Rad53^{AID*}$ degradation while maintaining G1 arrest.
6. At 30 min after auxin addition, transfer 1 mL of the culture to a 1.5-mL microcentrifuge tube, harvest cells by centrifugation at 8000 × *g* for 1 min, resuspend in 1 mL 70% ethanol, and keep at 4°C until later processing for cell cycle analysis by flow cytometry (see Section 3.4). At the same time, transfer another 1 mL aliquot of the culture into a microcentrifuge tube and keep on ice for plating in order to determine viability. These are the "G1 samples."
7. Transfer 20 mL of the remaining culture into a 50-mL conical tube and harvest cells at 3000 × *g*, at room temperature, for 2 min. Wash cells by resuspending in 20 mL of sterile water containing doxycycline + auxin and repeating the centrifugation step.

8. Resuspend cells in 20 mL water containing doxycycline + auxin and spread the cell suspension evenly onto two water-agar square plates (10 mL of cells on each plate) by gently shaking the plates.
9. Expose both plates in parallel to UV light (254 nm) at a total energy dose of $20\,J/m^2$ using a DNA cross-linker.
10. Collect irradiated cells into a fresh 50-mL conical tube wrapped in aluminum foil to prevent exposure to light that might result in photolyase activation (Sancar, 2000). Wash each water-agar plate with 10 mL water + doxycycline + auxin and combine with the collected cells in the same tube.
11. Pellet cells by centrifugation at $3000 \times g$ for 2 min (remove the foil temporarily during centrifugation if necessary). Resuspend in 20 mL of YPD + doxycycline + auxin and take 1 mL of the culture for cell cycle analysis and 1 mL culture for plating as $t = 0$ h samples. Incubate the rest of the culture in a shake incubator, in the dark, at 30°C.
12. Before the next time point, plate the G1 and $t = 0$ h samples. To perform colony counting as a way to assess cell survival rates, prepare serial dilutions of the samples using 96-well plates or individual microcentrifuge tubes, and plate 50 μL of appropriate dilutions to yield 100–200 colonies per round plate, in triplicate. This is estimated by first measuring the OD_{600} values for each sample (assume that one OD_{600} corresponds to ca. 2×10^7 cells at 100% viability). The necessary dilution factor has to be adjusted according to the survival rate of the specific strain to allow reliable colony counting. Spreading the cell evenly on plates is achieved using sterile disposable inoculating loops, glass beads, or similar methods. Undamaged G1 samples serve as reference and are only plated onto YPD, while $t = 0$ h samples are plated onto YPD containing auxin with (–Rad18) or without doxycycline (+Rad18) to either maintain repression or allow reactivation of *RAD18* gene transcription. Importantly, cells need to be plated from less diluted samples in the presence of doxycycline since viability in the presence of DNA damage is expected to be decreased when the cells lack Rad18.
13. At the following time points (e.g., $t = 2$ h and $t = 4$ h), harvest 1 mL culture for cell cycle analysis and 1 mL for determining viability and proceed with dilutions and plating as in the previous step.
14. Incubate plates at 30°C for 2–3 days and count colonies of the three technical replicates per strain and time point. Calculate the average survival for each condition and determine the percentage of surviving cells relative to the undamaged G1 samples. Repeat the experiment independently at least twice more.

3.2.3 Scheme B: Assessing the effects of checkpoint signaling on damage bypass after completion of S phase

In this procedure, recovery of viability after UV damage is assessed upon reactivation of damage bypass as above; however, in this case Rad18 is induced under conditions where checkpoint signaling is abolished by depletion of Rad53 only after cells have reached the G2/M phase (Fig. 3B).

Proceed as described earlier for Scheme A (Section 3.2.2) with the following modification:

- In step 4, incubate the cultures for 2 h in the presence of alpha-factor.
- Omit step 5.
- At 2 h after alpha-factor addition, take G1 samples as described in step 6.
- In steps 7, 8 and 10, washes are performed with water containing doxycycline, but not auxin.
- In step 11, you do not need to collect $t=0$ h samples.
- In step 13, incubate cultures for 4 h after release into S phase, collect samples for cell cycle analysis (see Section 3.4) and viability ($t=4$ h) and add auxin to the rest of the cultures to induce $Rad53^{AID*}$ degradation in the G2/M phase of the cell cycle. Collect additional samples at $t=6$ h and $t=8$ h.

3.2.4 Scheme C: Assessing the effects of checkpoint reactivation during S phase on damage bypass

In this procedure, recovery of viability after UV damage is assessed under conditions of simultaneous reactivation of damage bypass (by induction of Rad18) and checkpoint signaling (by terminating $Rad53^{AID*}$ degradation) (Fig. 3C). A transient loss of Rad53 is achievable because the auxin-inducible degron system is reversible and resynthesis of Rad53 is rapid.

Proceed as described earlier for Scheme A (Section 3.2.2) with the following modifications:

- In step 2, prepare six YPD plates and three YPD + auxin plates per strain. This number of plates refers to one time point. If taking more time points, prepare three additional YPD + auxin plates and three YPD plates per additional time point.
- In step 12, undamaged G1 samples are plated onto YPD + auxin, while $t=0$ h, $t=2$ h, and $t=4$ h samples are plated onto both YPD and YPD + auxin.

3.2.5 Notes

- During preparation of YPD plates containing auxin and/or doxycycline, YPD-agar must be cooled down to 55–60°C prior to addition of the compound. Plates can be stored at 4°C for up to 3 days before use.

- DNA damage can also be induced by other means, such as treatment with the alkylating agent methyl methanesulfonate (MMS). In this case, 0.02%–0.08% MMS is added to the culture during the synchronization step, at 1.5 h after addition of alpha-factor. Cells are then incubated in the presence of MMS for 30 min, washed twice with water + doxycycline + auxin and then released into S phase.

3.3 Preparation and Western blot analysis of total cell extracts

To determine if both regulated gene expression by means of the Tet-promoter and proper proteolytic control of AID*-tagged proteins are occurring, immunoblot analysis should be performed with antibodies against the relevant proteins or epitope tags. There are several methods for extracting total yeast proteins; here, we briefly describe a procedure for extraction using trichloroacetic acid (TCA) precipitation.

3.3.1 Materials and equipment

- NaOH/BME: 1.85 *M* NaOH and 7.5% (v/v) β-mercaptoethanol. Prepare from a stock solution of 2 *M* NaOH by addition of β-mercaptoethanol immediately before use.
- TCA solution: 55% (w/v) trichloroacetic acid in water.
- LDS sample buffer (Thermo Fisher Scientific).
- 1.5 mL microcentrifuge tubes and table-top microcentrifuge.

3.3.2 Procedure

1. Samples of ~4×10^7 cells are harvested by centrifugation at 16,000 × *g* for 15 s in a microcentrifuge.
2. The cell pellet is resuspended in 500 μL of water. Add 75 μL of NaOH/BME solution, mix thoroughly, and incubate for 15 min on ice.
3. Add 75 μL of TCA solution, mix, and incubate for a further 10 min on ice.
4. After centrifugation at 16,000 × *g* for 10 min, the supernatant is aspirated and the pellet is resuspended in 40 μL 1 × LDS sample buffer, following incubation at 65°C for 10 min.
5. Proteins are then analyzed by SDS-polyacrylamide gel electrophoresis and Western blotting using an appropriate antibody.

3.4 Analysis of cell cycle profiles by flow cytometry

In order to monitor the progression through the cell cycle and determine the cell cycle profile at the time of plating, samples are taken and subjected to flow cytometry analysis.

3.4.1 Materials and equipment

- 70% (v/v) ethanol in water
- Sodium citrate buffer: 50 m*M*, pH 7.0
- RNase A: stock solution at 10 mg/mL in 10 m*M* Tris–HCl, pH 7.5, 15 m*M* NaCl
- Proteinase K: stock solution at 10 mg/mL in 40% glycerol (v/v), 10 m*M* Tris–HCl, pH 7.5, 1 m*M* calcium acetate
- Propidium iodide: stock solution at 2 mg/mL in water
- Round-bottom polystyrene tubes suitable for flow cytometry
- Ultrasonic homogenizer or sonicator
- Flow cytometer

3.4.2 Procedure

1. Collect a 1-mL aliquot of yeast culture at each relevant time point, harvest cells by centrifugation, and resuspend in 1 mL 70% ethanol. Samples can be stored at 4°C for days.
2. Wash cells twice with 1 mL sodium citrate buffer and resuspend in 1 mL of sodium citrate buffer.
3. Add 8 μL RNase A solution (10 mg/mL) and incubate for 1 h at 50°C.
4. Add 8 μL proteinase K solution (10 mg/mL) and incubate for 1 h at 50°C.
5. Mix 250 μL of sample with 250 μL of a propidium iodide solution in sodium citrate buffer (final concentration of propidium iodide: 16 μg/mL) in round-bottom polystyrene tubes.
6. Sonicate samples using a Sonifier for 3 s, at 10% amplitude, to disperse cells. Exact settings will depend on the instrument used.
7. Analyze samples by flow cytometry using a flow cytometer according to the manufacturer's instructions. Collect data for at least 10,000 cells and visualize the propidium iodide signal as a histogram plot.

4. Limitations of the dual expression-control method

By using the *Tet-RAD18* and Rad53^{AID*} systems in combination, we were able to demonstrate that Rad53 function is required specifically during S phase to maintain DNA damage bypass competence, but not for the postreplicative ssDNA gap filling reaction in G2/M-arrested cells. Furthermore, we showed that a transient loss of Rad53 during replication is sufficient to irreversibly prevent damage bypass (Garcia-Rodriguez et al., 2018). This dual system can also be applied to other checkpoint factors.

In fact, we have observed that Ddc1, a component of the 9-1-1 checkpoint sensor clamp, is not only required to maintain bypass competence during S phase, but also in G2/M (unpublished observations), thus supporting previous findings that implicate the 9-1-1 complex in the damage bypass process itself (Karras et al., 2013). Apart from checkpoint factors, the role of other cellular components in DNA damage bypass, such as chromatin remodelers, can also be investigated using the same approach.

Importantly, some drawbacks are inherent in these conditional systems. For example, it has been shown that the addition of an AID-tag, or its truncated version, can reduce basal protein levels in a TIR1-independent manner (Morawska & Ulrich, 2013), and high levels of TIR1 can lead to auxin-independent degradation (Mendoza-Ochoa et al., 2019). Furthermore, proteins might not be degraded to a sufficient degree, or their degradation might require incubation with auxin for extended time. If recovery from degradation is required, slow protein synthesis might interfere with the timely reaccumulation of the protein, although this is generally not expected to be problematic, as continuous protein synthesis is maintained during auxin treatment. All these factors would prevent efficient use of the AID-based degron system. Hence, we recommend considering individual proteins on a case-by-case basis to assess depletion efficiency and to carefully compare basal protein levels of tagged and untagged targets.

Similar limitations in terms of response time, protein abundance, and reversibility apply to the use of the Tet-promoter system. Therefore, the expression of the gene of interest and the response to a range of doxycycline concentrations should be tested because Tet-regulated promoters can suffer either from "leakiness" in the "off" state of the promoter or from suboptimal protein levels in the "on" state (Mnaimneh et al., 2004). Another problem frequently experienced with conditional expression systems stems from so-called phenotypic lag, a time interval between the imposition of nonpermissive conditions and the appearance of a relevant null phenotype, which depends on the half-life of the expressed protein. Likewise, transcription and translation rates will be limiting factors for the reaccumulation of a protein upon promoter induction. Thus, the effects of gene perturbation might become apparent over a longer time frame than the process under investigation. It is worth noting that in our experimental setup we avoided any potential phenotypic lag in the *Tet-RAD18* system by shutting off the promoter 14h before the start of the experiment.

5. Outlook

An alternative to regulable promoters or inducible degrons can be found in the use of cell cycle-regulated alleles that restrict the presence of the protein of interest to a specific phase of the cell cycle. This approach is particularly useful in situations where the analysis of a single cell cycle is not required, but a "chronic" condition over several generations is desired. These alleles are constructed by placing the corresponding gene under the control of the entire suite of regulatory features of a cell cycle-regulated factor, including its promoter sequence and relevant portions of the ORF that mediate cell cycle-regulated degradation via a natural degron. For example, using this approach, restriction to M/G1 phase has been achieved by the promoter and N-terminal portions of the CDK inhibitor Sic1; an "S-phase tag" has been generated by utilizing the promoter and N-terminal degron element of the S phase cyclin Clb6; and G2/M-specific expression has been specified in an analogous fashion by imposing the regulatory elements of the mitotic cyclin Clb2 (Hombauer, Srivatsan, Putnam, & Kolodner, 2011; Johnson, Gali, Takahashi, & Kubota, 2016; Karras & Jentsch, 2010). Although these tags have the advantage of combining accurate expression at a specific phase of the cell cycle with low levels of "leakiness," they require an N-terminal modification of the target protein, which might impair its function.

In principle, the dual system described in this chapter can be applied to studies of human cells, as both the Tet-regulated conditional expression and the AID-degron system are functional in mammalian cells (Gossen & Bujard, 1992; Gossen et al., 1995; Nishimura et al., 2009). Since homologous recombination is less efficient in human cells and targeting of both alleles of diploid cells is required to achieve proper regulation, construction of appropriate cell lines for the use of these systems is more challenging. However, with the rapid advances in CRISPR–Cas9 technology, genetic manipulation of diploid human cell lines is moving into the realm of standard laboratory routine.

Acknowledgments

We thank Magdalena Morawska for her contribution to the experiments in the original paper. Research in H.D.U.'s lab is supported by grants from the European Research Council (ERC) and the German Research Association (DFG). N.G.R. is supported by the Marie Sklodowska-Curie Fellowship H2020-MSCA-IF-2017-794054 from the European Union's Horizon 2020 Research and Innovation Programme.

References

Belli, G., Gari, E., Aldea, M., & Herrero, E. (1998). Functional analysis of yeast essential genes using a promoter-substitution cassette and the tetracycline-regulatable dual expression system. *Yeast*, *14*, 1127–1138.

Belli, G., Gari, E., Piedrafita, L., Aldea, M., & Herrero, E. (1998). An activator/repressor dual system allows tight tetracycline-regulated gene expression in budding yeast. *Nucleic Acids Research*, *26*, 942–947.

Bergkessel, M., & Guthrie, C. (2013). Colony PCR. *Methods in Enzymology*, *529*, 299–309.

Botstein, D., & Fink, G. R. (2011). Yeast: An experimental organism for 21st century biology. *Genetics*, *189*, 695–704.

Cherest, H., Nguyen, N. T., & Surdin-Kerjan, Y. (1985). Transcriptional regulation of the MET3 gene of *Saccharomyces cerevisiae*. *Gene*, *34*, 269–281.

Daigaku, Y., Davies, A. A., & Ulrich, H. D. (2010). Ubiquitin-dependent DNA damage bypass is separable from genome replication. *Nature*, *465*, 951–955.

Davies, A. A., Huttner, D., Daigaku, Y., Chen, S., & Ulrich, H. D. (2008). Activation of ubiquitin-dependent DNA damage bypass is mediated by replication protein A. *Molecular Cell*, *29*, 625–636.

Dharmasiri, N., Dharmasiri, S., & Estelle, M. (2005). The F-box protein TIR1 is an auxin receptor. *Nature*, *435*, 441–445.

Dohmen, R. J., & Varshavsky, A. (2005). Heat-inducible degron and the making of conditional mutants. *Methods in Enzymology*, *399*, 799–822.

Dohmen, R. J., Wu, P., & Varshavsky, A. (1994). Heat-inducible degron: A method for constructing temperature-sensitive mutants. *Science*, *263*, 1273–1276.

Duina, A. A., Miller, M. E., & Keeney, J. B. (2014). Budding yeast for budding geneticists: A primer on the *Saccharomyces cerevisiae* model system. *Genetics*, *197*, 33–48.

Faden, F., Mielke, S., Lange, D., & Dissmeyer, N. (2014). Generic tools for conditionally altering protein abundance and phenotypes on demand. *Biological Chemistry*, *395*, 737–762.

Finley, D., Ozkaynak, E., & Varshavsky, A. (1987). The yeast polyubiquitin gene is essential for resistance to high temperatures, starvation, and other stresses. *Cell*, *48*, 1035–1046.

Friedberg, E. C. (2005). Suffering in silence: The tolerance of DNA damage. *Nature Reviews. Molecular Cell Biology*, *6*, 943–953.

Garcia-Rodriguez, N., Morawska, M., Wong, R. P., Daigaku, Y., & Ulrich, H. D. (2018). Spatial separation between replisome- and template-induced replication stress signaling. *The EMBO Journal*, *37*, e98369.

Gari, E., Piedrafita, L., Aldea, M., & Herrero, E. (1997). A set of vectors with a tetracycline-regulatable promoter system for modulated gene expression in *Saccharomyces cerevisiae*. *Yeast*, *13*, 837–848.

Gietz, D., St Jean, A., Woods, R. A., & Schiestl, R. H. (1992). Improved method for high efficiency transformation of intact yeast cells. *Nucleic Acids Research*, *20*, 1425.

Gossen, M., & Bujard, H. (1992). Tight control of gene expression in mammalian cells by tetracycline-responsive promoters. *Proceedings of the National Academy of Sciences of the United States of America*, *89*, 5547–5551.

Gossen, M., Freundlieb, S., Bender, G., Muller, G., Hillen, W., & Bujard, H. (1995). Transcriptional activation by tetracyclines in mammalian cells. *Science*, *268*, 1766–1769.

Havens, K. A., Guseman, J. M., Jang, S. S., Pierre-Jerome, E., Bolten, N., Klavins, E., et al. (2012). A synthetic approach reveals extensive tunability of auxin signaling. *Plant Physiology*, *160*, 135–142.

Hoege, C., Pfander, B., Moldovan, G. L., Pyrowolakis, G., & Jentsch, S. (2002). RAD6-dependent DNA repair is linked to modification of PCNA by ubiquitin and SUMO. *Nature*, *419*, 135–141.

Hombauer, H., Srivatsan, A., Putnam, C. D., & Kolodner, R. D. (2011). Mismatch repair, but not heteroduplex rejection, is temporally coupled to DNA replication. *Science*, *334*, 1713–1716.

Johnson, C., Gali, V. K., Takahashi, T. S., & Kubota, T. (2016). PCNA retention on DNA into G2/M phase causes genome instability in cells lacking Elg1. *Cell Reports*, *16*, 684–695.

Johnston, M., & Davis, R. W. (1984). Sequences that regulate the divergent GAL1-GAL10 promoter in *Saccharomyces cerevisiae*. *Molecular and Cellular Biology*, *4*, 1440–1448.

Karin, M., Najarian, R., Haslinger, A., Valenzuela, P., Welch, J., & Fogel, S. (1984). Primary structure and transcription of an amplified genetic locus: The CUP1 locus of yeast. *Proceedings of the National Academy of Sciences of the United States of America*, *81*, 337–341.

Karras, G. I., Fumasoni, M., Sienski, G., Vanoli, F., Branzei, D., & Jentsch, S. (2013). Noncanonical role of the 9-1-1 clamp in the error-free DNA damage tolerance pathway. *Molecular Cell*, *49*, 536–546.

Karras, G. I., & Jentsch, S. (2010). The RAD6 DNA damage tolerance pathway operates uncoupled from the replication fork and is functional beyond S phase. *Cell*, *141*, 255–267.

Kepinski, S., & Leyser, O. (2005). The Arabidopsis F-box protein TIR1 is an auxin receptor. *Nature*, *435*, 446–451.

Kubota, T., Nishimura, K., Kanemaki, M. T., & Donaldson, A. D. (2013). The Elg1 replication factor C-like complex functions in PCNA unloading during DNA replication. *Molecular Cell*, *50*, 273–280.

Maya, D., Quintero, M. J., de la, C., Munoz-Centeno, M., & Chavez, S. (2008). Systems for applied gene control in *Saccharomyces cerevisiae*. *Biotechnology Letters*, *30*, 979–987.

Mendoza-Ochoa, G. I., Barrass, J. D., Terlouw, B. R., Maudlin, I. E., de Lucas, S., Sani, E., et al. (2019). A fast and tuneable auxin-inducible degron for depletion of target proteins in budding yeast. *Yeast*, *36*, 75–81.

Mnaimneh, S., Davierwala, A. P., Haynes, J., Moffat, J., Peng, W. T., Zhang, W., et al. (2004). Exploration of essential gene functions via titratable promoter alleles. *Cell*, *118*, 31–44.

Moldovan, G. L., Pfander, B., & Jentsch, S. (2007). PCNA, the maestro of the replication fork. *Cell*, *129*, 665–679.

Morawska, M., & Ulrich, H. D. (2013). An expanded tool kit for the auxin-inducible degron system in budding yeast. *Yeast*, *30*, 341–351.

Nishimura, K., Fukagawa, T., Takisawa, H., Kakimoto, T., & Kanemaki, M. (2009). An auxin-based degron system for the rapid depletion of proteins in nonplant cells. *Nature Methods*, *6*, 917–922.

Nyberg, K. A., Michelson, R. J., Putnam, C. W., & Weinert, T. A. (2002). Toward maintaining the genome: DNA damage and replication checkpoints. *Annual Review of Genetics*, *36*, 617–656.

Pardo, B., Crabbe, L., & Pasero, P. (2017). Signaling pathways of replication stress in yeast. *FEMS Yeast Research*, *17*(2). https://doi.org/10.1093/femsyr/fow101.

Rogers, D. T., Lemire, J. M., & Bostian, K. A. (1982). Acid phosphatase polypeptides in *Saccharomyces cerevisiae* are encoded by a differentially regulated multigene family. *Proceedings of the National Academy of Sciences of the United States of America*, *79*, 2157–2161.

Rowley, A., Johnston, G. C., Butler, B., Werner-Washburne, M., & Singer, R. A. (1993). Heat shock-mediated cell cycle blockage and G1 cyclin expression in the yeast *Saccharomyces cerevisiae*. *Molecular and Cellular Biology*, *13*, 1034–1041.

Sancar, G. B. (2000). Enzymatic photoreactivation: 50 years and counting. *Mutation Research*, *451*, 25–37.

Sangsoda, S., Cherest, H., & Surdin-Kerjan, Y. (1985). The expression of the MET25 gene of *Saccharomyces cerevisiae* is regulated transcriptionally. *Molecular & General Genetics*, *200*, 407–414.
Thomas, B. J., & Rothstein, R. (1989). Elevated recombination rates in transcriptionally active DNA. *Cell*, *56*, 619–630.
Ulrich, H. D. (2009). Regulating post-translational modifications of the eukaryotic replication clamp PCNA. *DNA Repair (Amst)*, *8*, 461–469.
Varshavsky, A. (1996). The N-end rule: Functions, mysteries, uses. *Proceedings of the National Academy of Sciences of the United States of America*, *93*, 12142–12149.
Wishart, J. A., Hayes, A., Wardleworth, L., Zhang, N., & Oliver, S. G. (2005). Doxycycline, the drug used to control the tet-regulatable promoter system, has no effect on global gene expression in *Saccharomyces cerevisiae*. *Yeast*, *22*, 565–569.
Zimmerman, E. S., Schulman, B. A., & Zheng, N. (2010). Structural assembly of cullin-RING ubiquitin ligase complexes. *Current Opinion in Structural Biology*, *20*, 714–721.
Zou, L., & Elledge, S. J. (2003). Sensing DNA damage through ATRIP recognition of RPA-ssDNA complexes. *Science*, *300*, 1542–1548.

CHAPTER SEVEN

Engineered disulfide crosslinking to measure conformational changes in the 26S proteasome

Randi G. Reed, Robert J. Tomko Jr.*
Department of Biomedical Sciences, Florida State University College of Medicine, Tallahassee, FL, United States
*Corresponding author: e-mail address: robert.tomko@med.fsu.edu

Contents

Abstract

The 26S proteasome is a multisubunit ATP-dependent peptidase complex mediating most regulated protein degradation in eukaryotes. The proteasome undergoes several coordinated conformational changes during catalysis that activate it for substrate processing and functionally couple distinct enzymatic activities during substrate degradation. Understanding the impact of substrate interactions and individual ATP binding events on these conformational changes is currently a major bottleneck in the study of proteasome function. Here, we describe a simple biochemical reporter based on engineered disulfide crosslinking for measuring the conformational distribution of the *Saccharomyces cerevisiae* 26S proteasome. We demonstrate its use to investigate the impact of ATP analogs and proteasome inhibitors on proteasome conformational equilibria. This reporter allows simultaneous and rapid comparison of multiple treatments or conditions on the steady-state conformational distribution of the proteasome and can be readily extended to the study of other multisubunit complexes for which multiple conformational states are known at near-atomic resolution.

Methods in Enzymology, Volume 619
ISSN 0076-6879
https://doi.org/10.1016/bs.mie.2018.11.006
© 2019 Elsevier Inc.
All rights reserved.

1. Introduction

The 26S proteasome is a 2.5-MDa ATP-dependent peptidase complex responsible for most regulated intracellular degradation in eukaryotes (Bard, Goodall, et al., 2018; Finley, Chen, & Walters, 2016; Tomko & Hochstrasser, 2013). It consists of a barrel-shaped, proteolytic core particle (CP) capped on one or both ends by a 19S regulatory particle (RP) (Fig. 1A). The RP can be further subdivided into two subcomplexes, the lid and the base. The lid consists of Rpn3, 5–9, 11, 12, and Rpn15/Sem1 subunits. The base consists of non-ATPase subunits Rpn1, 2, and 13, and a hexameric ring of six ATPase subunits, Rpt1–Rpt6. Rpn1, Rpn13, and an additional subunit, Rpn10, serve as receptors for incoming substrates. During substrate processing, the Rpn11 lid subunit removes the polyubiquitin targeting

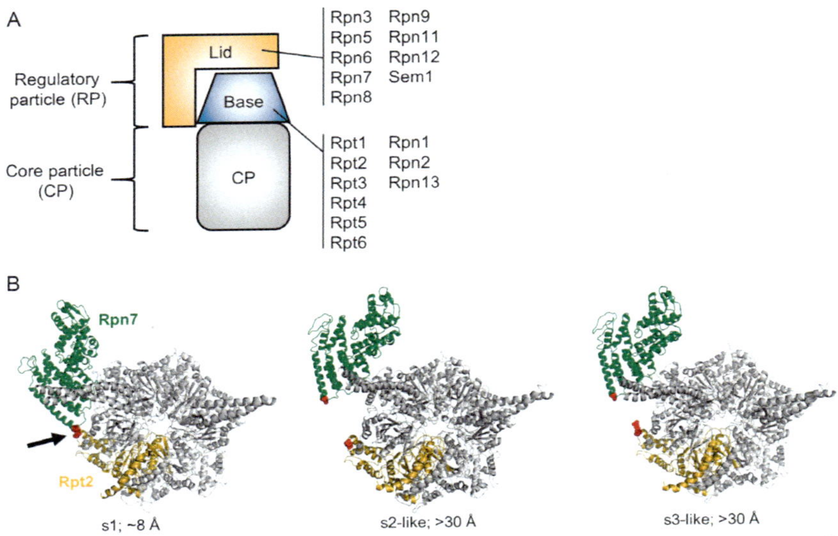

Fig. 1 (A) Cartoon depiction of the 26S proteasome. The subunit composition of the lid and base subcomplexes is shown. (B) Rpn7-D123 and Rpt2-R407 are juxtaposed in the s1 conformation of the yeast 26S proteasome (*red spheres* indicated by *arrow*) (PDB 4CR2), but are distant in the s2–s6 states. In the *middle panel*, the s2 state (PDB 4CR3) is shown, but the s5 state is highly similar. In the *right panel*, the s3 state (PDB 4CR4) is shown, but the s4 and s6 states are highly similar. The approximate distances in angstroms between the Rpn7-D123 and Rpt2-R407 alpha carbons are shown. *Panel (B) is modified from Eisele, M. R., Reed, R. G., Rudack, T., Schweitzer, A., Beck, F., Nagy, I., et al. (2018). Expanded coverage of the 26S proteasome conformational landscape reveals mechanisms of peptidase gating.* Cell Reports, *24(5), 1301–1315.e5. doi:10.1016/j.celrep.2018.07.004 with permission from Elsevier.*

signal (Verma et al., 2002; Worden, Dong, & Martin, 2017; Yao & Cohen, 2002). The base unfolds the substrate and translocates it through the central pore of the ATPase ring to the peptidase active sites of the CP, where it is cleaved into short peptides.

An onslaught of recent cryo-electron microscopy (cryo-EM) analyses has demonstrated that 26S proteasomes can adopt at least six conformational states, herein referred to as s1–s6 (Asano et al., 2015; Aufderheide et al., 2015; Bashore et al., 2015; Chen et al., 2016; de la Pena, Goodall, Gates, Lander, & Martin, 2018; Dong et al., 2018; Eisele et al., 2018; Guo et al., 2018; Haselbach et al., 2017; Huang, Luan, Wu, & Shi, 2016; Luan et al., 2016; Matyskiela, Lander, & Martin, 2013; Schweitzer et al., 2016; Sledz et al., 2013; Unverdorben et al., 2014; Wehmer et al., 2017; Zhu et al., 2018) (reviewed in Bard, Goodall, et al., 2018). These states can be broadly grouped into three categories: s1, s2-like, and s3-like. The s1 state appears to be the ground or resting state of the proteasome, and it is believed to be the major substrate-accepting state. In the s2-like states, the lid subcomplex undergoes a rotation that positions the deubiquitinase subunit Rpn11 directly over the pore of the ATPase ring of the base for substrate deubiquitination (Matyskiela et al., 2013; Sledz et al., 2013). Although these states appear primed for deubiquitination, a mismatch between the pore of the ATPase ring and the central channel of the CP suggests that they may not be fully competent for substrate degradation (Eisele et al., 2018; Unverdorben et al., 2014). Finally, in the s3-like states, the ATPase ring undergoes movements that align its pore with the central channel of the CP, providing an unobstructed passageway for processing and delivery of the substrate into the peptidase sites (Chen et al., 2016; de la Pena et al., 2018; Dong et al., 2018; Eisele et al., 2018; Luan et al., 2016; Matyskiela et al., 2013; Sledz et al., 2013; Unverdorben et al., 2014; Wehmer et al., 2017). Thus, the s3-like states represent actively processing proteasomes (de la Pena et al., 2018; Eisele et al., 2018; Matyskiela et al., 2013). Although these cryo-EM structures have been groundbreaking in terms of visualizing the substrate processing mechanism of the proteasome, the events and signals that promote each of these rearrangements remain poorly understood (Bard, Bashore, Dong, & Martin, 2018; Bard, Goodall, et al., 2018; Eisele et al., 2018).

The large-scale repositioning of the lid relative to the base that occurs as the proteasome shifts from the resting s1 state to the s2/s3-like states results in the formation and breakage of specific molecular contacts between lid and base subunits (for example, see Fig. 1B). Such dynamic contacts can serve as reporters to discriminate the resting s1 state from the s2/s3-like states, permitting measurement of the fraction of resting vs activated proteasomes. Engineered disulfide crosslinking has previously been used to assay static interactions between adjacent subunits within the proteasome (Tomko, Funakoshi, Schneider, Wang, & Hochstrasser, 2010; Velichutina, Connerly, Arendt, Li, & Hochstrasser, 2004). Here, we apply this approach to dynamic subunit–subunit contacts to measure the relative fraction of s1 proteasomes. We demonstrate the use of this method for assessing the impact of ATP analogs and proteasome inhibitors on the conformational distribution of 26S proteasomes in yeast whole cell extracts (WCE). This approach is simple, rapid, and can be readily implemented for studying the conformational dynamics of other large protein complexes that adopt multiple conformational states.

2. Assessing the impact of ATP analogs on 26S proteasome conformational distribution

We have found that conformation-specific crosslinking performed in WCE accurately reports on the conformation of 26S proteasomes with little to no influence from proteasome assembly intermediates. This is because the abundances of assembly intermediates are typically very low compared to the abundance of 26S proteasomes. For initial analyses, we prefer yeast strains harboring cysteine substitutions at Rpn7-D123 and Rpt2-R407 for crosslinking because these mutations have no obvious impact on proteasome structure, assembly, or function, and the spacing of these residues increases substantially between the s1 and s2-/s3-like states (Fig. 1B). However, other substitutions can also be used if the cysteine substitutions are well tolerated and are appropriately juxtaposed in the s1 state but not the s2- or s3-like states. We have also successfully utilized Rpn6 (T203C)–Rpt6(G387C) and Rpn7(A236C)–Rpt6(R339C) substitution pairs (Eisele et al., 2018; unpublished observations). These pairs do not obviously compromise proteasome structure or function and respond to ATP analogs similarly to the Rpn7(D123C)–Rpt2(R407C) pair. Site-directed mutagenesis to introduce cysteine substitutions is not covered here.

2.1 Materials

Unless listed below, all reagents are purchased from Millipore Sigma.

- Zymolyase buffer (ZB): 1.2 *M* D-sorbitol, 50 m*M* Tris–HCl, pH 8.0, and 500 μ*M* $MgCl_2$ dissolved in deionized water. This reagent can be stored at room temperature indefinitely (Table 1).
- 1 *M* dithiothreitol (DTT): dissolve in deionized water. This reagent can be stored in small aliquots for at least a year at −20°C.

Table 1 Concentrations of key reagents used during extract preparation and crosslinking

Reagent	Stock concentration	Final concentration
For spheroplasting		
DTT	1 *M*	30 m*M*
Zymolyase 20T	20 mg/mL	0.8 mg/mL
For cell lysis		
ATP	100 m*M*	2 m*M*
ATPγS	100 m*M*	2 m*M*
AMP-PNP	100 m*M*	2 m*M*
For crosslinking		
$CuCl_2$	10 m*M*	250 μ*M*
NEM	200 m*M*	10 m*M*
EDTA	100 m*M*	10 m*M*
DTT (reduced control only)	1 *M*	20 m*M*

- 20 mg/mL Zymolyase 20T (MP Biomedicals): dissolve Zymolyase 20T in ZB. This reagent can be stored in small aliquots at −20°C for at least a year, and working stocks can be kept refrigerated for at least 3 months.
- Lysis buffer (LB): 50 m*M* HEPES–NaOH, pH 7.5, 150 m*M* NaCl, and 5 m*M* $MgCl_2$. This reagent can be stored at room temperature indefinitely. Nucleotides are added from a concentrated stock immediately before use.
- Protein concentration assay reagent (such as Bio-Rad Protein Assay Reagent or similar).
- 100 m*M* adenosine triphosphate disodium salt hydrate (ATP): dissolve in 50 m*M* Tris–HCl, pH 7.5. Store as single-use aliquots at −80°C. This reagent is stable for at least 6 months.
- 100 m*M* adenosine 5′-(3-thiotriphosphate) tetralithium salt (ATPγS): dissolve in 50 m*M* Tris–Cl, pH 7.5. Store as single-use aliquots at −80°C. This reagent is stable for at least 6 months.
- 100 m*M* adenosine 5′-(β,γ-imido)triphosphate lithium salt hydrate (AMP-PNP): dissolve in 50 m*M* Tris–Cl, pH 7.5. Store as single-use aliquots at −80°C. This reagent is stable for at least 6 months.
- 10 m*M* $CuCl_2$ prepared in deionized water. This reagent can be stored at room temperature indefinitely.

- 200 m*M* *N*-ethylmaleimide (NEM) prepared in 95% ethanol. This reagent can be stored at −20°C for up to 1 week.
- 100 m*M* EDTA disodium salt dihydrate prepared in deionized water, pH 8.0. This reagent can be stored at room temperature indefinitely.
- 5 × nonreducing loading buffer: 300 m*M* Tris–Cl, pH 6.8, 10% SDS, 50% glycerol, and 0.04% bromophenol blue. This reagent can be stored at −20°C indefinitely.
- Congenic yeast strains harboring appropriate single or double cysteine substitutions, such as RTY2091 (*RPN7-D123C-6xGly-V5:kanMX6*), RTY2099 (*RPT2-R407C:natMX4*), and RTY2112 (*RPN7-D123C-6xGly-V5:kanMX6 RPT2-R407C:natMX4*) (Eisele et al., 2018).
- Antibodies against V5 tag (Life Technologies) and glucose-6-phosphate dehydrogenase (G6PD) (Millipore Sigma).

2.2 Equipment

- Absorbance spectrophotometer and appropriate cuvettes
- SDS-PAGE electrophoresis equipment (such as Bio-Rad Mini-PROTEAN3 system or similar)
- Electrotransfer apparatus (such as Bio-Rad Mini Trans-Blot or similar)
- Chemiluminescence imaging station (such as Bio-Rad Chemidoc MP or similar)
- Refrigerated microcentrifuge accepting 1.5 mL Eppendorf tubes
- Heat blocks or water baths set at 25°C and 100°C

2.3 Procedure

2.3.1 Cell growth, spheroplasting, and lysis

1. Inoculate 3 mL of YPD medium with each reporter strain from a colony on a freshly streaked YPD plate and incubate overnight at 30°C with aeration.
2. The following morning, use the overnight culture to inoculate 15 mL of YPD at an optical density of $OD_{600} \approx 0.2$. Incubate at 30°C with aeration.
3. Monitor cell growth via OD_{600}. At $OD_{600} \approx 1.0$, harvest 10 mL of cell culture in a 15-mL disposable conical tube. Centrifuge cells at 4000 × *g* for 2 min at room temperature.
4. Aspirate the supernatant and resuspend the cell pellet in 1 mL of deionized water. Transfer the cell suspension to a 1.5-mL centrifuge tube. Centrifuge the cells at 10,000 × *g* for 30 s to pellet cells.

5. Repeat the wash in step 4.
6. Aspirate the supernatant and resuspend the cells in 100 μL ZB supplemented with 3 μL of 1 *M* DTT. Incubate at room temperature for 15 min.
7. Pellet cells at 10,000 × *g* for 30 s at room temperature and discard the supernatant.
8. Resuspend the pellet in 100 μL ZB and add 4 μL of 20 mg/mL Zymolyase 20T. Mix, and incubate at 30°C for 30 min with gentle agitation to keep cells in suspension.
9. Gently pellet the spheroplasts by centrifuging at 1200 × *g* for 1 min at room temperature (see Note 1). Resuspend the pellet in 200 μL ZB. Break up the pellet by gently stirring with a wide-bore pipette tip and slowly pipetting up and down. Pellet the cells at 1200 × *g* for 1 min at room temperature. Discard the supernatant.
10. Add 150 μL of ice-cold LB containing 2 m*M* of the desired nucleotide (see Note 2). Stir the pellet with a pipette tip to partially resuspend the cells.
11. Lyse the spheroplasts by vortexing three times for 30 s at top speed, leaving on ice for 1 min in between vortexings.
12. Pellet cell debris and unbroken cells at 21,000 × *g* for 10 min at 4°C and transfer the supernatant to a fresh 1.5-mL centrifuge tube.
13. Perform a protein concentration assay on a small aliquot of the cleared WCE following manufacturer's instructions to determine the protein concentration. Adjust all sample concentrations to the least concentrated sample using LB containing the appropriate nucleotide (see Note 3).

2.3.2 Crosslinking of WCE

1. Transfer 50 μL of the normalized WCE from Section 2.3.1, step 13 to a fresh Eppendorf tube containing 2.5 μL of 200 m*M* NEM. Add 5.25 μL of 100 m*M* EDTA, mix well, and store sample at −80°C until analysis. This sample will serve as the noncrosslinked control.
2. Transfer 50 μL of the remaining WCE from Section 2.3.1, step 13 to a fresh tube. Add 1.25 μL of 10 m*M* $CuCl_2$ to the lysate and mix gently but thoroughly to initiate crosslinking (see Note 4). Immediately place tubes in a 25°C water bath to maintain temperature (see Note 5).
3. After 10 min, add 2.5 μL of 200 m*M* NEM and 5.25 μL of 100 m*M* EDTA to terminate further crosslinking (see Note 6). Mix well.

4. For a control to demonstrate the reversibility of disulfide formation, add 1 μL of 1 *M* DTT to 50 μL of crosslinked WCE. Incubate sample for 10 min at room temperature before adding 2.5 μL of 200 m*M* NEM and 5.25 μL of 100 m*M* EDTA as in step 3 above.
5. Freeze samples at −80°C until analysis or proceed directly to Section 2.3.3.

2.3.3 Nonreducing SDS-PAGE immunoblotting and quantitation of crosslinking

1. Thaw frozen samples on ice if necessary.
2. Add 17.5 μL of 5 × nonreducing SDS loading buffer to each sample. Boil for exactly 3 min (see Note 7).
3. Load 15 μL of each sample onto a 10% Tris–glycine SDS-polyacrylamide gel. If reduced and nonreduced samples will be loaded on the same gel, it is best to leave an empty lane between them to avoid diffusion of DTT into nonreduced sample lanes during electrophoresis.
4. Subsequent steps of gel electrophoresis, protein electrotransfer to poly-vinylidene difluoride (PVDF) membranes (Millipore Sigma), and immu-nodetection follow standard procedures. Perform immunoblotting against the V5 epitope on Rpn7 and against G6PD as a loading control.
5. Using image quantitation software, measure the volume of the crosslinked and uncrosslinked Rpn7 signal for each lane (see Note 8).
6. The volume of the crosslinked band divided by the sum of the crosslinked and uncrosslinked volumes, multiplied by 100, yields the percent subunit crosslinking.

3. Assessing the impact of proteasomal peptidase inhibitor bortezomib on 26S proteasome conformational distribution

Some proteasomal peptidase inhibitors bias the conformational distribution of 26S proteasomes toward the activated states (Haselbach et al., 2017). The methodology in Section 2 can be readily modified to assess the impact of such inhibitors on the conformational distribution of the 26S proteasome. We have accomplished this as described below via pre-treatment of WCE with the desired inhibitor(s) prior to crosslinking and analysis, but in principle cells could instead be treated with inhibitor prior to generation of cell extracts. By performing the crosslinking in the presence of the inhibitor and either ATP or a hydrolysis-resistant analog such as ATPγS or AMP-PNP, inhibitor bias toward or away from the s1 conformation can be assessed. Concurrent measurement of peptidase activity inhibition in an aliquot of the treated lysates using a fluorogenic substrate such as *N*-succinyl-leu-leu-val-tyr-7-amino-4-methylcoumarin (suc-LLVY-AMC) serves to confirm the inhibitor is bound to proteasomes under the crosslinking conditions (see Note 9). We demonstrate this approach using the FDA-approved proteasome inhibitor bortezomib as an example.

3.1 Materials

- Materials described in Section 2.1
- 10 m*M* bortezomib (Selleckchem): dissolve in anhydrous dimethylsulfoxide (DMSO) and store in single-use aliquots at −80°C. This reagent is stable for at least 6 months
- 2.5 m*M* suc-LLVY-AMC (R&D Systems): dissolve in anhydrous DMSO and store in single-use aliquots at −80°C. This reagent is stable for at least a year
- Black polystyrene 384-well plates (Corning Life Sciences)

3.2 Equipment

- Equipment described in Section 2.2
- Temperature-controlled multiwell fluorescence plate reader capable of reading 384-well plates and equipped with appropriate optics for imaging AMC fluorescence (BioTek Synergy H1 or similar)

3.3 Procedure

3.3.1 Disulfide crosslinking of vehicle- or bortezomib-treated WCE

1. Overnight yeast cultures are grown as in Section 2.3.1, but are diluted into 30 mL of culture instead of 15 mL. One culture for DMSO vehicle and one for bortezomib treatment should be prepared.
2. Cell harvest is performed as in Section 2.3.1, except 20 mL of culture is harvested instead of 10 mL.
3. Spheroplasting and lysis are performed as in Section 2.3.1, except the reagent volumes listed in steps 6, 8, 9, and 10 are doubled.
4. Add 3 μL of DMSO or bortezomib (final concentration 100 μ*M*) to 300 μL of the appropriate WCE and mix gently but thoroughly. Incubate on ice for 5 min.

5. Set ~160 μL of the treated WCE aside at 4°C for measurement of peptidase activity (Section 3.3.2).
6. Using the remaining WCE, perform crosslinking and analysis as in Sections 2.3.2 and 2.3.3.

3.3.2 Confirming proteasome inhibition via fluorogenic substrate cleavage assay

1. Transfer 50 μL of uncrosslinked, vehicle- or inhibitor-treated WCE from Section 3.3.1, step 5 to each of three wells of a 384-well black polystyrene microplate.
2. Add 1 μL of 2.5 m*M* suc-LLVY-AMC to each well. Mix thoroughly by pipetting up and down without introducing bubbles.
3. Measure fluorescence resulting from cleavage of suc-LLVY-AMC at excitation 360 nm, emission 460 nm at 30°C every minute for 60 min using the multiwell fluorescence plate reader.
4. Average the three readings for each sample and plot the average fluorescence intensity in arbitrary units vs time in minutes for the first 20 time points. Calculate the slope of the line to establish the peptidase rate in arbitrary fluorescence units per minute (see Note 10). Inhibition of peptidase activity by ≈90% is anticipated (some minimal fluorescence liberation by other cellular peptidases is typical).

4. Summary and conclusion

Cryo-EM has been enormously powerful in documenting the conformational states of multisubunit complexes. However, it requires access to extremely expensive equipment and the preparation of highly pure samples. Further, data processing is time- and effort-intensive, resulting in limited throughput that hinders detailed comparative mechanistic studies. The crosslinking approach described here circumvents each of these limitations and thus provides a powerful means to rapidly and efficiently compare many samples in parallel. Further, because it relies on methods and equipment that are already available in most molecular biology labs, it can be broadly implemented to study any multisubunit protein complex for which high resolution structures of multiple conformational states are known.

An additional advantage of this approach is that, due to the precise amino acid spacing necessary for disulfide bond formation (~5.9 Å between cysteine α carbons), smaller conformational changes can be measured than are possible with several other methods. For example, Förster resonance energy transfer is widely used to measure protein conformational changes, but is restricted to movements on the order of ~20–100 Å (Heyduk, 2002). In contrast, the Rpn6(T203C)–Rpt6(G387C) crosslinker pair described above discriminates a movement of ~9 Å between the s1 and s2/s3-like states (Eisele et al., 2018). Theoretically, any pair of residues whose α carbons are spaced ~5–6 Å apart in one conformation and >10–15 Å in another can constitute a reporter pair upon cysteine substitution. In practice, these distance constraints are influenced by the local flexibilities of the protein regions in which the residues are located, and this should be considered when choosing sites for cysteine substitution.

Typical results for ATP analogs (Fig. 2A) and for bortezomib treatment (Fig. 2B and C) are shown. Crosslinking efficiency of the Rpn7(D123C)–Rpt2(R407C) reporter typically ranges from <3% in the presence of high concentrations of ATPγS or AMP-PNP to approximately 30% in the presence of ATP and is highly reproducible. Incomplete crosslinking in the presence of ATP is likely due to a combination of incomplete conformational switching, imperfect alignment or spacing of cysteines for disulfide formation (see below), or spontaneous oxidation of cysteines prior to crosslinking. Generally, our efforts to further enhance the efficiency via extended crosslinking time courses or use of other oxidants have not been effective.

Although the utility of this crosslinking-based approach has been established (Eisele et al., 2018), there are two potential limitations that should be carefully considered before use. First, the assay is not time-resolved. Specifically, it reports on all proteasomes cycling through the s1 state over the course of crosslinking. Thus, fast rearrangements may be difficult to capture and quantitate accurately via this approach. For this reason, we have had the greatest success using this approach in combination with chemicals or mutations that arrest particular enzymatic activities of the proteasome, as they often trap particular conformational states or distributions.

The second consideration is that disulfide formation is dependent on precise juxtaposition of cysteine residues. Although this means that crosslinking is highly selective for a particular conformation or subunit juxtaposition, it also means that small perturbations to the positioning of an engineered cysteine-containing subunit may alter the crosslinking efficiency due to relatively small changes in the positions of the engineered cysteines. The resultant loss or gain of crosslinking could thus be misidentified as global changes to the conformational state. For this reason, we recommend using multiple pairs of engineered cysteines, ideally present in different subunits and regions of the multisubunit complex, to validate the initial results whenever possible (Eisele et al., 2018).

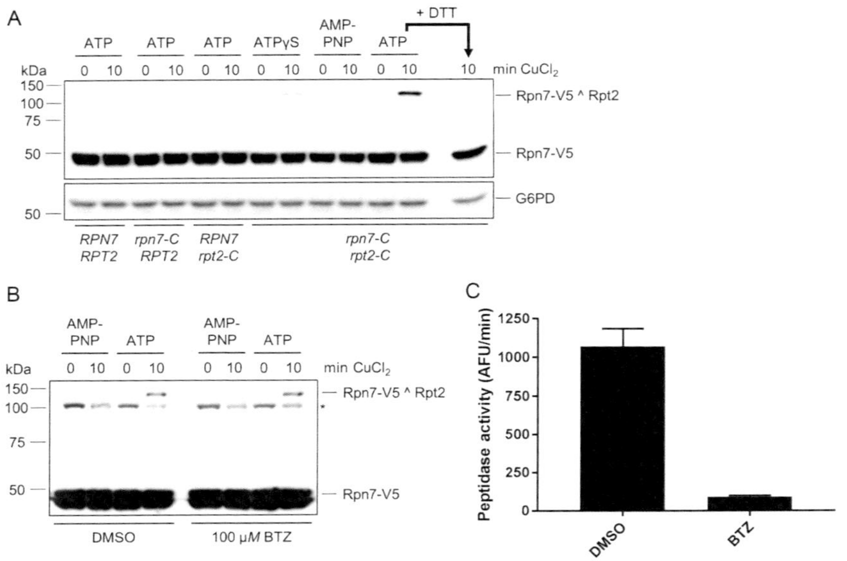

Fig. 2 (A) Conformation-dependent crosslinking of Rpn7 and Rpt2. Crosslinking requires both engineered cysteines and is regulated by nucleotide. Crosslinking was induced in the presence of 2 m*M* of the indicated nucleotide. For the last lane, the WCE was pre-incubated with DTT to reduce disulfide crosslinks prior to loading. (B) Bortezomib (BTZ) treatment does not influence Rpn7-Rpt2 crosslinking. Extracts from DMSO- or BTZ- (100 μ*M*) treated cells were crosslinked as in (A) in the presence of 2 m*M* ATP or AMP-PNP as shown. *Asterisk*, nonspecific cross-reacting band occasionally seen with V5 antibody. (C) Measurement of proteasomal peptidase activity in extracts from (B). Cell extract (20 μg) from DMSO- or BTZ-treated cells was incubated with 50 μ*M* suc-LLVY-AMC for 60 min at 30°C and fluorescence from liberated AMC was measured. The rate of fluorescence liberation is shown. *AFU*, arbitrary fluorescence units. Error bars indicate standard deviations (*N* = 5). *Panel (A) is modified from Eisele, M. R., Reed, R. G., Rudack, T., Schweitzer, A., Beck, F., Nagy, I., et al. (2018). Expanded coverage of the 26S proteasome conformational landscape reveals mechanisms of peptidase gating.* Cell Reports, 24*(5), 1301–1315.e5. doi:10.1016/j.celrep.2018.07.004 with permission from Elsevier.*

5. Notes

1. The cells are no longer protected by the cell wall and are now very fragile. Handle gently to prevent premature cell lysis.
2. Samples should be kept on ice at all times from this step forward to limit proteolysis and spontaneous oxidation of cysteines.
3. To ensure reproducible crosslinking efficiency both between samples and from experiment to experiment, we routinely dilute our extracts to 4 mg/mL prior to crosslinking.

4. Although we prefer $CuCl_2$ as the mild oxidant to induce disulfide formation, many other commercially available oxidants can be used. Examples include tetrathionate and aqueous iodine, both of which have been used successfully to induce disulfide crosslinks in proteasome complexes (Snoberger, Brettrager, & Smith, 2018; Velichutina et al., 2004).
5. Some engineered cysteines are prone to spontaneous disulfide formation even in the absence of exogenous oxidant, which can confound interpretation of crosslinking results. This spontaneous oxidation can be minimized by incubating the samples on ice instead of 25°C during crosslinking. We have generally found that there is no need to extend the crosslinking time to accommodate for the decrease in temperature.
6. Typically, maximal crosslinking occurs in less than 5 min, and that nonspecific crosslinking increases over time. For this particular crosslink pair, 10 min consistently provides near-maximal site-specific crosslinking efficiency with minimal nonspecific crosslinking. If desired, a crosslinking time course can be used to optimize the results for this and other cysteine pairs.
7. Despite the presence of NEM and EDTA that should prevent further oxidation of free protein thiols, we have found that extensive boiling, as well as repeated boiling, of nonreducing SDS-PAGE samples leads to additional nonspecific disulfide formation over time, possibly due to incomplete reaction of NEM with free thiols. For this reason, we carefully control the boiling time and make aliquots of crosslinked samples prior to freezing if multiple analyses are planned.
8. It is critical to use images in which the signals for both the uncrosslinked and crosslinked subunits are unsaturated to ensure accurate calculation of crosslinking efficiency. Most modern biomedical imaging systems have very large dynamic ranges and automatically detect and alert the user to signal saturation. For this reason, such imagers are preferred over autoradiography film.
9. Bortezomib preferentially inhibits the chymotryptic-like activity of the proteasome. The fluorogenic substrate suc-LLVY-AMC reports on this chymotryptic-like activity. If inhibitors of the trypsin-like or caspase-like activities of the proteasome will be assayed instead, fluorogenic substrates suitable for those activities should be substituted for suc-LLVY-AMC.
10. It is important that peptidase assays are performed under conditions where substrate is in great molar excess of proteasomes to ensure a linear relationship between fluorescence liberation and time. If substantial deviation from linearity is observed over the first 20 min of the assay, then the concentration of the extracts used in the assay should be reduced.

Acknowledgments

We thank members of the Tomko laboratory for critical reading of the manuscript. Work from our laboratory that led to the development of this approach was supported by NIH grant 1RO1GM118600 to R.J.T.

References

Asano, S., Fukuda, Y., Beck, F., Aufderheide, A., Forster, F., Danev, R. et al. (2015). Proteasomes. A molecular census of 26S proteasomes in intact neurons. *Science*, *347*(6220), 439–442. https://doi.org/10.1126/science.1261197.

Aufderheide, A., Beck, F., Stengel, F., Hartwig, M., Schweitzer, A., Pfeifer, G., et al. (2015). Structural characterization of the interaction of Ubp6 with the 26S proteasome. *Proceedings of the National Academy of Sciences of the United States of America*, *112*(28), 8626–8631. https://doi.org/10.1073/pnas.1510449112.

Bard, J. A. M., Bashore, C., Dong, K. C., & Martin, A. (2018). Deconvolution of substrate processing by the 26S proteasome reveals a selective kinetic gateway to degradation. *bioRxiv*. https://doi.org/10.1101/359695.

Bard, J. A. M., Goodall, E. A., Greene, E. R., Jonsson, E., Dong, K. C., & Martin, A. (2018). Structure and function of the 26S proteasome. *Annual Review of Biochemistry*, *87*, 697–724. https://doi.org/10.1146/annurev-biochem-062917-011931.

Bashore, C., Dambacher, C. M., Goodall, E. A., Matyskiela, M. E., Lander, G. C., & Martin, A. (2015). Ubp6 deubiquitinase controls conformational dynamics and substrate degradation of the 26S proteasome. *Nature Structural & Molecular Biology*, *22*(9), 712–719. https://doi.org/10.1038/nsmb.3075.

Chen, S., Wu, J., Lu, Y., Ma, Y. B., Lee, B. H., Yu, Z., et al. (2016). Structural basis for dynamic regulation of the human 26S proteasome. *Proceedings of the National Academy of Sciences of the United States of America*, *113*(46), 12991–12996. https://doi.org/10.1073/pnas.1614614113.

de la Pena, A. H., Goodall, E. A., Gates, S. N., Lander, G. C., & Martin, A. (2018). Substrate-engaged 26S proteasome structures reveal mechanisms for ATP-hydrolysis-driven translocation. *Science*. https://doi.org/10.1126/science.aav0725. [Epub ahead of print].

Dong, Y., Zhang, S., Wu, Z., Li, X., Wang, W. L., Zhu, Y., et al. (2018). Visualizing ATP-dependent substrate-processing dynamics of the human 26S proteasome at near-atomic resolution. *bioRxiv*. https://doi.org/10.1101/419051.

Eisele, M. R., Reed, R. G., Rudack, T., Schweitzer, A., Beck, F., Nagy, I., et al. (2018). Expanded coverage of the 26S proteasome conformational landscape reveals mechanisms of peptidase gating. *Cell Reports*, *24*(5), 1301–1315.e5. https://doi.org/10.1016/j.celrep.2018.07.004.

Finley, D., Chen, X., & Walters, K. J. (2016). Gates, channels, and switches: Elements of the proteasome machine. *Trends in Biochemical Sciences*, *41*(1), 77–93. https://doi.org/10.1016/j.tibs.2015.10.009.

Guo, Q., Lehmer, C., Martinez-Sanchez, A., Rudack, T., Beck, F., Hartmann, H., et al. (2018). In situ structure of neuronal C9orf72 poly-GA aggregates reveals proteasome recruitment. *Cell*, *172*(4), 696–705.e12. https://doi.org/10.1016/j.cell.2017.12.030.

Haselbach, D., Schrader, J., Lambrecht, F., Henneberg, F., Chari, A., & Stark, H. (2017). Long-range allosteric regulation of the human 26S proteasome by 20S proteasome-targeting cancer drugs. *Nature Communications*, *8*, 15578. https://doi.org/10.1038/ncomms15578.

Heyduk, T. (2002). Measuring protein conformational changes by FRET/LRET. *Current Opinion in Biotechnology*, *13*(4), 292–296. https://doi.org/10.1016/S0958-1669(02)00332-4.

Huang, X., Luan, B., Wu, J., & Shi, Y. (2016). An atomic structure of the human 26S proteasome. *Nature Structural & Molecular Biology*, *23*(9), 778–785. https://doi.org/10.1038/nsmb.3273.

Luan, B., Huang, X., Wu, J., Mei, Z., Wang, Y., Xue, X., et al. (2016). Structure of an endogenous yeast 26S proteasome reveals two major conformational states. *Proceedings of the National Academy of Sciences of the United States of America*, *113*(10), 2642–2647. https://doi.org/10.1073/pnas.1601561113.

Matyskiela, M. E., Lander, G. C., & Martin, A. (2013). Conformational switching of the 26S proteasome enables substrate degradation. *Nature Structural & Molecular Biology*, *20*(7), 781–788. https://doi.org/10.1038/nsmb.2616.

Schweitzer, A., Aufderheide, A., Rudack, T., Beck, F., Pfeifer, G., Plitzko, J. M., et al. (2016). Structure of the human 26S proteasome at a resolution of 3.9 A. *Proceedings of the National Academy of Sciences of the United States of America*, *113*(28), 7816–7821. https://doi.org/10.1073/pnas.1608050113.

Sledz, P., Unverdorben, P., Beck, F., Pfeifer, G., Schweitzer, A., Forster, F., et al. (2013). Structure of the 26S proteasome with ATP-gammaS bound provides insights into the mechanism of nucleotide-dependent substrate translocation. *Proceedings of the National Academy of Sciences of the United States of America*, *110*(18), 7264–7269. https://doi.org/10.1073/pnas.1305782110.

Snoberger, A., Brettrager, E. J., & Smith, D. M. (2018). Conformational switching in the coiled-coil domains of a proteasomal ATPase regulates substrate processing. *Nature Communications*, *9*(1), 2374. https://doi.org/10.1038/s41467-018-04731-6.

Tomko, R. J., Jr., Funakoshi, M., Schneider, K., Wang, J., & Hochstrasser, M. (2010). Heterohexameric ring arrangement of the eukaryotic proteasomal ATPases: Implications for proteasome structure and assembly. *Molecular Cell*, *38*(3), 393–403. https://doi.org/10.1016/j.molcel.2010.02.035.

Tomko, R. J., Jr., & Hochstrasser, M. (2013). Molecular architecture and assembly of the eukaryotic proteasome. *Annual Review of Biochemistry*, *82*, 415–445. https://doi.org/10.1146/annurev-biochem-060410-150257.

Unverdorben, P., Beck, F., Sledz, P., Schweitzer, A., Pfeifer, G., Plitzko, J. M., et al. (2014). Deep classification of a large cryo-EM dataset defines the conformational landscape of the 26S proteasome. *Proceedings of the National Academy of Sciences of the United States of America*, *111*(15), 5544–5549. https://doi.org/10.1073/pnas.1403409111.

Velichutina, I., Connerly, P. L., Arendt, C. S., Li, X., & Hochstrasser, M. (2004). Plasticity in eucaryotic 20S proteasome ring assembly revealed by a subunit deletion in yeast. *The EMBO Journal*, *23*(3), 500–510. https://doi.org/10.1038/sj.emboj.7600059.

Verma, R., Aravind, L., Oania, R., McDonald, W. H., Yates, J. R., 3rd, Koonin, E. V., et al. (2002). Role of Rpn11 metalloprotease in deubiquitination and degradation by the 26S proteasome. *Science*, *298*(5593), 611–615. https://doi.org/10.1126/science.1075898.

Wehmer, M., Rudack, T., Beck, F., Aufderheide, A., Pfeifer, G., Plitzko, J. M., et al. (2017). Structural insights into the functional cycle of the ATPase module of the 26S proteasome. *Proceedings of the National Academy of Sciences of the United States of America*, *114*(6), 1305–1310. https://doi.org/10.1073/pnas.1621129114.

Worden, E. J., Dong, K. C., & Martin, A. (2017). An AAA motor-driven mechanical switch in Rpn11 controls deubiquitination at the 26S proteasome. *Molecular Cell*, *67*(5), 799–811.e8. https://doi.org/10.1016/j.molcel.2017.07.023.

Yao, T., & Cohen, R. E. (2002). A cryptic protease couples deubiquitination and degradation by the proteasome. *Nature*, *419*(6905), 403–407. https://doi.org/10.1038/nature01071.

Zhu, Y., Wang, W. L., Yu, D., Ouyang, Q., Lu, Y., & Mao, Y. (2018). Structural mechanism for nucleotide-driven remodeling of the AAA-ATPase unfoldase in the activated human 26S proteasome. *Nature Communications*, *9*(1), 1360. https://doi.org/10.1038/s41467-018-03785-w.

CHAPTER EIGHT

Assays for ubiquitin-like protein ligation and proteasome function in archaea

Xian Fu[a], Zachary Adams[a], Julie Maupin-Furlow[a,b,*]

[a]Department of Microbiology and Cell Science, Institute of Food and Agricultural Sciences, University of Florida, Gainesville, FL, United States
[b]University of Florida, Genetics Institute, Gainesville, FL, United States
*Corresponding author: e-mail address: jmaupin@ufl.edu

Contents

Abstract

Ubiquitin-like protein (Ubl) ligation is common to diverse archaea and targets many cellular pathways, including those associated with sulfur mobilization, and also tags proteins as substrates for degradation by the proteasome. Here we highlight protocols to assay proteasome function and Ubl ligation in archaea. A chase assay is described to monitor the impact of proteasome function on the stability of Ubl-modified proteins in the cell. A method to reconstitute Ubl ligation using a purified E1-like enzyme (UbaA), Ubl (SAMP2), methionine sulfoxide reductase A (MsrA), and cell lysate of an *ΔmsrA ΔubaA Δsamp1–3* mutant is also described. MsrA is found to have the surprising ability to stimulate the formation of Ubl bonds. *Haloferax volcanii*, a halophilic archaeon originally isolated from the Dead Sea, serves as the model organism for these protocols.

Methods in Enzymology, Volume 619
ISSN 0076-6879
https://doi.org/10.1016/bs.mie.2018.12.036
© 2019 Elsevier Inc.
All rights reserved.

1. Introduction

Archaea can target proteins for destruction by proteasomes using a ubiquitin-like protein (Ubl) ligation system (Fu et al., 2016; Humbard et al., 2010) (Fig. 1). Archaeal Ubls differ in primary sequence but share a common β-grasp fold and C-terminal diglycine motif that can be activated by E1-like enzymes (EC:2.7.7.) (Miranda et al., 2014, 2011). A Ubl~E1 thioester intermediate is formed that commits the Ubl to protein ligation instead of sulfur mobilization (Hepowit et al., 2016). Methionine sulfoxide reductase (MsrA) stimulates the Ubl ligation system by a mechanism that is independent of its methionine sulfoxide (MetO) reductase activity (Fu et al., 2017). The Ubl modification can transiently inactivate enzymes (e.g., Ubl-MoaE of molybdopterin biosynthesis) (Cao, Hepowit, & Maupin-Furlow, 2015)

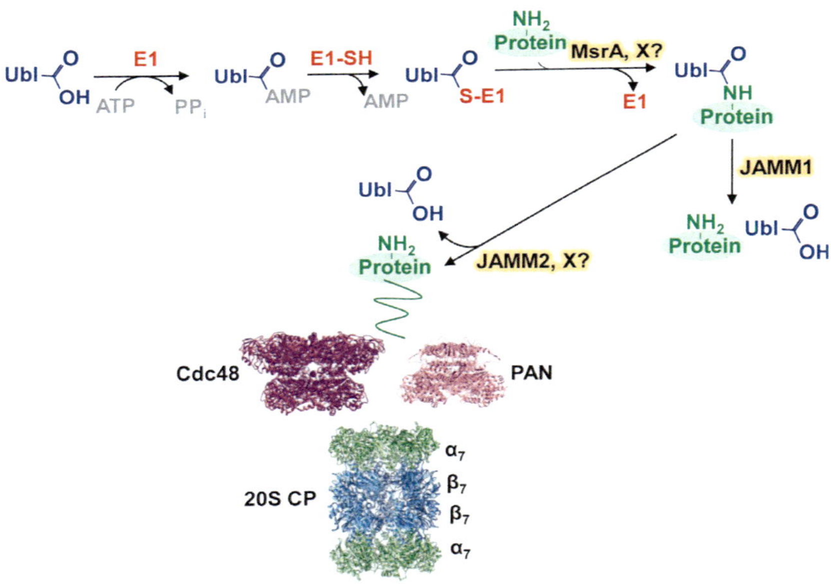

Fig. 1 Ubiquitin-like protein (Ubl) ligation and proteasome systems of Archaea. The Ubl protein (*blue*) is activated at its C-terminus by an E1-like enzyme (*red*). Adenylation (Ubl~AMP) and thioester (Ubl~E1) intermediates are formed. The Ubl is transferred from the Ubl~E1 to the target protein (*green*) by a mechanism that is stimulated by MsrA. The Ubl is linked by an amide (isopeptide) bond to the target protein. JAMM1 can cleave the Ubl-modified protein at the isopeptide linkage. The activity of JAMM2, AAA ATPases (Cdc48 and PAN), and 20S core particles is required for turnover of Ubl-modified proteins. X?, additional factors may be required.

or target proteins (e.g., TATA-binding protein TBP2) for proteolysis by proteasomes (Fu et al., 2016). The archaeal proteasome system is composed of a network of 20S core particles (CPs) and AAA ATPases homologous to the eukaryotic Rpt1–6 proteasome subunits and Cdc48/p97 segregase (PAN and VAT, respectively) (Barthelme, Chen, Grabenstatter, Baker, & Sauer, 2014; Forouzan et al., 2012). The ATPases open the 20S CP gates into the CP interior and facilitate the unfolding and translocation of the target protein into the central proteolytic chamber of the 20S CP (Kim, Snoberger, Schupp, & Smith, 2015; Rabl et al., 2008). Associated with the archaeal Ubl-proteasome system are JAMM/MPN+ metalloproteases of at least two distinct types (Hepowit et al., 2012). JAMM1 can cleave (iso) peptide bonds between the C-terminus of the Ubl and substrate and can reactivate enzymes such as Ubl-MoaE (Cao et al., 2015). By contrast, JAMM2 is inactive in vitro, most likely due to the absence of a key α-helix that is used by JAMM1 to bind the Ubl (Cao et al., 2017). This missing α-helix may be compensated by a protein partner(s), based on the finding that JAMM2 is needed in vivo for proteasome-mediated proteolysis (Fu et al., 2016). Thus, JAMM2 is proposed to release Ubls from target proteins prior to proteolysis.

Here we describe protocols for assays of proteasome function and Ubl ligation in archaea. *Haloferax volcanii* ATCC 29605, of the phylum *Euryarchaeota*, serves as the model system for these protocols as derivatives of this DS2 strain are amenable to genetics and protein biochemistry.

2. Methods for assay of proteasome function in archaea

To assay for proteasome function in degrading Ubl-tagged proteins, the abundance of a suspected target protein can be monitored by a chase assay that measures its rate of disappearance after shutoff of protein synthesis in proteasome mutant strains compared to wild-type cells. Chemical inhibitors can also be used to perturb proteasome activity. This section is based on Fu et al. (2016).

2.1 Monitoring the levels of proteasome targets in archaea

To monitor the levels of Ubl-modified proteins in *H. volcanii* for the chase assay, one can produce the target protein with an engineered C-terminal StrepII tag and the Ubl with an N-terminal Flag tag from an expression cassette. This approach frees the C-terminus of the Ubl to form (iso)peptide linkages with the target protein and frees the N-terminus of the target protein for possible regulation by the N-end rule, a mechanism of substrate targeting that requires a particular subset of N-terminal residues. The selection of the target protein is guided by previous detection of the protein in a Ubl-modified form. For this example, tandem mass spectrometry (LC–MS/MS) analysis revealed the TATA-binding protein TBP2 to be covalently linked to the Ubl SAMP2 by isopeptide bonds (i.e., Ubl linked to substrate lysine side chains) (Humbard et al., 2010).

TBP2-StrepII and Flag-SAMP2 are expressed on a bicistronic message from a gene cassette that uses an unregulated promoter (P2*rrnA*). Alternative promoters (Gregor & Pfeifer, 2005; Large et al., 2007) and epitope tags (Dilks, Gimenez, & Pohlschroder, 2005; Gimenez, Dilks, & Pohlschroder, 2007; Karadzic et al., 2012; Martin et al., 2018; McMillan, Hepowit, & Maupin-Furlow, 2016) are available for use in this halophilic archaeon as needed. A shuttle plasmid that includes selection markers and replication origins for *Escherichia coli* (ampicillin resistance and ColE1, respectively) and *H. volcanii* (novobiocin resistance and pHV2, respectively) is used to carry the cassette. The plasmid DNA is prepared from an *E. coli* dam^- dcm^- strain that is deficient in DNA methylation, to ensure high transformation efficiency. During transformation, the *H. volcanii* cells are exposed to EDTA which strips off the S-layer of the cell wall, generating spheroplasts, and allows for polyethylene (PEG)-mediated uptake of the plasmid DNA into the cell (Dyall-Smith, 2009). The transformed cells are analyzed for steady state levels of TBP2 and Ubl SAMP2 by immunoblotting analysis as described in Section 2.4. Cells carrying an empty vector are included as a control to ensure antibody specificity.

2.2 Generation of proteasome mutant strains

Genes predicted to impact proteasome function based on comparative genomics or other methods are deleted from the *H. volcanii* genome using a markerless deletion strategy (Cao et al., 2015; Fu et al., 2016; Miranda et al., 2011; Zhou, Kowalczyk, Humbard, Rohatgi, & Maupin-Furlow, 2008) (Fig. 2). The open reading frame (ORF) along with ~500 bp of 5′ and 3′ flanking sequences is isolated from the *H. volcanii* H26 genome by PCR with primers that encode the desired restriction sites. The PCR product is ligated into the multiple cloning site of pTA131 (Allers, Ngo, Mevarech, & Lloyd, 2004) (see later) using restriction enzymes such as *Hin*dIII and *Bam*HI. The resulting "predeletion" plasmid is subsequently used as DNA template for inverse PCR to generate an in-frame deletion of the gene while avoiding disruption of adjacent ORFs.

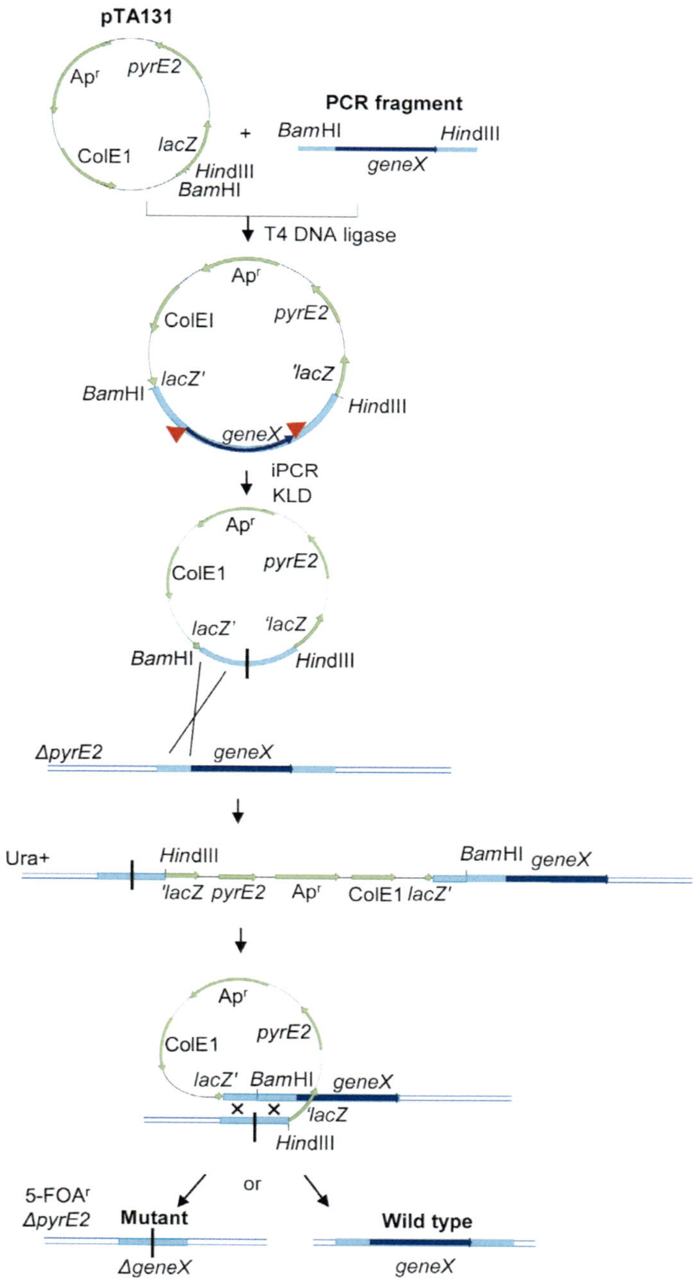

Fig. 2 See legend on next page.

The "deletion" plasmid is generated from this PCR product using KLD Enzyme Mix (NE Biolabs) that combines T4 polynucleotide kinase to phosphorylate the 5′ ends of the PCR product, T4 DNA ligase to form circular DNA, and DpnI to cleave the methylated DNA template. The deletion plasmid is based on pTA131 (Allers et al., 2004), a vector that carries an origin of replication for *E. coli* (but not *H. volcanii*) and harbors *pyrE2*, a gene that codes for orotate phosphoribosyl transferase, an enzyme needed for uracil biosynthesis in *H. volcanii* H26 (*ΔpyrE2*) (Bitan-Banin, Ortenberg, & Mevarech, 2003). The deletion plasmid is integrated into the *H. volcanii* genome by selecting for growth on casamino acid medium (CA) minus uracil (Dyall-Smith, 2009) at 42°C, within the optimal growth temperature range of 40–45°C reported for this archaeon (Mullakhanbhai & Larsen, 1975; Robinson et al., 2005). Isolated colonies are screened for plasmid integration by PCR using primers that anneal 500 bp 5′ and 3′ to the target gene.

Subsequently, integrant-positive cells are subject to counterselection against the *pyrE2* gene in the integrated plasmid by inoculating into 4 mL ATCC 974 medium supplemented with 5-fluoroorotic acid at 50 μg mL^{-1} (5-FOA medium) (last step in Fig. 2). Cultures are grown with rotary shaking in borosilicate glass tubes (13 × 100 mm) at 42°C for 15 h. Serial dilutions of the cultures are plated for isolating single colonies on 5-FOA medium; plates at dilutions with cleanly separated colonies are then patched onto the same 5-FOA medium. Clones of interest are identified by PCR with Taq or Phusion DNA polymerase using primers 700 bp upstream and downstream, respectively, of the ORF. Mutant strains are further purified by restreaking on the same medium at least twice for isolated colonies. DNA sequencing of the PCR product is used to confirm the deletion.

If the gene appears to be essential for growth (i.e., no clones with the desired gene deletion are identified), a conditional allele can be generated (Large et al., 2007). The 20S proteasome β subunit gene is essential,

Fig. 2 Markerless deletion of a gene (*geneX*) predicted to impact proteasome function. A DNA fragment carrying *geneX* and 500 bp 5′ and 3′ of the ORF is generated by PCR from *H. volcanii* genomic DNA. The DNA fragment is ligated into the multiple cloning site of pTA131. This "predeletion" plasmid is used as a template for inverse PCR (iPCR) using primers indicated by *red arrows*. The iPCR product is converted to the "deletion" plasmid using DNA polynucleotide kinase, T4 ligase, and DpnI (KLD). The deletion plasmid is transformed into an H26 *ΔpyrE2* strain. Integrants, generated via homologous recombination, are selected by growth in the absence of uracil. The counterselective agent 5-fluoroorotic acid (5-FOA) is used to select for loss of the plasmid from the genome, which can be resolved in the two ways shown. Wild type and mutant strains are distinguished at this stage by PCR screening.

for example (Zhou et al., 2008). For generating a conditional allele of the gene of interest, the native promoter in the *H. volcanii* chromosome is replaced with a tryptophan-inducible promoter (P*tnaA*) that can be repressed by removal of tryptophan from the medium (Large et al., 2007). For details on creating such a conditionally lethal strain, please see Large et al. (2007) and Zhou et al. (2008).

H. volcanii strains are stored short term (1–2 weeks) at room temperature in the dark (in plastic bags to maintain moisture) or long term in 20% (v/v) glycerol stocks at −80°C. *H. volcanii* strains are not stored at 4°C (haloarchaea are sensitive to cold shock). To generate the glycerol stocks, the strains are grown to stationary phase (OD_{600} 2.0–3.0) in ATCC974 medium and sterilely mixed 1:1 with 40% (v/v) glycerol. The stocks are transferred to −80°C. To initiate growth, the cells are inoculated by taking a small lump of frozen cells from the 20% (v/v) glycerol stock using a sterilized wooden stick or toothpick. The frozen cells are streaked for isolation from this toothpick onto ATCC974 medium supplemented with 2% (w/v) agar in a culture plate. Plates are transferred to zip-lock plastic bags with wet Kimwipes to prevent loss of moisture; the plates are incubated at 42°C until visible colonies form.

2.3 Assay of protein stability in proteasome mutant strains

To assess proteasome system function, the stability of proteins targeted for Ubl modification is assayed by chase assay in wild type and proteasomal system mutant strains. The TBP2-StrepII Flag-Ubl expression cassette described in Section 2.1 will be used as the example here. The expression plasmid is transformed into *H. volcanii* H26 (the parent strain used to generate mutations by homologous recombination) and mutant strains including in-frame deletions of the genes encoding the E1-like UbaA (Miranda et al., 2011), JAMM1/2 (Cao et al., 2015; Fu et al., 2016), Cdc48b/c (Fu et al., 2016), and PAN1/2 (Zhou et al., 2008) proteins. The cells are grown to log phase in 4 mL ATCC 974 medium (an LB medium for salt-loving archaea) using 13 × 100 mm culture tubes with rotary shaking (200 rpm). Log phase is determined based on an OD_{600} of 0.4–0.7 units using a UV–Vis spectrophotometer at 600 nm. To bring the cells into the same exponential growth phase, the cells are diluted to an OD_{600} of 0.02 into 5.1 mL ATCC 974 medium and again grown to log phase. The cells are centrifuged (10,000 × *g* for 6 min at 25°C) and resuspended in 4.2 mL of a buffered salt solution (4.1 *M* NaCl, 150 m*M* $MgCl_2 \cdot 6H_2O$, 140 m*M*

$MgSO_4{\cdot}7H_2O$, 100 m*M* KCl, and 20 m*M* Tris–HCl, pH 7.5) to maintain cell integrity while removing the carbon/energy source.

The cells are treated with actinomycin D (20 μg mL^{-1}) to inhibit DNA-primed RNA synthesis and anisomycin (50 μg mL^{-1}) to block ribosomal-based protein synthesis. Stock solutions of the antibiotics are made by dissolving the actinomycin D at 10 mg mL^{-1} in acetone and the anisomycin at 20 mg mL^{-1} in methanol. The stock solutions are vortexed immediately prior to use and are added to the cultures to initiate the chase immediately following removal of cell aliquots to provide initial time points. The combination of these two inhibitors is found necessary to prevent the synthesis of new transcript and protein during the chase assay. Culture aliquots are removed at intervals between 0 and 2 h and stored on ice prior to harvesting by centrifugation (10,000 × *g* for 6 min at 4°C). Cell pellets are suspended in lysis buffer (50 m*M* Tris–HCl, pH 6.8, 2 *M* urea, 2% (wt/vol) SDS, and 0.1 mg mL^{-1} EDTA-free protease inhibitor cocktail (Roche)) with glass beads (Chemglass Life Sciences). Cells are lysed with glass disruptor beads (0.1 mm in diameter) by vortexing vigorously 5–7 times with cycles of 1 min on and 1 min off/on ice. Cell debris and glass beads are removed by centrifugation (16,800 × *g* for 20 min at 4°C). Supernatants are transferred into fresh 1.5-mL microcentrifuge tubes on ice. The concentration of protein in the cell lysate is measured by bicinchoninic acid (BCA) protein assay according to the supplier's instructions, using bovine serum albumin (BSA) as the protein standard (Thermo Scientific).

Cell lysate is mixed 1:1 with 2 × SDS reducing sample buffer (100 m*M* Tris–HCl, pH 6.8, 4% (wt/vol) SDS, 20% (vol/vol) glycerol, 0.6 mg mL^{-1} bromophenol blue, and 5% (vol/vol) β-mercaptoethanol) and boiled for 10 min. Equivalent amounts of cell lysate (in the range of 5–20 μg per lane) for each time point are separated by SDS-PAGE (12%) in parallel gels. To assess equivalent protein loading, gel 1 is stained with Coomassie blue (Welinder & Ekblad, 2011). Proteins in gel 2 are analyzed by quantitative immunoblotting (see Section 2.4). In this way, the relative steady state levels (Fig. 3A) and the degradation rates (Fig. 3B–D) of the target protein in the various strains can be determined.

2.4 Immunoblotting analysis and quantification of Ubl-proteasome system targets

Proteins are electroblotted from the unstained SDS-PAGE gels onto a 0.45 μm polyvinylidene difluoride (PVDF) membrane (Amersham) at 4°C for 3 h at 90 V or 13–14 h at 30 V using a mini trans-blot module according to the supplier's instructions (Bio-Rad). The membrane is placed upright in

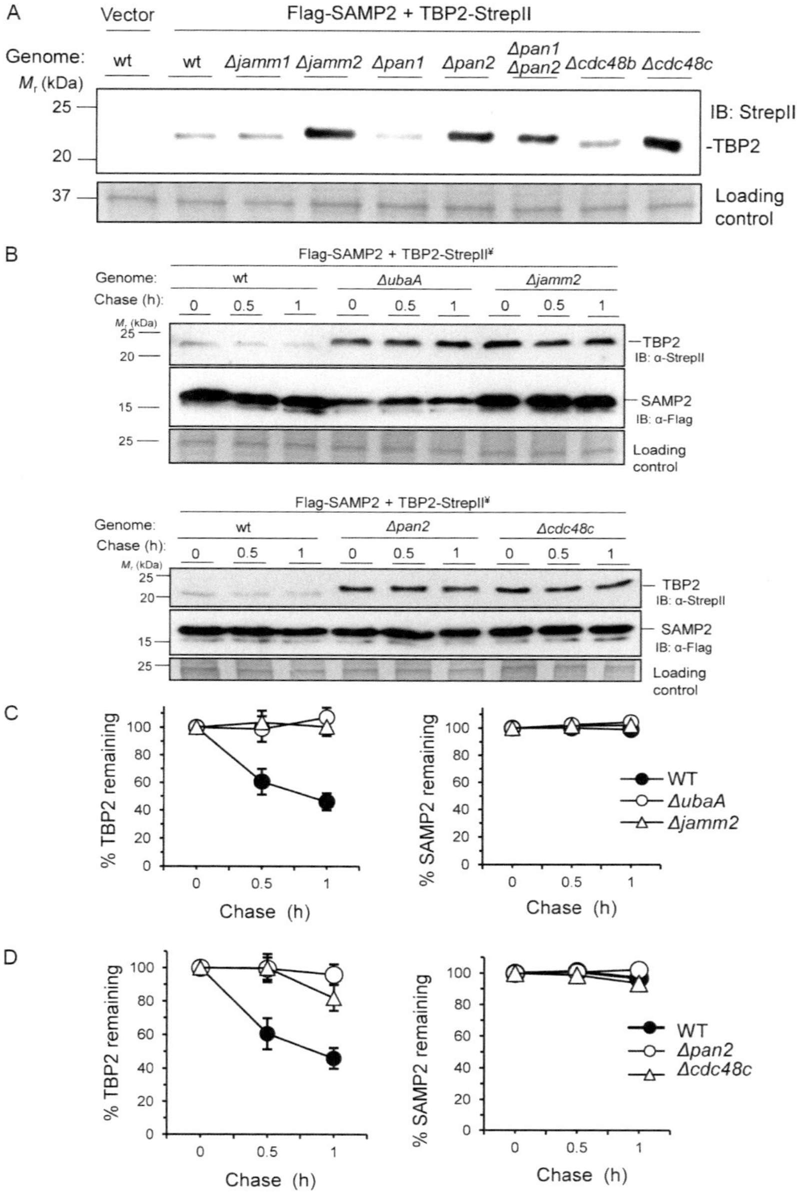

Fig. 3 Proteasome system function evaluated in vivo by monitoring the stability of a protein target (TBP2) of the Ubl modification pathway. In this example, the E1-like UbaA, JAMM2, and the AAA ATPases (PAN2 and Cdc48c) are found to be important in the turnover of TBP2. TBP2 steady state levels (A) and turnover monitored by chase assay

(Continued)

an 18 × 10 cm plastic container and briefly rinsed with 25 mL of TBS (50 m*M* Tris–Cl, pH 7.5, 150 m*M* NaCl). The membrane is blocked for 2 h at room temperature or overnight at 4°C in 25 mL TBS supplemented with 0.05% (v/v) Tween 20 and 5% (w/v) nonfat dry milk. The primary antibody is added to 10 mL of the blocking buffer in the following dilutions: 1:5000 of the Strep-tag II monoclonal antibody (in mouse) (QIAGEN, catalog number: 34850) to detect MsrA-StrepII or TBP2-StrepII 1:10,000 of the HRP-conjugated His6 monoclonal antibody (Proteintech, catalog number: HRP-66005) to detect His6-UbaA; and 1:10,000 of the alkaline phosphatase-linked anti-Flag M2 monoclonal antibody (Sigma-Aldrich, catalog number: A9469-2MG) to detect Flag-His6-SAMP2. The membrane is incubated with the primary antibody solution for 1 h at room temperature with medium speed rocking. The membrane is rinsed 3 × 15–30 min at room temperature with 25–50 mL of TBS supplemented with 0.05% (v/v) Tween 20 (TBST) using the high setting of the rocker. A secondary antibody of goat antimouse IgG (whole molecule)-alkaline phosphatase-linked antibody (Sigma-Aldrich, catalog number: A5153-1ML) is used to detect the anti-Strep-tag II antibody. The secondary antibody is diluted 1:10,000 in 10 mL of blocking solution and incubated with the membrane for 1 h with medium rocking followed by rinsing of the membrane 5 × in TBST buffer as described earlier. The antibody:protein complexes are visualized on the PVDF membrane by chemiluminescence using CDP-Star (Thermo Scientific) for the alkaline phosphatase-linked antibodies and ECL Prime (GE Healthcare) for the HRP-linked antibodies according to supplier's guidelines. The membranes are exposed to X-ray film (Thermo Fisher Scientific) for 1 min for the anti-His antibody and for 10 min for the anti-Flag and anti-Strep antibodies, with longer exposure times as needed. The intensities of the protein bands are quantified on the developed X-ray film using ImageJ (Schneider, Rasband, & Eliceiri, 2012). Future recommendations to enhance sensitivity in determining the turnover rate of the target protein are to label the proteome

Fig. 3—Cont'd (B) in wt (wild type) and ubiquitin-like (Ubl) proteasome system mutant strains. Log-phase cells were treated with 20 μg mL^{-1} actinomycin D and 50 μg mL^{-1} anisomycin for the indicated times and collected for the chase assay. (C–D) TBP2 and SAMP2 intensities were quantified by ImageJ for the chase assay. Results are expressed as the percent change from time zero, which was set at 100%. The *H. volcanii* strains expressed TBP2-StrepII and Flag-SAMP2. TBP2 and SAMP2 abundance was determined by anti-StrepII and anti-Flag antibody immunoblotting analysis (IB), respectively. Equal loading of total protein was confirmed by Coomassie blue (CB) staining. Data represent mean results ± standard errors of the mean (SEM) from three independent experiments. For details, see text and Fu et al. (2016).

using ^{35}S methionine, inhibit protein synthesis, immunoprecipitate the target protein in time intervals, and monitor the level of radioactivity in the samples over time. A control strain devoid of the tag used to immunoprecipitate the target protein would be used to assess off-target signals.

3. Methods for reconstitution of archaeal ubiquitin-like protein ligation

3.1 Expression and purification of Ubl ligation system components

The E1-like enzyme UbaA mediates the ligation of Ubl to itself and MoaE (Hepowit et al., 2016) but requires additional factors, such as MsrA, to ligate the Ubl to other protein targets (Fu et al., 2017). To reconstitute Ubl ligation, the E1-like UbaA, Ubl SAMPs, and MsrA are purified (Fu et al., 2017; Hepowit et al., 2016). Cell lysate of *H. volcanii* mutant strains deficient in Ubl ligation is generated and used as a source of protein substrates (Fu et al., 2017). To maintain the stability of the *H. volcanii* proteins, 2 *M* NaCl is included in the purification buffers, as this archaeon uses a "salt-in-cytoplasm" strategy for osmoregulation (Roesser & Muller, 2001). To cleave the Ubl from the target protein, the metalloprotease JAMM1 is purified and used as a control (Hepowit et al., 2012). Chromatography steps are done at room temperature as haloarchaeal proteins are typically thermotolerant. Samples are stored at 4°C. Protein concentration is determined by the BCA protein assay (Thermo Scientific) using BSA (Thermo Scientific) as a protein standard. Homogeneity of the proteins is determined by SDS-PAGE and mass spectrometry analysis (e.g., LC–MS/MS or ESI-TOF MS).

3.1.1 E1-like enzyme UbaA purification

The E1-like enzyme UbaA (HVO_0558) is expressed with an N-terminal His6-tag from the plasmid pJAM933 in *H. volcanii* NH02. To enhance the yield of His6-UbaA, NH02 is used since it is deficient in Ubl modification (*Δsamp1/2/3*) and does not express UbaA from the genome (*ΔubaA*). The strain is grown to stationary phase (OD_{600} of 3.0–4.0) at 42°C in ATCC974 medium with novobiocin (Nv, $0.2\,\mu g\,mL^{-1}$) (4 × 1 L in 2.8-L Fernbach flasks). Rotary shaking at 200 rpm is used to aerate the cultures. The cells are harvested by centrifugation (4500 × *g*, 10 min, 15°C). The cells are resuspended at $5\,mL\,g^{-1}$ wet weight of cell pellet in Tris-salt buffer (2 *M* NaCl, 50 m*M* Tris–HCl, pH 7.4) with 40 m*M* imidazole. Cells are lysed by 4–6 passages through a chilled French Press cell in a French Press at 20,000 psi, with the lysate collected on ice. The lysate is clarified by centrifugation (12,000 × *g*, 20 min, 4°C) and sequential filtration using 0.8 and 0.45 μm pore size surfactant-free cellulose acetate (SFCA) filters (Nalgene). Sample is applied to a HisTrap HP Ni^{2+}-affinity column (1 mL bed volume, GE Healthcare) preequilibrated in Tris-salt buffer with 40 m*M* imidazole and washed with 5–10 column volumes of the same buffer at a flow rate of 0.5 mL min^{-1}. Proteins are eluted in Tris-salt buffer containing 500 m*M* imidazole and pooled. Samples are dialyzed against Tris-salt buffer and concentrated to 0.5–2.0 mg mL^{-1} protein by centrifugal filtration using an Ultra-4 centrifugal filter unit with Ultracel-3 membrane (Merck) in an HL-8 swinging bucket rotor at 4°C and 4000 × *g*. Samples are further purified by gel filtration chromatography by applying the samples (0.5 mL per run) to a Superdex 75 10/300 GL column (GE Healthcare) equilibrated in Tris-salt buffer at 0.3 mL per min. Protein samples purified by gel filtration are further concentrated to 10–15 mg mL^{-} for future assay.

3.1.2 Ubl SAMP2 purification

The Ubl (SAMP2, HVO_0202) is ectopically expressed with an N-terminal Flag-His6-tag from the plasmid pJAM1132 in *E. coli* Rosetta (DE3). The *E. coli* strain is grown to log phase (OD_{600} of 0.6–0.8) at 37°C in 25 mL LB medium with kanamycin (Km, 50 μg mL^{-1}) with rotary shaking at 220 rpm. This culture serves to inoculate 500 mL LB + Km medium (in 2.8-L Fernbach flasks) to an OD_{600} of 0.01; this larger culture is similarly grown to log phase. Isopropyl β-D-1-thiogalactopyranoside (IPTG, 0.4 m*M*) is added to the culture to induce protein expression. The culture is shifted to 25°C and incubated for 12 h with rotary shaking at 200 rpm. The cells are harvested, and the protein is purified by Ni^{2+}-affinity chromatography and gel filtration chromatography and concentrated, as described in Section 3.1.1.

3.1.3 MsrA purification

To prevent Ubl modification, the *H. volcanii* MsrA (HVO_A0230) is expressed in *E. coli* Rosetta (DE3) from plasmid pJAM3200. The MsrA is fused to a C-terminal StrepII tag to facilitate purification. The *E. coli* strain is used to produce MsrA-StrepII as described earlier for the Ubl SAMP2. The cells are resuspended at 5 mL g^{-1} wet weight of cell pellet in Tris-salt buffer (50 mM Tris–Cl, pH 7.4, 2 M NaCl) supplemented with 1 mg mL^{-1} EDTA-free protease inhibitor cocktail (Roche). The cells are lysed by French press, and the cell lysate is centrifuged and filtered, as described for the E1-like UbaA. The sample is applied to a StrepTrap HP column (5 mL bed volume, GE Healthcare) equilibrated with Tris-salt buffer at a flow rate of 0.5 mL min^{-1}. The column is washed with 140 mL Tris-salt buffer at a flow rate of 1.2 mL min^{-1} to remove unbound proteins. Proteins are eluted in Tris-salt buffer supplemented with 5 mM D-desthiobiotin at a flow rate of 0.5 mL min^{-1}. Fractions are pooled, dialyzed against Tris-salt buffer, concentrated, further purified by gel filtration chromatography, and concentrated, as described in Section 3.1.1.

3.1.4 JAMM1 metalloprotease purification

JAMM1 (HVO_2505) is ectopically expressed with an N-terminal His6-tag in *E. coli* Rosetta (DE3) from the plasmid pJAM991. Fresh transformants are grown at 37°C in LB medium supplemented with ampicillin (Amp, 0.1 mg mL^{-1}) (0.5- to 1-L cultures) in 2.8-L Fernbach flasks with rotary shaking at 200 rpm. At log phase (OD_{600} of 0.4–0.6), IPTG (0.4 mM) and $ZnCl_2$ (100 μM) are added, and cultures are shifted to 25°C for 4.5–12 h. Cells are harvested by centrifugation (10–15 min at 9200 × *g* and 4°C) and washed in ice-chilled Tris-low salt buffer (20 mM Tris–HCl, pH 7.5, 150 mM NaCl). Cells are stored at −80°C until lysis. Cells are resuspended in 5 mL g^{-1} wet weight of cell pellet in lysis buffer (20 mM Tris–HCl, pH 7.5, 150 mM NaCl, 40 mM imidazole, 1 mM PMSF) and lysed thrice in a French Press (20,000 psi). An equal volume of dilution buffer (20 mM Tris–HCl, pH 7.5, 4 M NaCl) is added to the cell lysate. The sample is clarified by centrifugation (twice for 10–20 min at 9200 × *g* and 4°C) and filtration (SFCA filters, 0.2 μm). Sample is applied to a HisTrap HP Ni^{2+} column preequilibrated in Tris-salt buffer with 40 mM imidazole and washed with the same buffer at a flow rate of 0.5 mL min^{-1}. JAMM1 is eluted using Tris-salt buffer containing 500 mM imidazole. Purified protein fractions are pooled and buffer-exchanged twice by dialysis into HEPES-salt buffer (20 mM HEPES buffer, pH 7.5, 2 M NaCl) for storage at 4°C as needed. Prior to assay, JAMM1 is purified by gel filtration chromatography using a Superdex 75 10/300 GL column (GE Healthcare) equilibrated in Tris-salt buffer at 0.3 mL per min. Peak fractions from the gel filtration chromatography step are then concentrated to 10–15 mg mL^{-1} by centrifugal filtration, as described previously.

3.1.5 Cell lysate preparation

Cell lysate is derived from *H. volcanii* LR03 (*Δsamp1 Δsamp2 Δsamp3 ΔmsrA ΔubaA*). The *H. volcanii* strain is grown to stationary phase (OD_{600} of 3.0–4.0) in ATCC974 medium supplemented with Nv and 25 m*M* DMSO (0.5 L in 2.8-L Fernbach flask) with rotary shaking at 200 rpm. Cells are harvested by centrifugation (3500 × *g*, 10 min at 15°C). Cell pellets are suspended in 8 mL of Tris-salt buffer-7.5 supplemented with 10 m*M* $MgCl_2$, 100 μ*M* bortezomib (LC Laboratories) and 30 μg mL^{-1} DNase I from bovine pancreas (Sigma). Cells are lysed in a French Press (20,000 psi). Cell debris is removed by centrifugation (12,000 × *g*, 20 min, 4°C). Cell lysate is filtered (0.2 μm SFCA, Nalgene) and dialyzed 3 times against Tris-salt buffer at 4°C. As an example, 7 mL of cell lysate is loaded into the dialysis tubing (3.5 kDa MWCO, SnakeSkin prewet according to the manufacturer's instructions, Thermo Scientific). The sample (in the dialysis tubing) is placed in 4 L of buffer with stirring for 1.5–2 h at 4°C. Repeat this last step two more times.

3.2 Reconstitution of Ubl-protein ligation

Ubl ligation is reconstituted in 100 μL reaction volumes distributed in 0.5 mL Eppendorf tubes. The final components in the reactions are: cell lysate (80 μL, 20–30 mg mL^{-1}), 5 μ*M* MsrA-StrepII, 10 μ*M* Flag-His-SAMP2, 5 μ*M* His-UbaA, 50 m*M* Tris–Cl, pH 7.5, 2 m*M* $MgCl_2$, 4 m*M* ATP, 25 m*M* DMSO, 0.5 m*M* DTT, and 2 *M* NaCl. Components of the reaction are omitted to assess their impact. The reactions are incubated at 45°C for 0–18 h. After assay, the salts are removed by application of the reactions to a Zeba Spin Desalting Column (7K MWCO) according to the supplier (Thermo Scientific). The desalted reaction products are separated by SDS-PAGE (12%) and analyzed by immunoblotting using anti-StrepII, anti-His6, and anti-Flag antibodies as described in Section 2.4 and shown in Fig. 4.

3.3 Cleavage of the Ubl modification

To cleave the Ubl from its target protein, HvJAMM1 (5 μ*M*) is added to Ubl-modified proteins (Fig. 5). The samples are incubated for 2–4 h at 45–50°C. EDTA (50 m*M*) is included to inactivate the metalloprotease as a negative control. HvJAMM1 can cleave Ubl-modified proteins that are generated in the cell (Fig. 5A) or are reconstituted in vitro (Fig. 5B). The cleaved products are desalted and analyzed by immunoblotting as in Section 3.2. To generate the Ubl-modified proteins in vivo to serve as

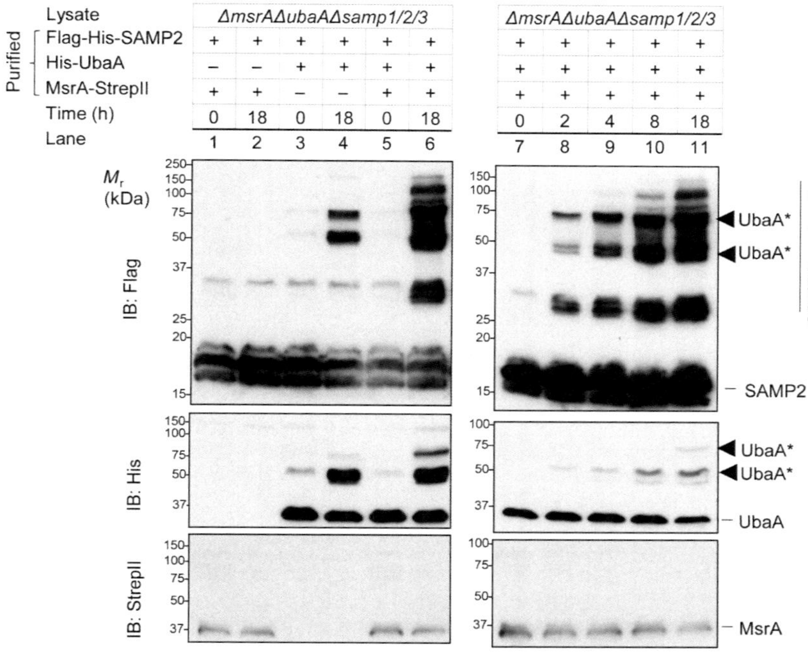

Fig. 4 Reconstitution of ubiquitin-like protein ligation. Reaction components are MsrA-StrepII (5 μ*M*), Flag-His-SAMP2 (10 μ*M*), His-UbaA (5 μ*M*), cell lysate of LR03 (H26 *ΔmsrA ΔubaA Δsamp1–3* mutant) (80 μL, 20–30 mg mL^{-1}), 50 m*M* Tris–Cl, pH 7.5, 2 m*M* $MgCl_2$, 4 m*M* ATP, 25 m*M* DMSO, 0.5 m*M* DTT, and 2 *M* NaCl. Components of the reaction are omitted to assess their impact. The time dependence of the reaction is demonstrated in lanes 7–11. In this example, the impact of MsrA on the reaction is examined by comparing lanes 4 and 6. *For details, see text and Fu et al. (2017).*

substrates for cleavage by HvJAMM1, *H. volcanii* H26-pJAM949 expressing Flag-SAMP2 can be grown to stationary phase in 100 mL of ATCC 974 medium supplemented with Nv (0.2 μg mL^{-1}) and 100 m*M* DMSO. The cells are harvested by centrifugation (15 min at 9200 × *g* and 4°C), washed twice with 20 mL of HEPES-salt buffer (20 m*M* HEPES, pH 7.5, 2 *M* NaCl), and resuspended in the same buffer at 1 g wet wt cells per 1 mL. Cells are lysed by sonication (3 × 30–60 s at 140 W on ice). The lysate is clarified by centrifugation (as earlier) and its residual protease activity is inactivated by boiling for 20 min. The boiled lysate is chilled, clarified by centrifugation (as earlier), and adjusted to 5 mg protein per mL with HEPES-salt buffer. These samples containing Ubl-modified proteins formed in vivo can be stored at −80°C until use.

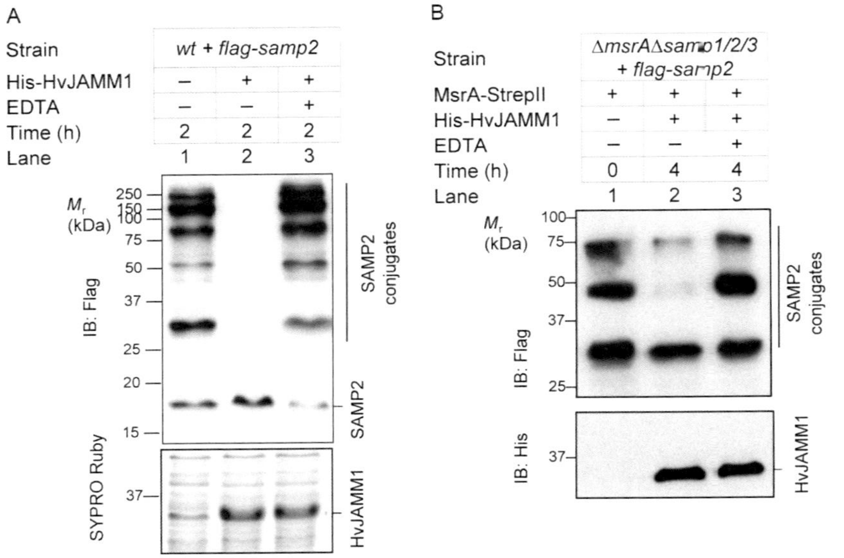

Fig. 5 HvJAMM1-mediated cleavage of ubiquitin-like protein ligation products formed in vivo (A) and in vitro (B). (A) Reactions (10 μL) included 5 μg of heat-treated Flag-SAMP conjugate enrichment incubated for 2 h at 50°C in assay buffer (20 m*M* HEPES, pH 7.5, 2 *M* NaCl, 50 μ*M* $ZnCl_2$) with no additional effector (lane 1), 10 μ*M* HvJAMM1 (lane 2), or 10 μ*M* HvJAMM1 plus 50 m*M* EDTA (lane 3). (B) Purified MsrA-StrepII (5 μ*M*) incubated in assay buffer (50 m*M* Tris–Cl, pH 7.5, 2 m*M* $MgCl_2$, 4 m*M* ATP, 25 m*M* DMSO, 0.5 m*M* DTT, and 2 *M* NaCl) with 1.6 mg mL^{-1} cell lysate of strain LR02 (H26 Δ*msrA* Δ*samp1/2/3*) expressing Flag-SAMP1 from plasmid pJAM947 for 18 h at 45°C. The reactions were subsequently mixed with no effector (lane 1), 5 μ*M* HvJAMM1 (lane 2), or 10 μ*M* HvJAMM1 plus 50 m*M* EDTA (lane 3). Proteins were separated by reducing 12% SDS-PAGE and analyzed by immunoblotting (IB: anti-Flag and anti-His) and total protein stain (SYPRO Ruby) as indicated on the *left*. Migration of molecular mass standards (Mr, kDa) indicated on *left*. Migration of SAMP2, SAMP2 conjugates, and HvJAMM1 indicated on *right*. The Ubl conjugates formed in vitro in Fig. 5B are less abundant than when reconstituted with purified SAMP and E1 (UbaA) (Fig. 4) but are highly dependent on MsrA and DMSO, as the level of E1 and its automodification are limited. For details, see text and Fu et al. (2017) and Hepowit et al. (2012).

4. Conclusions

Assays for Ubl-protein ligation and proteasome function have proved fruitful in understanding how archaea may target proteins for energy-dependent proteolysis. The nanocompartmentalized structure of proteasomes and Ubl tagging make the Ubl-proteasome system energy dependent and provide a means to regulate the destructive forces of proteolysis. Finding that TBP2 may be phosphorylated at Ser2 prior to Ubl ligation provides an extra layer of regulation for substrate targeting in the archaeal cell. The reconstitution of Ubl ligation with E1-UbaA, SAMP2 and MsrA provides an opportunity to fractionate and identify the additional cofactors and/or substrates of the cell lysate that may be needed for this in vitro reaction. The recent development of SILAC (stable isotope labeling with amino acids in cell culture) for quantitative analysis of the *H. volcanii* proteome in a multiplex platform (McMillan et al., 2018) offers another avenue for discovery of novel factors associated with the Ubl-protein ligation system of archaea.

Acknowledgments

Funds for this project were awarded to J.M.-F. through the Bilateral NSF/BIO-BBSRC program (NSF 1642283), the U.S. Department of Energy, Office of Basic Energy Sciences, Division of Chemical Sciences, Geosciences and Biosciences, Physical Biosciences Program (DOE DE-FG02-05ER15650) and the National Institutes of Health (NIH R01 GM57498).

References

Allers, T., Ngo, H. P., Mevarech, M., & Lloyd, R. G. (2004). Development of additional selectable markers for the halophilic archaeon *Haloferax volcanii* based on the *leuB* and *trpA* genes. *Applied and Environmental Microbiology, 70*, 943–953.

Barthelme, D., Chen, J. Z., Grabenstatter, J., Baker, T. A., & Sauer, R. T. (2014). Architecture and assembly of the archaeal Cdc48*20S proteasome. *Proceedings of the National Academy of Sciences of the United States of America, 111*, E1687–E1694.

Bitan-Banin, G., Ortenberg, R., & Mevarech, M. (2003). Development of a gene knockout system for the halophilic archaeon *Haloferax volcanii* by use of the *pyrE* gene. *Journal of Bacteriology, 185*, 772–778.

Cao, S., Engilberge, S., Girard, E., Gabel, F., Franzetti, B., & Maupin-Furlow, J. A. (2017). Structural insight into ubiquitin-like protein recognition and oligomeric states of JAMM/MPN+ proteases. *Structure, 25*, 823–833. e826.

Cao, S., Hepowit, N., & Maupin-Furlow, J. A. (2015). Ubiquitin-like protein SAMP1 and JAMM/MPN+ metalloprotease HvJAMM1 constitute a system for reversible regulation of metabolic enzyme activity in archaea. *PLoS One, 10*, e0128399.

Dilks, K., Gimenez, M. I., & Pohlschroder, M. (2005). Genetic and biochemical analysis of the twin-arginine translocation pathway in halophilic archaea. *Journal of Bacteriology, 187*, 8104–8113.

Dyall-Smith, M. (2009). In M. Dyall-Smith (Ed.), *The halohandbook: protocols for halobacterial genetics version 7.2*. https://haloarchaea.com/halohandbook/.

Forouzan, D., Ammelburg, M., Hobel, C. F., Ströh, L. J., Sessler, N., Martin, J., et al. (2012). The archaeal proteasome is regulated by a network of AAA ATPases. *The Journal of Biological Chemistry, 287*, 39254–39262.

Fu, X., Adams, Z., Liu, R., Hepowit, N. L., Wu, Y., Bowmann, C. F., et al. (2017). Methionine sulfoxide reductase A (MsrA) and its function in ubiquitin-like protein modification in archaea. *MBio, 8*, e01117–e01169.

Fu, X., Liu, R., Sanchez, I., Silva-Sanchez, C., Hepowit, N. L., Cao, S., et al. (2016). Ubiquitin-like proteasome system represents a eukaryotic-like pathway for targeted proteolysis in archaea. *MBio, 7*, e00316–e00379.

Gimenez, M. I., Dilks, K., & Pohlschroder, M. (2007). *Haloferax volcanii* twin-arginine translocation substates include secreted soluble, C-terminally anchored and lipoproteins. *Molecular Microbiology, 66*, 1597–1606.

Gregor, D., & Pfeifer, F. (2005). In vivo analyses of constitutive and regulated promoters in halophilic archaea. *Microbiology*, *151*, 25–33.

Hepowit, N. L., de Vera, I. M., Cao, S., Fu, X., Wu, Y., Uthandi, S., et al. (2016). Mechanistic insight into protein modification and sulfur mobilization activities of noncanonical E1 and associated ubiquitin-like proteins of archaea. *The FEBS Journal*, *283*, 3567–3586.

Hepowit, N. L., Uthandi, S., Miranda, H. V., Toniutti, M., Prunetti, L., Olivarez, O., et al. (2012). Archaeal JAB1/MPN/MOV34 metalloenzyme (HvJAMM1) cleaves ubiquitin-like small archaeal modifier proteins (SAMPs) from protein-conjugates. *Molecular Microbiology*, *86*, 971–987.

Humbard, M., Miranda, H., Lim, J., Krause, D., Pritz, J., Zhou, G., et al. (2010). Ubiquitin-like small archaeal modifier proteins (SAMPs) in *Haloferax volcanii*. *Nature*, *463*, 54–60.

Karadzic, I., Maupin-Furlow, J., Humbard, M., Prunetti, L., Singh, P., & Goodlett, D. R. (2012). Chemical cross-linking, mass spectrometry, and *in silico* modeling of proteasomal 20S core particles of the haloarchaeon *Haloferax volcanii*. *Proteomics*, *12*, 1806–1814.

Kim, Y. C., Snoberger, A., Schupp, J., & Smith, D. M. (2015). ATP binding to neighbouring subunits and intersubunit allosteric coupling underlie proteasomal ATPase function. *Nature Communications*, *6*, 8520.

Large, A., Stamme, C., Lange, C., Duan, Z., Allers, T., Soppa, J., et al. (2007). Characterization of a tightly controlled promoter of the halophilic archaeon *Haloferax volcanii* and its use in the analysis of the essential *cct1* gene. *Molecular Microbiology*, *66*, 1092–1106.

Martin, J. H., Rawls, K. S., Chan, J. C., Hwang, S., Martinez-Pastor, M., McMillan, L. J., et al. (2018). GlpR is a direct transcriptional repressor of fructose metabolic genes in *Haloferax volcanii*. *Journal of Bacteriology*, *200*(17), e00244-18.

McMillan, L. J., Hepowit, N. L., & Maupin-Furlow, J. A. (2016). Archaeal inorganic pyrophosphatase displays robust activity under high-salt conditions and in organic solvents. *Applied and Environmental Microbiology*, *82*, 538–548.

McMillan, L. J., Hwang, S., Farah, R. E., Koh, J., Chen, S., & Maupin-Furlow, J. A. (2018). Multiplex quantitative SILAC for analysis of archaeal proteomes: A case study of oxidative stress responses. *Environmental Microbiology*, *20*, 385–401.

Miranda, H. V., Antelmann, H., Hepowit, N., Chavarria, N. E., Krause, D. J., Pritz, J. R., et al. (2014). Archaeal ubiquitin-like SAMP3 is isopeptide-linked to proteins via a UbaA-dependent mechanism. *Molecular & Cellular Proteomics*, *13*, 220–239.

Miranda, H., Nembhard, N., Su, D., Hepowit, N., Krause, D., Pritz, J., et al. (2011). E1- and ubiquitin-like proteins provide a direct link between protein conjugation and sulfur transfer in archaea. *Proceedings of the National Academy of Sciences of the United States of America*, *108*, 4417–4422.

Mullakhanbhai, M. F., & Larsen, H. (1975). *Halobacterium volcanii* spec. nov., a dead sea halobacterium with a moderate salt requirement. *Archives of Microbiology*, *104*, 207–214.

Rabl, J., Smith, D. M., Yu, Y., Chang, S. C., Goldberg, A. L., & Cheng, Y. (2008). Mechanism of gate opening in the 20S proteasome by the proteasomal ATPases. *Molecular Cell*, *30*, 360–368.

Robinson, J. L., Pyzyna, B., Atrasz, R. G., Henderson, C. A., Morrill, K. L., Burch, A. M., et al. (2005). Growth kinetics of extremely halophilic *Archaea* (family *Halobacteriaceae*) as revealed by Arrhenius plots. *Journal of Bacteriology*, *187*, 923–929.

Roesser, M., & Muller, V. (2001). Osmoadaptation in bacteria and archaea: Common principles and differences. *Environmental Microbiology*, *3*, 743–754.

Schneider, C. A., Rasband, W. S., & Eliceiri, K. W. (2012). NIH image to imageJ: 25 Years of image analysis. *Nature Methods*, *9*, 671–675.

Welinder, C., & Ekblad, L. (2011). Coomassie staining as loading control in Western blot analysis. *Journal of Proteome Research*, *10*, 1416–1419.

Zhou, G., Kowalczyk, D., Humbard, M., Rohatgi, S., & Maupin-Furlow, J. (2008). Proteasomal components required for cell growth and stress responses in the haloarchaeon *Haloferax volcanii*. *Journal of Bacteriology*, *190*, 8096–8105.

CHAPTER NINE

Structural mass spectrometry approaches to study the 20S proteasome

Gili Ben-Nissan, Shay Vimer, Mark Tarnavsky, Michal Sharon*
Department of Biomolecular Sciences, Weizmann Institute of Science, Rehovot, Israel
*Corresponding author: e-mail address: michal.sharon@weizmann.ac.il

Contents

Abstract

The 20S proteasome is a large multisubunit proteolytic machine that is central to intracellular protein degradation. It is found in all three kingdoms of life and is ubiquitous in archaea and eukaryotes. Since its discovery, much effort employing a diverse array of structural biology methods has been applied to help understand its structure/function relationships. Here, we will specifically focus on the application of native mass spectrometry (MS) approaches for structural investigations of the 20S proteasome. Native MS is a method that examines intact protein assemblies, without disturbing the noncovalent interactions that govern the overall structure. This method is ideally suited to revealing the intrinsic heterogeneity of a given sample and provides insight into the composition, stoichiometry, subunit architecture, and topology of the protein assembly. Initially, we

Methods in Enzymology, Volume 619
ISSN 0076-6879
https://doi.org/10.1016/bs.mie.2018.12.029
© 2019 Elsevier Inc.
All rights reserved.

describe native MS-oriented protocols for the isolation of endogenous 20S proteasomes from yeast, rat liver, and human cells. We then highlight the applicability of native MS methodologies, using different instrumental platforms, for structural investigations of the complex. In particular, by means of proteasome biology, we highlight the different approaches used to analyze both intact complexes—their natural heterogeneity and interactions with substrates and regulators—and their individual constituent subunits.

Abbreviations

ATD arrival time distribution
CCS collision cross section
CID collisional-induced dissociation
CIU collisional-induced unfolding
GFP green fluorescent protein
HCD higher-energy collision dissociation
IM-MS ion mobility–mass spectrometry
MS mass spectrometry
nESI nano-electrospray ionization
PA projection approximation
PBS phosphate-buffered saline
PTM posttranslational modification
QTOF quadrupole time-of-flight
SID surface-induced dissociation
TEAA triethylammonium acetate
TOF time-of-flight

1. Introduction

The ability to sense and adapt to environmental changes involves constant adjustment of the cellular proteome. Various mechanisms for controlling protein half-life exist, with proteasome-mediated degradation being of critical importance (Glickman & Ciechanover, 2002; Goldberg, 2003). Through this cellular route, misfolded, damaged, and short-lived regulatory proteins are removed.

Proteolysis occurs within the 20S proteasome, which is a barrel-shaped particle. To degrade folded and ubiquitin-modified proteins, the 20S proteasome must associate with a 19S regulatory complex, forming the 26S proteasome. ATPases in the 19S subcomplex trigger structural rearrangements that lead to the opening of the sealed gate structure of the 20S complex (Kunjappu & Hochstrasser, 2014; Schmidt & Finley, 2014). These subunits also contribute to substrate unfolding and the delivery of the unfolded protein substrate into the 20S proteolytic chamber. Analysis of proteasome complexes distribution in different human cell lines revealed that 30% of the 20S complexes are associated with the 19S regulatory particle (Fabre et al., 2013, 2014). Approximately 20% of the remaining 20S proteasome pool is associated with other proteasome activators, such as the cytosolic PA28αβ (or 11S), as well as the nuclear PA28γ and PA200 regulators, that enhance the proteolytic capacity of the complex in an ATP- and ubiquitin-independent manner (DeMartino & Slaughter, 1993; Ma, Slaughter, & DeMartino, 1992; Sadre-Bazzaz, Whitby, Robinson, Formosa, & Hill, 2010).

At least half of the 20S proteasome pool is thought to function alone in its free form (Brooks et al., 2000; Fabre et al., 2013, 2014), leading to an ATP- and ubiquitin-independent mode of degradation. The substrates of this degradation route contain unstructured regions that enable their access into the restricted entry pore of the complex (reviewed in Ben-Nissan & Sharon, 2014). These can be proteins that have become partially or completely unfolded due to aging, mutations or oxidation or native proteins containing large unstructured segments (>30 amino acids in length), referred to as intrinsically disordered regions (IDRs), or proteins with entirely disordered sequences (intrinsically disordered proteins, IDPs) (Ben-Nissan & Sharon, 2014). Many key regulatory proteins, such as the tumor suppressors p53, p73, and retinoblastoma protein, the proto-oncoprotein c-Fos, the cell cycle regulators p27 and p21, and the neurodegenerative diseases related proteins tau and α-synuclein, have been shown to be substrates of the 20S proteasome. Here, we will share our lab's experience in purifying and characterizing the 20S proteasome by structural mass spectrometry.

The 20S proteasome is found in all the known domains of life, ranging from archaea through humans (Humbard & Maupin-Furlow, 2013). It is composed of 28 subunits, 14 α-type and 14 β-type, forming a ~700-kDa assembly with a highly conserved cylindrical structure of four rings, each containing seven subunits ($\alpha_7\beta_7\beta_7\alpha_7$) (Kunjappu & Hochstrasser, 2014; Voges, Zwickl, & Baumeister, 1999). The β-subunits which form the two inner rings contain the active proteolytic sites. The main difference between the prokaryotic and eukaryotic 20S proteasomes lies in their complexity (Fig. 1). Prokaryotic 20S proteasomes (e.g., from *Thermoplasma acidophilum*) (Zwickl, Lottspeich, & Baumeister, 1992) are generally composed of 14 identical α-subunits and 14 identical β-subunits, whereas canonical eukaryotic proteasomes include 7 different α- and 7 different β-subunits, each present at a specific position in each ring. Of the latter subunits, three, β_1, β_2, and β_5, are catalytically active; they are responsible for the caspase-like (or peptidylglutamyl-peptide), trypsin-like, and chymotrypsin-like

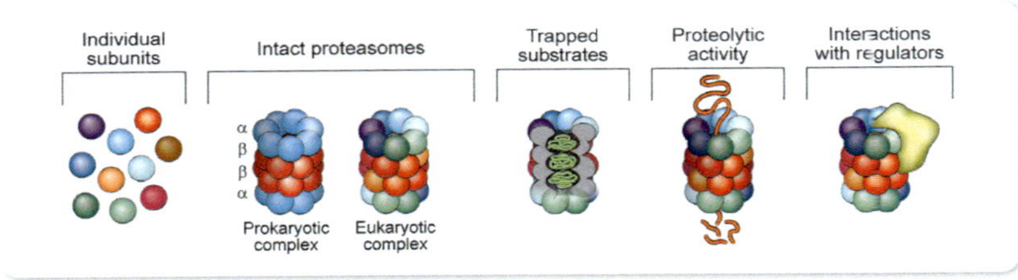

Fig. 1 Structural analysis of the 20S proteasome by native MS includes both examination of the heterogeneity of individual α- and β-subunits, as well as analysis of intact proteasomes. Using native MS, we can compare particles from different organisms, study the interactions between the 20S proteasome and its substrates, track proteolytic activity, and monitor interactions with regulators.

hydrolyzing functions, respectively. In mammals, 20S proteasome complexity is further enhanced by three γ-interferon inducible subunits, which are incorporated into the proteasome during the immune response, forming the immunoproteasome (Groettrup, Kirk, & Basler, 2010). Additionally, tissue-specific 20S proteasome particles with specific subunit replacements are found in the testis and thymus (Kniepert & Groettrup, 2014). Overall, the 20S proteasome architecture creates a compartment whose proteolytic active sites are restricted to its interior, so that only proteins entering this chamber are degraded.

Crystal structures of the eukaryotic 20S proteasome have been solved for a number of organisms, including yeast (Groll et al., 1997), bovine (Unno et al., 2002), mouse (Huber et al., 2012), and human (Harshbarger, Miller, Diedrich, & Sacchettini, 2015). Moreover, advances in high-resolution cryo-electron microscopy methods yielded structures of the 20S proteasome and its associated 19S regulatory particle (Eisele et al., 2018; Huang, Luan, Wu, & Shi, 2016; Schweitzer et al., 2016). Nuclear magnetic resonance has contributed to our understanding of the gating mechanism (Huang, Perez, & Kay, 2017; Latham, Sekhar, & Kay, 2014), and allosteric communication within the 20S proteasome complex (Ruschak & Kay, 2012). Here, however, we will focus on and present protocols involving native MS, an alternative structural biology method for investigation of the complex. Although native MS does not deliver structures at atomic resolution, its advantage lies in its ability to present the relative abundance of the different populations in a given assembly, within a single spectrum (Chen et al., 2016; Liko, Allison, Hopper, & Robinson, 2016; Lossl, van de Waterbeemd, & Heck, 2016; Sharon, 2013). This ability, together with the technique's high resolving power, enables native MS to reveal the various subpopulations of endogenous 20S proteasomes, and determine their associations with protein partners such as substrates and regulators (Fig. 1).

In native MS applications, intact protein assemblies are introduced into the mass spectrometer while preserving the weak noncovalent interactions between protein subunits and associated biomolecules (Chen et al., 2016; Liko et al., 2016; Lossl et al., 2016; Sharon, 2013). These types of experiments provide information on the composition, stoichiometry, subunit architecture, and topological organization of an assembly. The high sensitivity of native MS and, consequently, low quantity of sample required, constitutes another advantage, along with the speed of analysis, which enables the monitoring of reactions in a time-resolved manner. In addition, recent technological advances in high-resolution mass spectrometers afford increased mass accuracy and better separation between substoichiometric complexes, posttranslationally modified entities, and assemblies bound to small molecules.

In general, the study of large protein complexes has mainly been done with hybrid mass spectrometers equipped with a nano-electrospray ionization (nESI) source. These instruments, which combine a quadrupole mass filter with an orthogonal time-of-flight (TOF) analyzer (QTOF instruments), were specifically modified to enable the transmission and detection of high mass ions (Chernushevich & Thomson, 2004; Sobott & Robinson, 2002) (Fig. 2A). Recently, analysis of large assemblies, which benefits from improved mass accuracy, has also become possible by means of modified Orbitrap instruments (Rose, Damoc, Denisov, Makarov, & Heck, 2012) (Fig. 2B). These instrumental platforms enable multiple workflows for in-depth structural characterization of protein assemblies (Fig. 2C–G), as we describe later.

Over the past decade, we have investigated the 20S proteasome using a number of native MS approaches (Moscovitz et al., 2015, 2012; Olshina, Ben-Nissan, & Sharon, 2018; Sharon et al., 2006; Sharon, Witt, Glasmacher, Baumeister, & Robinson, 2007; Solomon et al., 2017; Tsvetkov et al., 2012; Witt et al., 2006). Here, we describe how this methodology can be applied to various 20S proteasome systems. First, we provide detailed purification protocols for the isolation of endogenous 20S proteasomes from rat livers, yeast, and human cells. Second, we discuss analysis of the intact 20S proteasome complex from different sources (archaea, rat, yeast, and human), using different instrumental platforms. We then demonstrate the multilevel analysis of subunits that comprise the 20S proteasome, and how the biological heterogeneity of individual subunits

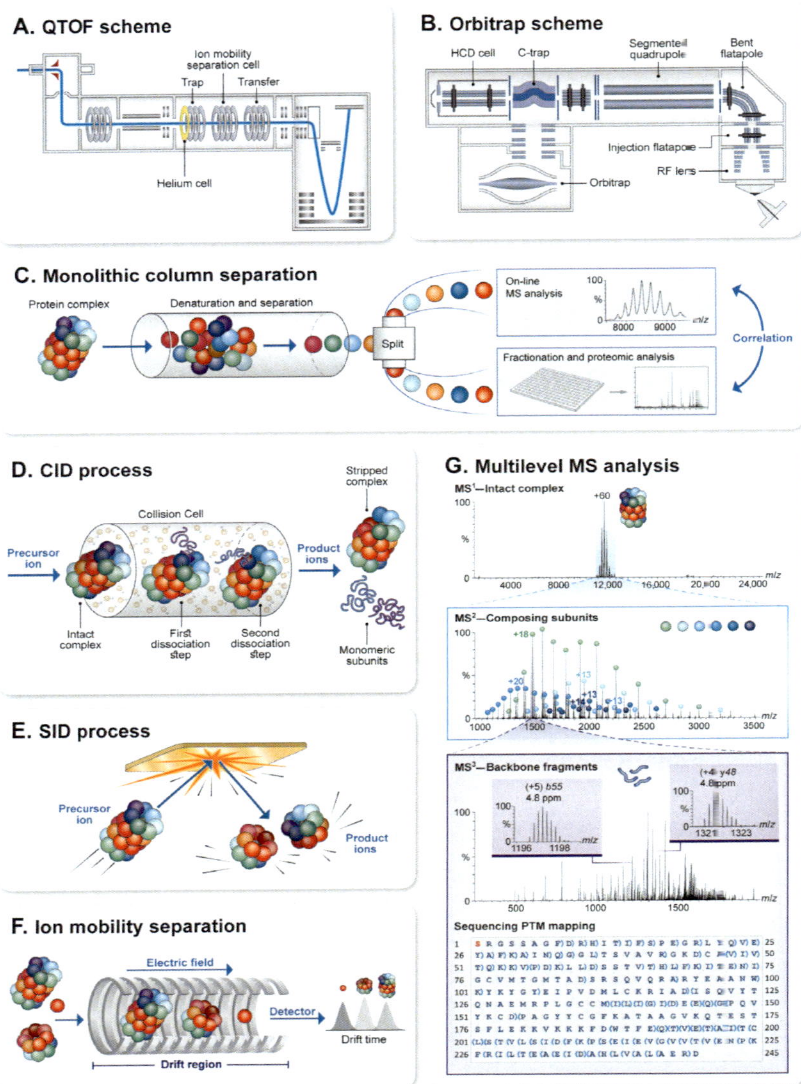

Fig. 2 Native MS offers multiple methodologies for in-depth structural characterization of the 20S proteasome. Two popular MS platforms for native analysis of large protein complexes are QTOF- (A) and Orbitrap-based instruments (B). Subunit heterogeneity can be studied under denaturing (C) or native conditions (D–G). In CID ions are accelerated into a gas-filled collision cell wherein the multicollision events generate dissociation products. SID can be applied to the study of intermolecular interface strength and assembly pathways (E). Coupling MS with ion mobility measurements can reveal information on the overall shape and compactness of the particles (F). Hybrid instruments equipped with multiple collision cells can perform multilevel analyses and examine the intact complexes, as well as separate them into the subunits comprising them and perform detailed, top-down sequencing and identification of posttranslational modifications (G).

can be revealed. We also demonstrate how native MS can be coupled with ion mobility (IM-MS) measurements for determining the overall shape and stability of the complex, and finally, we show how substrate and/or regulator interactions with the 20S proteasome can be characterized.

2. 20S proteasome purification protocols for native MS analysis

In this section, we describe the protocols developed in our lab for biochemical purification of 20S proteasomes from rat livers, and the FLAG-affinity purification of the complex from *Saccharomyces cerevisiae* and HEK293 cells (Fig. 3). The purification protocol of the recombinant archaeal 20S proteasome was published by the Lewis lab (Ruschak, Religa, Breuer, Witt, & Kay, 2010). Our protocols are grounded in the fundamental requirements necessary to generate complexes suitable for native MS analyses, namely, using proteins within a micromolar concentration range, in a volatile solution and avoiding the use of chemicals that remain bound to the proteins during the last steps of the purification procedure.

Classic biochemical purification procedures typically employ neutral, water-based solutions that contain a physiological concentration of salt and a pH-stabilizing buffer (Konermann, 2017). Commonly, we use solutions that are based on 150 m*M* NaCl and contain 10–50 m*M* of buffers such as phosphate buffer or Tris, at a pH close to 7. We avoid HEPES-based buffers since we found that they tend to adhere to the proteins and broaden the peaks in our measurements. Similarly, we omit glycerol and detergents such as NP-40, Tween 20, and Triton X-100 from the last steps of purification procedures.

If the purification protocol does not include a final step that transfers the sample into a volatile solution, it is necessary to exchange the buffer prior to the MS measurement. Ammonium acetate is the volatile solution most commonly used in native MS experiments, due to the tendency of its constituents to evaporate efficiently during nESI. It also serves as a stabilizing background electrolyte, and thus can be used at a concentration similar to that of the salts in the original buffer, typically around 150 m*M*. It is noteworthy that ammonium acetate is not a buffer at pH 7, but diminishes pH alterations occurring in positive mode nESI by buffering the solution within the range of 4.75 ± 1 (Konermann, 2017).

Typically, at the end of a successful purification, we concentrate the 20S proteasome sample to about 1–5 μ*M*, divide it into small aliquots (2–20 μL), snap-freeze in liquid nitrogen, and store at −80°C.

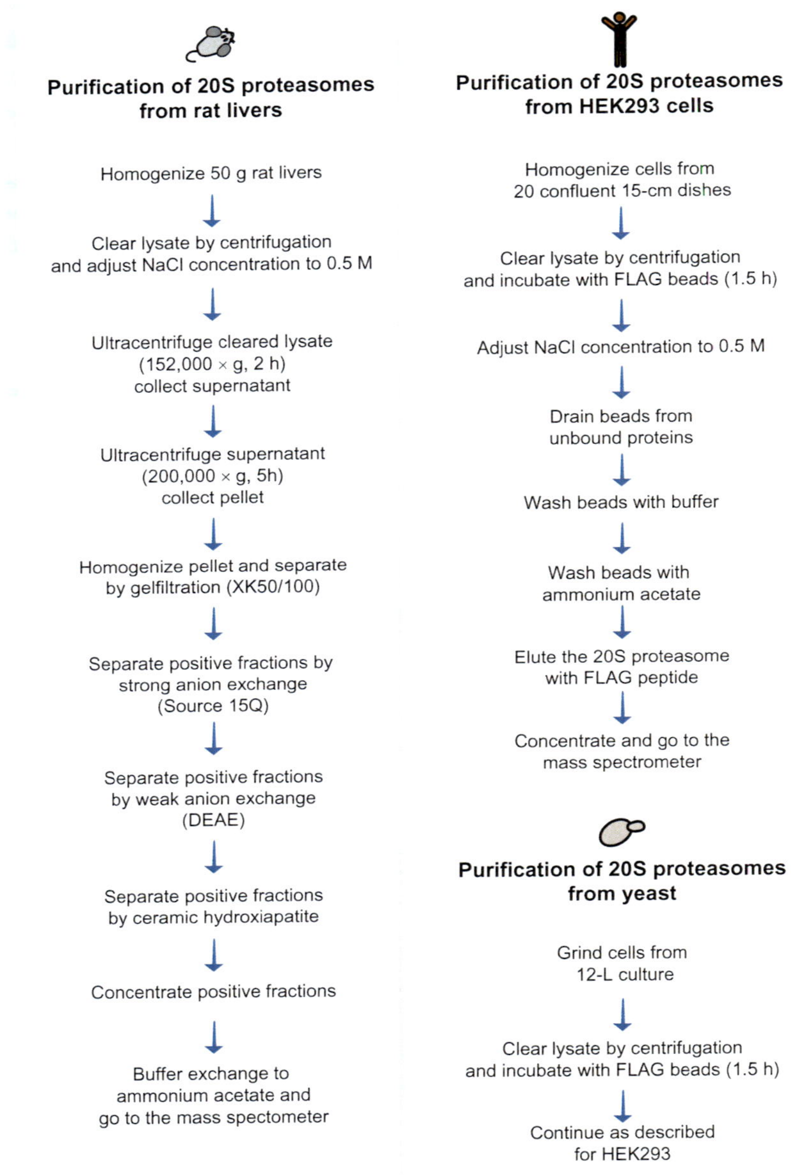

Fig. 3 Workflows for the purification of the 20S proteasome for native MS analysis. The scheme highlights the major steps in the biochemical purification of 20S proteasomes from rat livers, and FLAG-affinity purification of 20S proteasomes from yeast and human cells.

2.1 Purification of 20S proteasomes from rat livers

2.1.1 Special equipment

The list below describes special equipment required for the protocol that is used in our lab, but any other equipment with comparable specifications can be used.

- Eppendorf Bench Top Refrigerated Centrifuge (5810R) with a Fixed-Angle Rotor (cat. # F-34-6-38).
- Beckman Coulter Optima L-90K Ultracentrifuge, with a Type 45 Ti Fixed-Angle Titanium Rotor.
- 12 Beckman Coulter polycarbonate bottles with cap assemblies, 38 × 102 mm, for 70 mL (cat. # 355622).
- Kinematica Polytron tissue homogenizer with a 12 mm dispersing aggregate (for 2–250 mL of sample) (PT-DA 2112/EC).
- Overhead stirrer RW 20 (IKA cat. # 3593000) attached to a plate stand (IKA cat. # 31601000).
- Large Volume (100 mL) Potter-Elvehjem Glass Tube tissue grinder with PTFE Pestle (Kimble cat. # 885520-0100).
- Two types of Monoject™ Blunt Cannulas; lavender cannulas, gauge 21, needle length 5 cm (Covidien cat. # 8881202371), and orange cannulas, gauge 23, needle length 2.5 cm (Covidien cat. # 8881202397).
- AKTA protein purification system, equipped with a UV detector and fraction collector (GE Healthcare).
- XK 50/100 column (GE Healthcare) packed with 1.8 L 4% Rapid Run™ Agarose Bead Standard 50–150 μm for gel filtration (GF) (Agarose Bead Technologies cat. # 4RRS-1000).
- HiScale 26/20 column (GE Healthcare) packed with 50 mL of the strong anion exchange resin Source 15Q resin (GE Healthcare cat. # 17-0947-01).
- HiTrap® DEAE Fast Flow column (GE Healthcare cat. # 17-5055-01) packed with 5 mL of the weak anion exchange resin, DEAE.
- Tricorn 10/10 column packed with 8 mL of type I ceramic hydroxiapatite (CHT) (Bio-Rad cat. # 1582000) consisting of mixed mode functional groups of both positively and negatively charged groups.
- Amicon Ultra centrifugal filter units, Ultra-15, MWCO 30 kDa (Merck cat. # UFC903024).
- Two black 96-well plates with a flat bottom (Thermo cat. # 237108).
- Tecan Infinite M Plex plate reader for fluorescence reading.
- Micro BioSpin 6 buffer exchange columns (Bio-Rad cat. # 7326221).

2.1.2 Buffers and reagents

- 4–6 Rat livers (about 50 g).
- Lysis buffer (0.5 L): Tris–HCl 50 m*M*, pH 7.5, Sucrose 250 m*M*, EDTA 1 m*M*, DTT 1 m*M*, Benzonase endonuclease (2500 U).
- Buffer A (10 L): Tris–HCl 25 m*M*, pH 7.5, EDTA 1 m*M*, DTT 1 m*M*.
- Buffer B (4 L): Tris–HCl 25 m*M*, pH 7.5, NaCl 1 *M*, EDTA 1 m*M*, DTT 1 m*M*.
- PBS (phosphate-buffered saline) (1 L): 810 mL K_2HPO_4 1 *M* (pH 7.4) mixed with 190 mL KH_2PO_4 1 *M* (pH 7.4).
- PBS-A (1 L): 10 mL PBS in 990 mL double-distilled water (DDW) diluted to make a 10 m*M* solution at pH 7.4.
- PBS-B (1 L): 400 mL PBS in 990 mL DDW diluted to make a 10 m*M* solution at pH 7.4.
- NaCl 5 *M* (1 L)
- SDS 1% (50 mL).
- Fluorogenic peptide substrate: 50 m*M* Suc-Leu-Leu-Val-Tyr-AMC (Bachem cat. # I-1395.0100), dissolved in DMSO.
- Activity buffer: Buffer A containing 150 μ*M* fluorogenic peptide.
- SDS-activity buffer: Buffer A containing 150 μ*M* fluorogenic peptide and 0.027% SDS.
- $MgCl_2$ 1 *M* (200 μL).

2.1.3 General notes prior to starting the purification process

- Use fresh livers, obtained immediately after sacrifice of the rats. Place each of 2–3 livers in a 50-mL tube, and snap-freeze in liquid nitrogen. Store frozen livers at −80°C.
- The quality of purified 20S proteasomes from rat livers is highly dependent on the animals' overall condition. Insofar as is possible, aim to purify 20S proteasomes from a single batch of livers, obtained from animals that were exposed to similar growth conditions.
- DTT, Benzonase, and fluorogenic peptide should be added to the different buffer solutions immediately prior to use.
- All solutions (including water) need to be filtered through a 0.2 μm PES filter before introducing them into the FPLC system.
- After using the different columns, wash them according to the manufacturer's instructions.
- The purification process takes 5 consecutive days.

- During the days prior to the purification, precondition the XK 50/100 column by washing it with 2 column volumes (CV) of DDW, followed by 2 CV of buffer A (without DTT and Benzonase). Restrict the pressure in the column to 0.22 MPa and the flow to 1 mL/min, so that the preconditioning will end during the first day of the week-long purification.
- The first day of purification takes 12–14 h. To save time, begin the purification process on that day by turning on and cooling the benchtop refrigerated centrifuge, and the ultracentrifuge.
- Operation of the glass tissue grinder is performed in the cold room, and takes about 30 min. Make sure you wear a lab coat during this procedure.
- After each purification step, we examine all the eluted fractions by means of an activity assay (Li, Tomko, & Hochstrasser, 2015). This assay involves a fluorogenic peptide that is a substrate of the 20S proteasome. Before cleavage, the peptide is not fluorescent, whereas after cleavage by the 20S proteasome, the fluorophore is activated and fluorescence can be measured. We then pool the positive fractions, and utilize them in the next purification step.

2.1.4 Details of purification procedure

Day 1

1. Place a 1-L glass beaker in ice.
2. Add 75 mL PBS and then 260 mL lysis buffer to the beaker.
3. Thaw the tube with the livers for up to 5 min in a container with warm water.
4. Just before placing the livers in the glass beaker containing the buffer, add DTT and Benzonase to the solution, and mix gently.
5. Remove livers from the water, shake them gently to remove excess water, and place them in the beaker.
6. While immersed in the buffer, cut the livers into small pieces, using a scalpel.
7. Assemble the tissue homogenizer with the dispersing aggregate and dip it in the buffer.
8. Turn on the homogenizer at its lowest speed and slowly increase the speed, to avoid splashing. Homogenize the livers at maximum speed for 5–10 s, then turn off the homogenizer, to enable the contents of the beaker to cool down. Repeat several times, until all the tissue is completely homogenized, and no tissue pieces are evident. If the

sample starts to foam, stop homogenization immediately and let the sample cool down again.

9. Place six 50-mL centrifuge tubes, the glass tissue grinder, and the pestle in ice.

10. In the cold room, connect the pestle to the overhead stirrer. Operate at 1000 rpm.

11. Pour about 100 mL of homogenate into the glass tube, and perform 10 slow rounds of grinding while pushing the glass tube up and down. Do not let the PTFE pestle rise above the level of the liquid, and do not introduce air into the tube. Hold the glass tube straight and firm; otherwise it can break.

12. Repeat this procedure with the rest of the homogenate.

13. After grinding, divide the homogenate among the six 50-mL centrifuge tubes, and centrifuge at 4°C for 15 min, at 1000 × *g*.

14. Collect the supernatant in a 500-mL cylinder, and add lysis buffer to a final volume of 360 mL. Add 40 mL of 5 *M* NaCl, to a final concentration of 0.5 *M* (to separate the 20S proteasome from the 19S particle), and DTT to a final concentration of 1 m*M*, and mix gently.

15. Divide the supernatant between six prechilled ultracentrifugation bottles. Use lysis buffer to fill the bottles completely, and balance them with the lids.

16. Centrifuge in the ultracentrifuge at 4°C for 2 h, at 152,000 × *g*

17. Collect the supernatant, and divide it between six new prechilled ultracentrifuge bottles. Use lysis buffer to fill the bottles completely, and balance them with the lids.

18. Centrifuge in the ultracentrifuge at 4°C for 5 h, at 200,000 × *g*.

19. Discard the liquid, and gently resuspend each pellet in about 2.5 mL of buffer A and combine the six tubes. The final volume of the resuspended pellets should not exceed 30 mL.

20. Further, homogenize the resuspended pellets, using a 30-mL plastic syringe and the blunt cannulas until all aggregates have been resuspended and the homogenate passes easily through the cannula. Start by passing the resuspended pellet through the gauge-21 cannula several times, and then move to the 23-gauge cannula.

21. Load the homogenized pellets onto the XK50/100 GF column. Elute the proteins with 3 L of buffer A, at a flow rate of 2 mL/min, and restrict the pressure to a maximum 0.22 MPa. Collect fractions of 13 mL, in 15-mL glass tubes.

Day 2

22. Prewash the Source 15Q column with 2 column volumes (CV) DDW, 2 CV buffer A, 2 CV buffer B, then again with 2 CV buffer A.
23. Perform activity assays to examine the proteolytic activity of the GF fractions, to determine which of the fractions contain the 20S proteasome. Prepare two black 96-well plates with flat bottoms. In one plate, pipette 40 μL activity buffer into each well (to monitor 26S proteasome activity); in the second plate, pipette 40 μL of SDS-activity buffer into each well. Low levels of SDS open the gated pore into the catalytic core of the "latent" 20S proteasome.
24. To each well, add 10 μL of the eluted proteins from the GF fractions (examine every alternate fraction). Use three wells in each plate as control measurements and supplement them with 10 μL DDW.
25. Cover the plates with aluminum foil, and incubate at 30°C for 30 min.
26. Add 150 μL of 1% SDS solution to each well, to stop the reaction.
27. Read the plate using a fluorescence plate reader. Set excitation to 385 nm, and emission to 460 nm.
28. For each plate, subtract the averaged blank measurements from each reading. Plot the calculated values of the fluorescence from the two plates, side-by-side, against fraction number. Fractions displaying higher fluorescence levels in the SDS-activity buffer reflect enrichment for the 20S proteasome, whereas fractions displaying the opposite trend indicate enrichment for the 26S proteasome.
29. Combine positive fractions containing the 20S proteasome, and load onto the Source 15Q column. Wash the column with 2 CV buffer A, 2 CV of 10% buffer B, and elute the proteins with 4 CV of a gradient between 10% and 30% buffer B, and another 5 CV of 30% buffer B, at a flow rate of 5 mL/min. Collect fractions of 13 mL, in 15-mL glass tubes. The 20S proteasome typically elutes in one peak centered at 30% buffer B.

Day 3

30. Prewash the HiTrap® DEAE Fast Flow column with 5 CV DDW, 5 CV buffer A, 5 CV buffer B, and 5 CV buffer A.
31. Examine the proteolytic capacity of the eluted fractions from the Source 15Q column, as described in steps 22–29 (due to the gradual enrichment in 20S proteasomes, use half the amount of fluorogenic peptide from now on).

32. Combine positive fractions, dilute the final volume 3 times with buffer A to reduce salt concentration, and enable binding of the proteins to the DEAE resin. Load the proteins onto the DEAE column. Wash the column with 6 CV buffer A, and elute the proteins in a step gradient of 5 CV of 10% buffer B, and 6 CV of 30%, 40%, 50%, 60%, and 70% buffer B, at a flow rate of 5 mL/min. Collect fractions of 1.2 mL in 96-well deep well plates.
33. Examine the proteolytic capacity of the eluted fractions, as described earlier. The 20S proteasome typically elutes between 10% and 30% buffer B.
34. Combine positive fractions and dilute the final volume 3 times with buffer A. The final volume is estimated to be about 70 mL.

Day 4

35. Prewash the CHT column with 5 CV DDW, 5 CV buffer PBS-A, 5 CV buffer PBS-B, and 5 CV buffer PBS-A.
36. Change the buffer of the positive fractions 3 times to PBS-A buffer using a 30 kDa cut-off concentrator (15 mL), according to the manufacturer's instructions.
37. After the third round of buffer exchange, collect the concentrated sample into a 50-mL tube, and add PBS-A buffer to a final volume of 20 mL. Add $MgCl_2$ to a final concentration of 10 m*M*.
38. Load the sample onto the CHT column. Wash the column with 5 CV buffer PBS-A, and elute the proteins in steps of 5 CV of 10%, 20%, 40%, 50%, and 60% buffer PBS-B, at a flow rate of 0.5 mL/min. Collect fractions of 1.2 mL in 96-deep well plates.

Day 5

39. Examine the proteolytic capacity of the eluted fractions, as described earlier. The 20S proteasome typically elutes around 40% buffer B.
40. Combine positive fractions and concentrate about 30 times, using a 30 kDa concentrator. The final concentration of the 20S proteasome should be around 2–4 μ*M*. Aliquot to 20 μL/tube, flash-freeze in liquid nitrogen, and store at −80°C.
41. 20S proteasome quality can be assessed by denaturing SDS-PAGE, followed by Coomassie blue staining, native PAGE, and in-gel activity assays (Li et al., 2015), and by degradation assays (Moscovitz et al., 2015).

On the day of native MS analysis, perform buffer exchange with 150 mM ammonium acetate using Bio-Rad spin columns, according to the manufacturer's instructions.

2.2 Purification of the 20S proteasome from *S. cerevisiae*

2.2.1 Special equipment

- Biospec bead beater (model HBB908-CE).
- Acid-washed glass beads (200 mL), 0.5 mm soda lime glass, Marienfield cat. # 4901000.
- Dry ice.
- 4 Poly-Prep® Chromatography Columns, Bio-Rad cat. # 7311550.

2.2.2 Buffers and reagents

- YPD medium (3 L): 20 g Bacto Peptone and 10 g yeast extract per liter, 2% Glucose.
- Lysis buffer (400 mL): Tris–HCl 50 m*M*, pH 7.5, NaCl 150 m*M*, Glycerol 10%, $MgCl_2$ 5 m*M*. Immediately before use, add phosphatase inhibitors—Na-o-vanadate 5 m*M*, Na-pyrophosphate 4 m*M*, and β-glycerophosphate 4 m*M*.
- Wash buffer (100 mL): Lysis buffer, supplemented with 0.2% NP-40.
- Ammonium acetate (150 mL): 0.5 *M*, pH 7.
- Glycine-HCl (10 mL): 100 m*M*, pH 3.5.
- ANTI-FLAG® M2 Affinity Gel, Sigma cat. # A2220 (2 mL packed resin). The resin can be reused 2–3 times. Regeneration is performed according to the manufacturer's instructions.
- FLAG peptide (0.8 mL): (DYKDDDDK) 95% purity, GL Biochem (Shanghai), Ltd. 10 mg/mL, dissolved in ammonium acetate.
- 1 L Glycerol 50%.

2.2.3 General notes prior to starting the purification process

- 20S proteasomes are affinity purified from the yeast strain RJD1144, generated in the lab of Raymond Deshaies, Caltech, California, USA, which contains a chromosomally FLAG-tagged β4 (PRE1) subunit (Verma et al., 2000).
- In several steps during the protocol (Section 2.2.4, steps 17 and 21) we divide the FLAG resin between several chromatography columns, in order to shorten the washing time.

2.2.4 Procedure

1. Transfer a single colony of yeast into 40 mL YPD medium, and grow in a shaking incubator overnight at 30°C.

2. Divide the culture between four 3-L flasks, each containing 700 mL YPD, and grow for 24 h.
3. Pellet the cells by centrifugation for 20 min at $5000 \times g$, at room temperature.
4. Resuspend each pellet in 10 mL DDW, transfer cells to a 50-mL tube, and centrifuge again.
5. Discard the liquid, and flash-freeze the cells in liquid nitrogen. Store the cells at 80°C.
6. On the day of purification, thaw the cell pellet for several minutes in a water bath heated to 30°C.
7. Resuspend the cells in 100 mL lysis buffer.
8. Meanwhile, fill the grinding chamber of the bead beater with about 200 mL glass beads.
9. Wet the beads with 200 mL lysis buffer, and stir to make sure that they are completely soaked in buffer.
10. Add the resuspended cells to the grinding chamber and assemble into the grinder, according to the manufacturer's instructions.
11. Fill the outer jacket with dry ice and pour 50% glycerol over it, until the outer jacket is almost full.
12. Operate the bead beater for 7–10 cycles of 1 min on, 1 min off. Add dry ice if required.
13. Remove the glycerol and disassemble the grinding chamber.
14. Collect the lysate from the grinding chamber using a 25-mL pipette.
15. Centrifuge the lysate for 20 min, at 4°C, at $35{,}000 \times g$.
16. Collect the supernatant, divide between four 50-mL tubes, and keep tubes on ice.
17. Pipette 1 mL of FLAG resin (~1.2 mL slurry), into two empty chromatography columns.
18. Drain the resin completely and precondition by washing each column with 5 mL lysis buffer, 5 mL glycine solution, and another 10 mL lysis buffer.
19. Drain the columns and close the bottoms.
20. Resuspend the resin in each column with 1 mL lysis buffer.
21. Collect the slurry from the columns using a blunt tip (we cut the edge of a 1 mL plastic tip with scissors), and divide it between the four tubes containing the lysates.
22. Rotate the tubes gently for 1.5 h at 4°C.
23. Gradually load the content of each tube onto an empty chromatography column and drain the flow-through.

24. Wash each column with 25 mL wash buffer.
25. Wash each column with 2.5 mL lysis buffer.
26. Close the bottom of the columns, and add 2.5 mL lysis buffer containing 0.5 *M* NaCl, to separate the 20S proteasome from the 19S particle.
27. Seal the tubes with the caps and rotate them gently for 1 h at 4°C.
28. Combine the resin from the 4 columns, load onto a single column, and drain the liquid.
29. Wash the resin with 4 mL of lysis buffer.
30. Wash the resin with 20 mL of 0.5 *M* ammonium acetate, pH 7.
31. Elute the 20S proteasome with 3.8 mL of ammonium acetate 0.5 *M*, supplemented with 200 μL FLAG peptide (final concentration of the peptide is 0.5 mg/mL).
32. Concentrate the eluted 20S proteasomes to about 100 μL, using a 30 kDa concentrator. The final concentration of the 20S proteasome should be around 2–4 μ*M*. Aliquot to 2–5 μL/tube, flash-freeze in liquid nitrogen, and store at −80°C.
33. 20S proteasome quality can be assessed as detailed in Section 2.1.4, step 41.

2.3 Purification of the 20S proteasome from HEK293 cells

2.3.1 Special equipment

- Kimble® Kontes® 15 mL All Glass Dounce Tissue Grinder (cat. # 885300-0015).
- 2 Poly-Prep® Chromatography Columns (Bio-Rad cat. # 7311550).

2.3.2 Buffers and reagents

- Dulbecco's Modified Eagle Medium (4.5 g/L D-Glucose) (DMEM + 5), supplemented with 10% fetal calf serum, penicillin–streptomycin, sodium pyruvate, nonessential amino acids (Biological Industries), and MycoZap (Lonza cat. # VZA-2032).
- Lysis buffer (100 mL): Tris–HCl 50 m*M*, pH 7.5, NaCl 150 m*M*, NP40 0.5%, $MgCl_2$ 5 m*M*. Immediately before use, add phosphatase inhibitors: Na-o-vanadate 5 m*M*, Na-pyrophosphate 4 m*M*, and β-glycerophosphate 4 m*M*.
- Wash buffer (50 mL): Lysis buffer, supplemented with 0.2% NP-40.
- Ammonium acetate (25 mL): 0.5 *M*, pH 7.
- Glycine-HCl (10 mL): 100 m*M*, pH 3.5.

- ANTI-FLAG® M2 Affinity Gel, Sigma cat. # A2220 (1 mL packed resin). The resin can be reused 2–3 times. Regeneration is performed according to the manufacturer's instructions.
- FLAG peptide (0.4 mL): (DYKDDDDK) 95% purity, GL Biochem (Shanghai), Ltd. 10 mg/mL, dissolved in ammonium acetate.

2.3.3 General notes prior to starting the purification process

- 20S proteasomes are affinity purified from HEK293 cells stably expressing the C-terminally FLAG-tagged β4-subunit, produced by the lab of Chaim Kahana, Weizmann Institute of Science, Israel.
- In several steps during the protocol (Section 2.2.4, steps 12 and 15) we divide the FLAG resin between two chromatography columns, in order to shorten the time of washing.

2.3.4 Procedure

1. Grow cells to confluency in 20 15-cm culture dishes, in DMEM + 5, at 37°C in a 5% CO_2 humidified atmosphere, and collect cells by trypsinization.
2. Wash the cells with sterile PBS and pellet by gentle centrifugation at 300 × *g* for 3 min, at room temperature.
3. Discard the liquid, snap-freeze the cells in liquid nitrogen, and store at −80°C.
4. On the day of purification, transfer the tubes to ice, and let them stand for 2–3 min.
5. Add 400 μL lysis buffer to each tube, gently resuspend the cells by pipetting the buffer up and down a few times with a pipettor.
6. Pool the resuspended cells in a 15-mL tube, and rotate gently for 10 min at 4°C.
7. Chill the 15 mL glass tissue grinder on ice.
8. Pour the cells into the tissue grinder, and grind by moving the pestle gently and slowly up and down for 10 strokes. Do not let the pestle rise above the level of the liquid, do not introduce air into the tube, and do not foam the contents. Keep the grinder in ice during the entire procedure.
9. Leave the lysate in the tissue grinder for 10 min on ice and grind again as in previous step.
10. Repeat step 9.
11. Centrifuge the lysate at 4°C for 10 min, at 14,000 × *g*, collect the supernatant and keep it on ice.

12. Pipette 1 mL of FLAG resin (~1.2 mL slurry), into two empty chromatography columns.
13. Drain the resin completely and precondition by washing each column with 5 mL lysis buffer, 5 mL glycine solution, and another 10 mL lysis buffer.
14. Drain the columns and close the bottoms.
15. Split the supernatant between the two columns and resuspend the resin gently.
16. Seal the top of the columns with the caps and rotate gently for 1.5 h at 4°C.
17. Drain the columns from the unbound proteins.
18. Wash the columns with 25 mL wash buffer.
19. Wash the columns with 2.5 mL lysis buffer.
20. Close the bottom of the columns and add 2.5 mL lysis buffer containing 0.5 *M* NaCl, to separate the 20S proteasome from the 19S particle.
21. Seal the columns with the caps and gently rotate for 1 h at 4°C.
22. Combine the resin from the 4 columns, load onto a single column, and drain the liquid.
23. Wash the resin with 4 mL of lysis buffer.
24. Wash the resin with 20 mL of 0.5 *M* ammonium acetate, pH 7.
25. Elute the 20S proteasome from each column with 3.8 mL of ammonium acetate 0.5 *M*, supplemented with 200 μL FLAG peptide (final concentration of the peptide is 0.5 mg/mL).
26. Concentrate the eluted 20S proteasomes to about 100 μL, using a 30 kDa concentrator. The final concentration of the 20S proteasome should be around 2–4 μ*M*. Aliquot to 2–5 μL/tube, flash-freeze in liquid nitrogen, and store at −80°C.
27. 20S proteasome quality can be assessed as detailed in Section 2.1.4, step 41.

3. Native mass spectrometry approaches to characterize the 20S proteasome

3.1 General notes

Native MS of large protein complexes is commonly performed using direct infusion of the sample through pulled glass capillaries that are coated with a conductive metal (Kirshenbaum, Michaelevski, & Sharon, 2010). In brief, we typically load 2 μL of sample into a gold-coated capillary and spin it for about 20 s in a minicentrifuge, using a designated capillary holder, to push the sample to the edge of the tip. We then inspect the capillary under a microscope (such as a Leica ES2). If an air bubble is trapped at the edge, we spin the capillary again until the bubble disappears. Since the pulled capillary tip is essentially blocked, we cut it manually under the microscope, using tweezers with a flat gripping surface (EMS #72919-2A). Detailed information about needle preparation, and sample spraying into a mass spectrometer, can be found in Kirshenbaum et al. (2010).

We aim to generate droplets which contain a low number of proteasome assemblies, in order to reduce the chances of artificial dimerization. In theory, this can be achieved by spraying a 1 pmol/μL protein solution from a capillary with an inner diameter of ~1 μm (Wilm & Mann, 1996). Since we cut the tip manually, we cannot precisely control the size of the orifice; therefore, we start by cutting a relatively long tip, and shorten it if required. Generally, we tend to cut longer tips for our QTOF's, whereas in the Orbitrap we use slightly shorter tips.

For the initiation of the spray, the capillary position needs to be carefully optimized. If the spray does not initiate, some backpressure may be applied. If the spray still not does initiate and no signal is obtained, it may be useful to inspect the edge of the tip under the microscope. Sometimes the tip is too long (especially if it is slightly bent, or wiggles) or the gold coating has peeled off. In the latter instance, the tip of the capillary needs to be recut, and shortened slightly. This technique also applies in instances when the signal decays before the end of the sample.

3.2 MS^1 analysis of the intact 20S proteasome

MS^1 constitutes the first step in protein analysis. This type of analysis measures the intact protein complex in order to study its integrity and overall structure. Instrumental parameters need to be specifically tuned to enable high-resolution measurement of large particles and remove all adducts and buffer molecules; at the same time, the analysis should be performed under relatively soft conditions to keep the complex intact. Such conditions include high pressure and gentle activation energies.

3.2.1 Orbitrap instrumental settings for MS^1 analysis

We typically adjust the instrumental parameters to the following: capillary voltage 1.7 kV, at a temperature of 180°C. Fore-vacuum pressure is set to 1.5 mbar, and the trapping gas pressure to 4.1, corresponding to HV pressure of 1.2×10^{-4} mbar, and UHV pressure of 3.4×10^{-10} mbar. We operate the source at a constant energy of 2 V in the flatapole bias and interflatapole lens (no trapping). Bent flatapole DC bias and gradient are set to 2.5 and 35 V, respectively, and the HCD cell is operated at energies between 10 and 50 V. Note that our Q Exactive Plus EMR Orbitrap (Thermo) is a hybrid instrument, which contains an ion trap for in-source dissociation of large protein complexes (Ben-Nissan et al., 2017). Thus, flatapole bias and interflatapole lens voltages will not apply for EMR Orbitrap-based instruments that are lacking the in-source trapping device.

3.2.2 QTOF instrumental settings for MS^1 analysis

Instrumental settings on our Synapt G2 instrument (Waters) for analysis of the 20S proteasome are as follows: the manual backing pump is set to 6–7 mbar, capillary voltage 1.4–1.8 kV, sampling cone 60 V, extraction cone 5 V, cone gas 50 L/h, and purge gas 100 L/h. The trap gas and helium cell are operated at 8 and 120 L/h, respectively, and the trap collision energy is set to low energies, typically around 10 V.

The QStar Elite instrument (Sciex) is operated at a capillary voltage of 1250 V, declustering potential 200 V, focusing potential 250 V, second declustering potential 25 V, and collision energy 25 V.

In general, the output of a mass spectrometry measurement is a spectrum, which is a pattern representing the mass-to-charge ratio (m/z) of the ions vs intensity. The number of charges a protein can accommodate in nESI is not only proportional to its mass, but also to the overall shape of the protein. A narrow distribution of low charges (relative to the mass) is indicative of a folded structure, whereas a broad distribution of high charge states reflects an extended structure (Ben-Nissan & Sharon, 2011). Using the settings described earlier, mass spectra of 20S proteasomes from different eukaryotic organisms (yeast, rat, and human) were measured (Fig. 4A). In all measurements, the proteasome appears as a narrow charge state series centered around 11,000–12,000 m/z, suggesting that the 20S proteasomes adopt a similar compact structure.

For comparison, we also examined the MS^1 spectrum of the prokaryotic 20S proteasome from the archaeal species *T. acidophilum*. The major difference between prokaryotic and eukaryotic 20S proteasomes lies in the diversification of the α- and β-subunits throughout evolution (Humbard & Maupin-Furlow, 2013), but despite this fundamental difference, the essential activities of both particles are similar. MS^1 measurement of the prokaryotic, reconstituted particle is comparable to that of eukaryotic 20S proteasomes. The narrower peaks of the archaeal 20S proteasome, however, reflect the low biological heterogeneity of this bacterial particle.

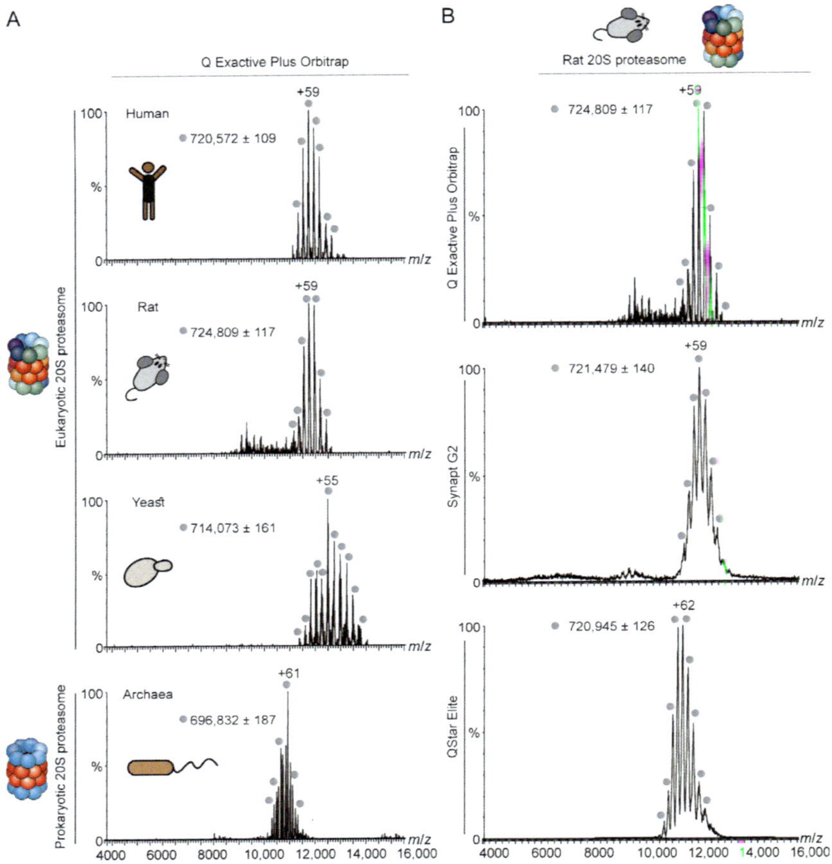

Fig. 4 Native MS analysis of the 20S proteasome. (A) Purified 20S proteasomes from four different organisms (human, rat, yeast, and archaea) were measured on a modified Q Exactive Plus Orbitrap. The three eukaryotic proteasomes are close in mass and charge distribution, indicating that they are markedly similar in overall shape. The prokaryotic proteasome displays a lower mass and additional charges, suggesting that its overall structure is more expanded. (B) The rat 20S proteasome was analyzed on two QTOF platforms (QStar Elite and Synapt G2), and on a high mass range Orbitrap instrument. In all measurements, the 20S proteasome appears as a charge state series around 11,000–12,000 *m/z*, with a mass of ~720 kDa.

Using different mass spectrometers, i.e., the QStar Elite and Synapt G2 QTOF platforms, and the high mass range Orbitrap instrument, spectra of the rat 20S proteasome were measured (Fig. 4B). The obtained charge series correspond in mass to an average of ~720 kDa. This measured mass is in close agreement with the theoretical mass, which ranges between

715,000 and 719,000 Da (based on masses of potential isoforms and proteoforms reported in the Uniprot database, and including major modifications such as removal of the initial methionine, N-terminal acetylation, and phosphorylation). Despite minor changes in charge distribution and basic differences in the resolution afforded by the instruments (Banerjee & Mazumdar, 2012; Krusemark, Frey, Belshaw, & Smith, 2009), the three spectra are essentially similar, demonstrating the stability of the rat 20S proteasome particle. Moreover, in all three measurements, the peaks of the different charge states are broad, reflecting the biological heterogeneity of this 28-subunit protein complex.

3.3 MS^2 analysis of the intact 20S proteasome

In addition to measuring the masses of the intact complex and inspecting the entire collection of complexes in a given sample, we can specifically isolate a particular species of interest in the quadrupole, and further characterize it in a process known as tandem MS. In this type of analysis, the isolated species is activated by energy impartation, until it dissociates into its composing subunits, revealing important layers of information as to its structural organization, assembly, and association of different subunits within the complex. To date, two major activation mechanisms have been utilized in native MS analyses of protein complexes. The more common activation method used in QTOF instruments is collisional-induced dissociation (CID). In CID, the ions are accelerated into a collision cell filled with gas (usually argon), and the multiple collision events lead to the generation of dissociation products (Fig. 2D) (Benesch, 2009). Orbitrap instruments are equipped with a beam-type, higher-energy collision dissociation (HCD) chamber, in which the energy deposition is typically higher than that found in conventional CID (Olsen et al., 2007). Surface-induced dissociation (SID) constitutes an alternative activation method that is based on a single high-energy event, in which the ions collide with an inert surface (Zhou & Wysocki, 2014) (Fig. 2E).

3.3.1 CID and HCD of the 20S proteasome

CID and HCD both result in a gradual dissociation of the isolated complex ions into their composite subunits. In such experiments, peripheral subunits, which are usually more weakly associated with the core of the particle, accumulate internal energy during the activation process, and gradually unfold within the collision cell (Fig. 2D) (Benesch, 2009). When sufficient energy is applied, the unfolded subunit will physically detach from the complex while carrying some of the charge with it, thus enabling its mass measurement, in conjunction with the remaining stripped complex. In CID, the charges of the fragment ions generated are divided in a manner asymmetrically proportional to their mass, generating highly charged monomer ions, and relatively lower charged stripped complexes (Aquilina, Benesch, Bateman, Slingsby, & Robinson, 2003). In HCD, the activation energy is typically higher, and dissociation occurs faster than in CID (Olsen et al., 2007), thus enabling more subunits to dissociate simultaneously, while competing for the same pool of charges from the intact particle.

We usually perform tandem MS experiments by elevating the collision energy in the QStar Elite to 120–150 V, and the HCD energy in the Orbitrap to 140–200 V. As exemplified in Fig. 5, tandem MS analysis of the rat 20S proteasome, performed on the QStar Elite and Orbitrap instruments (Fig. 5A and B, respectively), results in the dissociation of α- and β-subunits.

The high mass accuracy afforded by measurements of the dissociated subunits enables us to assign specific proteoforms to each existing subunit. In this example, in the Qstar Elite we could detect α6 and α7, both missing the initial methionine and acetylated. The specific proteoform identification of these two subunits was confirmed by top-down proteomic analysis, as demonstrated also in Section 3.4. Two populations of stripped complexes, missing either one or two α-subunits, appear at the right side of the *m/z* spectrum. In the Orbitrap, we could detect α2, α4, α6, and α7, all missing the initial methionine and acetylated; α3, missing the initial methionine, acetylated and phosphorylated (Bose, Brooks, Mason, & Rivett, 2001); the acetylated form of α5 and the mature β7-subunit. In addition stripped complexes missing one subunit could also be found.

Due to the higher activation energies available in the HCD-based platform, it is possible to observe dissociation of more subunits, making this analysis very powerful for revealing the major proteoforms of the 20S proteasome, within a given sample. At conditions modified for optimal dissociation (around HCD energy of 180 V), we could detect all seven α-subunits and even three β-subunits, which are buried within the two inner rings of the 20S proteasome, and are usually not detected in CID experiments (Fig. 6). By harnessing the power of the CID method, we have monitored the structural basis for proteasome activation and maturation (Witt et al., 2006) and identified the intermediate stages along the assembly pathway of the 20S particle (Sharon et al., 2007).

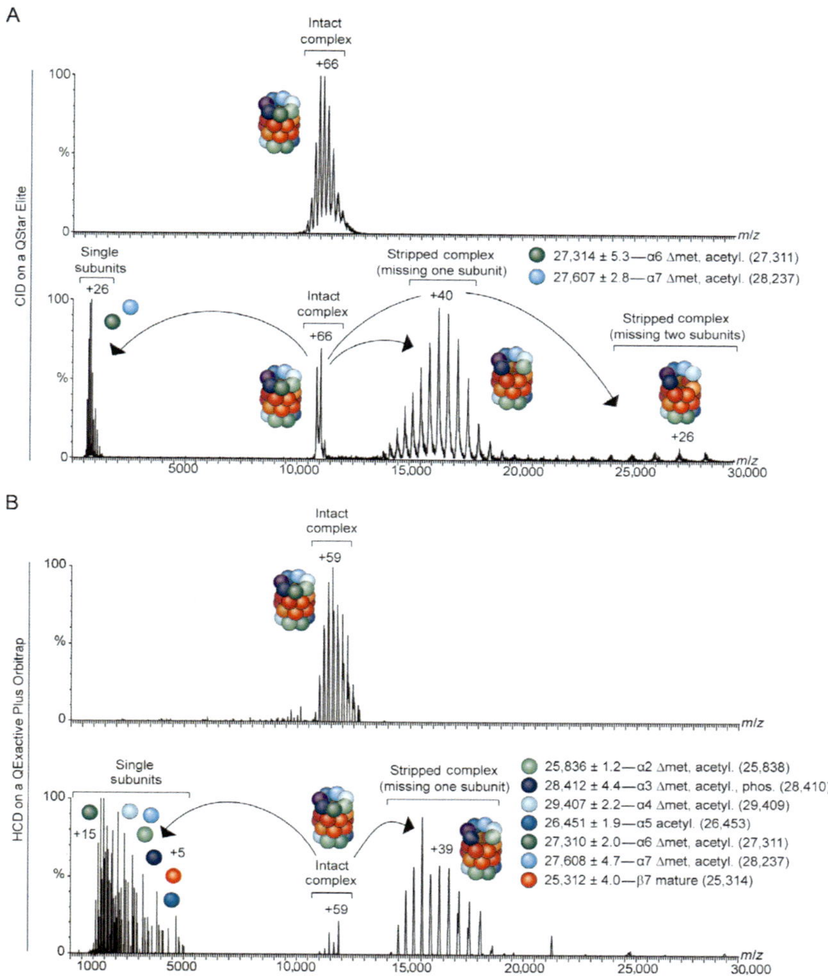

Fig. 5 Activation of the 20S proteasome results in dissociation of single subunits. The 20S proteasome was isolated in the quadrupole and collisionally activated by CID in the Qstar Elite instrument (A), and by HCD in the Orbitrap instrument (B). In both cases, monomeric subunits were detached from the complex, leaving behind stripped complexes lacking in the dissociated subunits. Values in brackets indicate the theoretical masses.

3.3.2 SID analysis of the 20S proteasome

SID is an alternative activation method that is based on a single collision event with an inert surface (Zhou & Wysocki, 2014). In SID experiments, the internal energy absorbed by the protein upon rapidly colliding with the

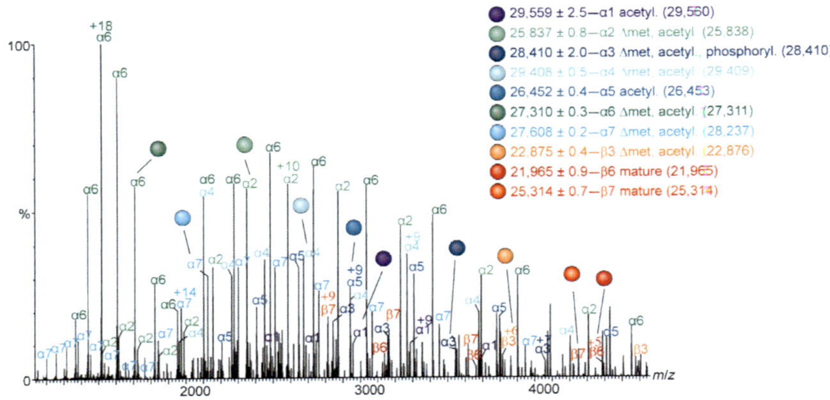

Fig. 6 The 20S proteasome can be dissociated almost fully at high HCD energies. Dissociation of the rat 20S proteasome at optimal conditions in the HCD cell results in the dissociation and detection of 10 subunits. The high mass accuracy enables us to assign the major proteoforms for each subunit. Measured masses for each of the subunits are shown. Values in brackets indicate the theoretical masses.

surface is high, due to the larger relative mass, yielding dissociation pathways with high energetic barriers, and with minimal unfolding and structural rearrangements (Wysocki, Joyce, Jones, & Beardsley, 2008) (Fig. 2E). Thus, SID is typically characterized by a more symmetric dissociation pathway, resulting in the formation of products carrying charges proportional to their mass (Zhou, Huang, & Wysocki, 2012; Zhou & Wysocki, 2014). This type of dissociation results not only in the ejection of monomers but also in the dissociation of the complex into relatively large subcomplexes, typically sharing smaller interfaces (Konijnenberg & Sobott, 2015). Our SID device, developed by the Wysocki group from the Ohio State University (Columbus, Ohio, USA), was installed in our Synapt G2 instrument between the Trap and IMS chambers. To operate the SID, we refer the readers to previously published settings (Sahasrabuddhe et al., 2018).

For efficient SID, we usually add to the sample a charge reducing agent, such as triethylammonium acetate (TEAA), which holds the protein in a more compact conformation. A typical buffer composition would consist of 150 mM ammonium supplemented with TEAA, at a ratio of 0.9/0.1 Ammonium Acetate/TEAA (v/v) (Ma, Loo, & Wysocki, 2015).

Fig. 7 shows the SID dissociation pattern of the rat 20S proteasome. After recording the full spectrum of the intact complex (Fig. 7A), the +43 charge state was isolated and subjected to SID fragmentation at 150 V (Fig. 7B). In this experiment, we used the following voltage setting on the different lenses

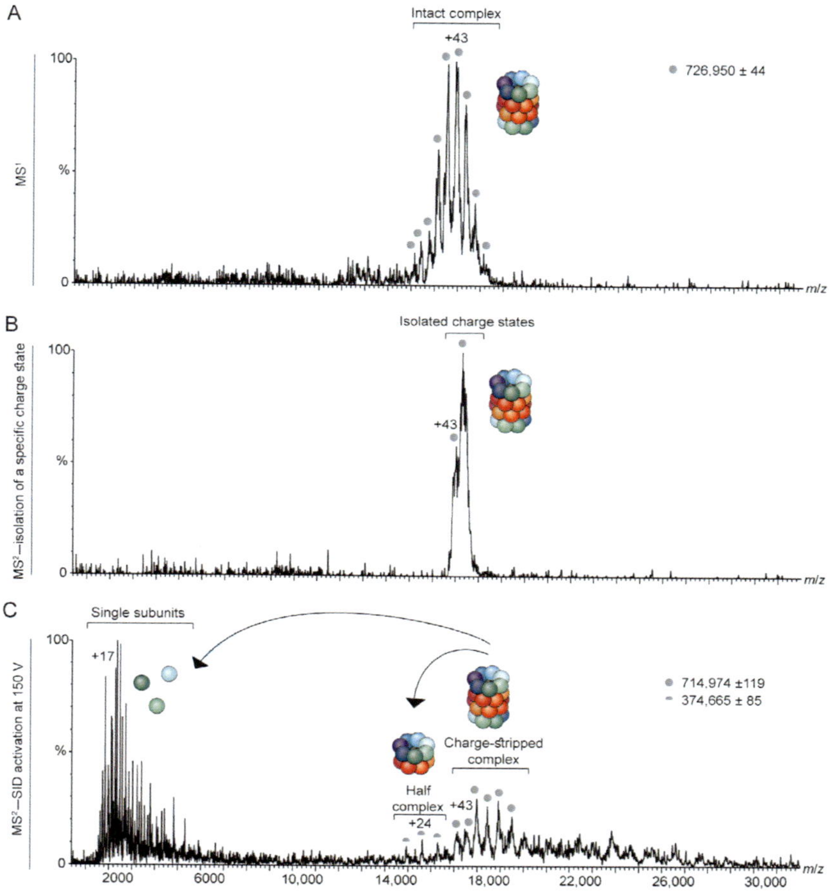

Fig. 7 SID can dissociate the 20S proteasome into half proteasomes. (A) The rat 20S proteasome was mixed with the charge-reducing agent TEAA and measured in a Syanpt G2 instrument which is equipped with a SID chamber, at instrumental conditions optimized for MS^1 analysis. The +43 charge state was isolated (B) and accelerated in the SID chamber at 150V (C). Collision with the surface resulted in dissociation of the 20S proteasome into half proteasomes. Single subunits were also detected.

of the SID device: entrance 1, 93V; entrance 2, −45.5V; front top, −145V; front bottom, 85V; mid bottom, −118.3V; surface, −50V; rear top, −182.6V; rear bottom, −75V; exit 1, −77.2V; exit 2, −75V. The Trap DC bias was set to 195V, resulting in a Trap DC exit of 56V. As a result, we could detect three α-subunits, α2, α6, and α7, each missing the initial methionine and acetylated. Remarkably, in addition to the dissociation of α-subunits, the 20S proteasome also dissociates into half proteasomes (composed of one α ring and one β ring), essentially reflecting the final assembly step of the endogenous complex (Ma et al., 2015; Murata, Yashiroda, & Tanaka, 2009).

3.3.3 Denaturing MS of the 20S proteasome

The profile of the major 20S proteasome proteoforms can be further studied under denaturing conditions. We have developed a liquid chromatography MS-based approach to expose the array of subunit variants that comprise protein complexes (Rozen et al., 2013) (Fig. 2C). In this approach, the intact complex is loaded onto a reverse phase monolithic column prepared in-house, and is then eluted by a gradient of a hydrophobic solution. This results in the stepwise unfolding and release of single subunits, which are then directly sprayed into the instrument, for accurate mass measurements. The eluted flow can be split and part of the flow collected in fractions, using a designated TriVersa NanoMate® robot (Advion) for subsequent proteomic analysis. Data from the intact mass measurements and the proteomic analysis can then be correlated for definite identification of all the proteins in the sample. In Fig. 8, we show an example of the human 20S proteasome,

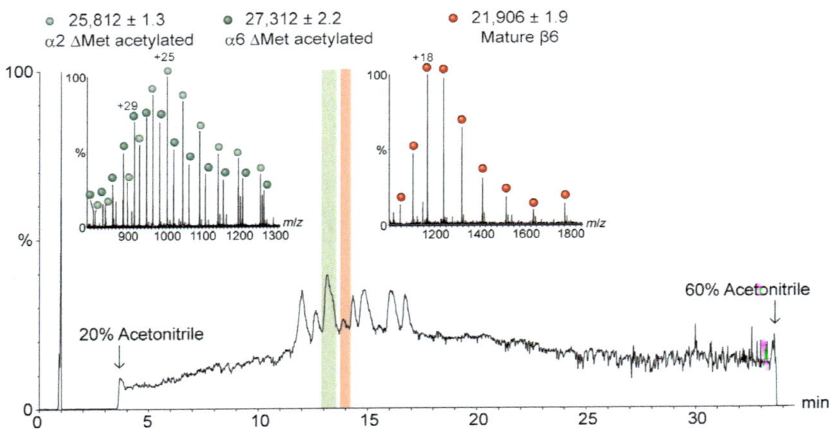

Fig. 8 Analysis of the 20S proteasome under denaturing conditions reveals all the underlying subunits. The human 20S proteasome was loaded onto a monolithic column and denatured with a gradient of 20%–60% acetonitrile over 30 min, at 60°C. The eluted subunits were sprayed directly into a Qstar XL instrument for accurate mass measurement. The total ion intensity chromatogram indicates that the different subunits elute in a series of peaks around 35% acetonitrile. Spectra measured at 13 and 14 min during the gradient (*green* and *red stripes*, respectively) are shown in the *insets*. At 13 min, both the α2- and α6-subunits eluted simultaneously, and at 14 min, the β6-subunit was eluted. Accurate mass measurements indicate the exact proteoform for each subunit.

separated over a gradient of 20%–60% acetonitrile at 60°C, during a period of 30 min, and analyzed by a QStar XL mass spectrometer. Acquisition parameters used in this type of measurement include declustering potential of 50 V, focusing potential 200 V, second declustering potential 25 V, and collision energy 20 V. As can be seen in the total ion chromatogram, the various subunits of the human 20S proteasome elute in a series of peaks around 30% acetonitrile. Inspection of the spectra recorded at the different elution times shows that some subunits elute separately, whereas others coelute. Due to the unfolded structure of the eluted subunits, they acquire a high number of charges and generate a broad series of narrow peaks, which together yield highly accurate mass measurements.

3.4 MS^3 analysis of the 20S proteasome

Another means to elucidate the biological heterogeneity of individual subunits of the 20S proteasome is to apply the MS^3 approach (Fig. 2G). Hybrid instruments equipped with a front-end trap (Belov et al., 2013; Ben-Nissan et al., 2017) can dissociate protein complexes into the subunits comprising them already at the source, enabling the subsequent isolation of single charge states of the dissociated subunits in the quadrupole, for top-down fragmentation and sequencing. Recently, Thermo has launched a Q Exactive UHMR Hybrid Quadrupole-Orbitrap mass spectrometer with such a capability built in (Fort et al., 2017).

Dissociation of a large particle such as the 20S proteasome requires high activation energies. We usually trap ions in cycles of 10 ms, at flatapole bias and interflatapole lens voltages of −220 and 15 V, respectively. Each trapping event is followed by a 200 μs release time, during which the inject flatapole bias and interflatapole lens voltages are set to 20 and 1 V, respectively. In this type of experiment, the other instrument settings need to be changed for detection of small proteins, in order to facilitate transmission and measurement of the dissociated subunits, and their associated subcomplexes. For this purpose, we reduce the trapping gas pressure to 1.5, which results in an HV pressure of 4.5×10^{-5} mbar and UHV pressure of 3.4×10^{-10} mbar. The bent flatapole DC bias and gradient are reduced to 1.8 and 15 V, respectively, and no HCD energy is applied.

Using these settings, we can efficiently dissociate the 20S proteasome at the front end, measure the mass of almost all the α-subunits, and assign their proteoforms (Fig. 9A). To validate the identity of the different subunits and pinpoint their PTMs, we can then isolate charge states of single subunits in

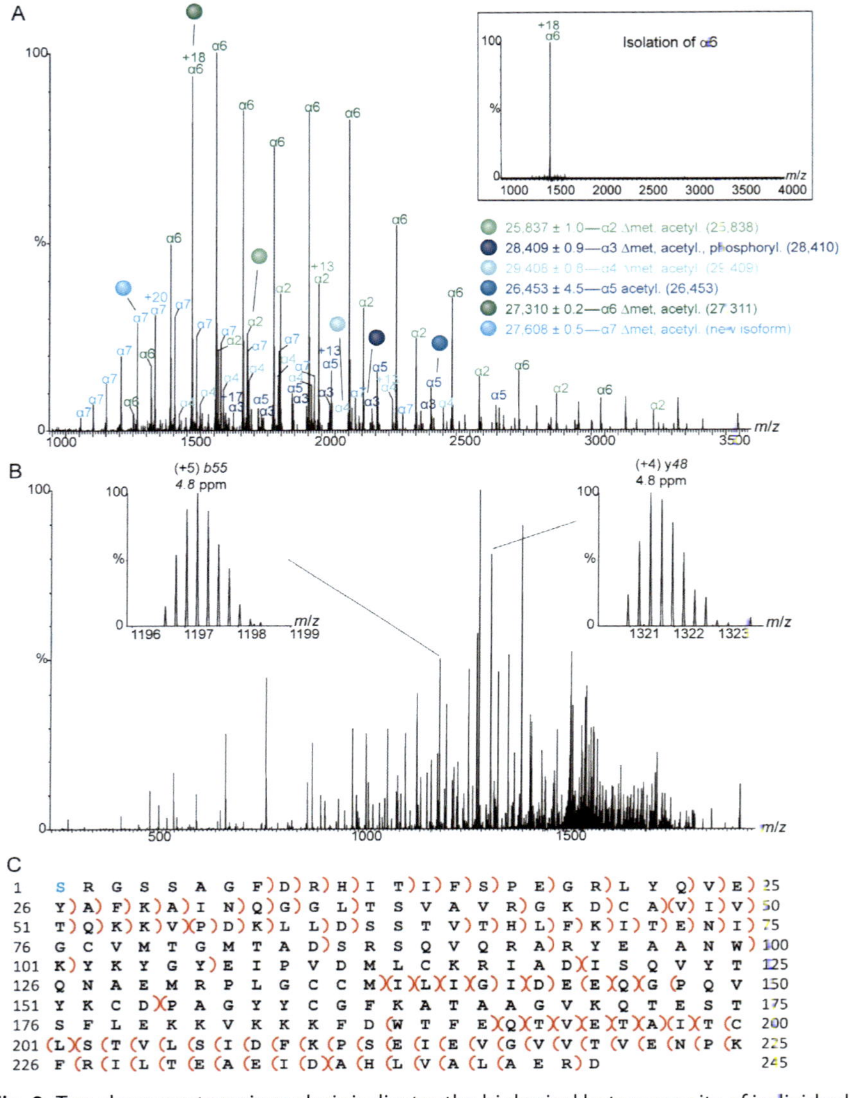

Fig. 9 Top-down proteomic analysis indicates the biological heterogeneity of individual 20S proteasome subunits. (A) The 20S proteasome was subjected to MS^2 activation in the front-end trap, which resulted in the dissociation of six α-subunits. Following dissociation, a single charge state of the α6-subunit was isolated in the quadrupole (*inset*), and subjected to MS^3 fragmentation in the HCD cell, yielding a collection of backbone fragments (B). *Insets* show spectra of two such fragments. (C) Accurate mass measurement of the generated fragments enabled us to determine the sequence of α6 at a coverage of 45%, and confirm that the subunit is missing the initial methionine and contains one acetylation, positioned on the N-terminal serine (labeled in *blue*). Brackets pointing to the *left* and *right* denote the identified *b*-ions and *y*-ions, respectively.

the quadrupole using a narrow isolation window of 5–10 *m*/*z* units (inset in Fig. 9A), and break them into backbone fragments in the HCD cell (Fig. 9B). HCD generates predominantly *b*- and *y*-type fragment ions (corresponding to the N-terminal and C-terminal fragments of a given polypeptide, respectively), which result from the breaking of the peptide amide bond (CO–NH).The HCD energy required for maximal backbone fragmentation varies, and can range between 20 and 80 V, depending on the protein measured. We therefore recommend that spectra be acquired from a range of HCD energies, and the top-down information combined into a comprehensive sequence coverage map. A common fragmentation spectrum may contain between several dozens to several hundreds of fragments (Fig. 9B). Enlarged examples of two such fragments are shown in the insets.

The literature offers various search engines for the assignment of top-down protein sequencing, such as Byonic™ (Protein Metrics, Inc.), MASH Suite Pro (Cai et al., 2016), and ProSight (developed by Proteinaceous, Inc., distributed by Thermo). In our lab, we use ProSight PC. In Fig. 9C, we show the sequence coverage obtained for the +18 charge state of α6, generated from the combined information measured at HCD energies of 30–55 V. The results confirm that α6 is missing the N-terminal methionine, and pinpoint the acetylation to the initial serine (at a sequence coverage of 45%), thus validating the proteoform assignment for this subunit in Figs. 5–7.

3.5 IM-MS analysis of the 20S proteasome

Additional structural information about the protein complex under study can be obtained by combining mass spectrometry with IM, using QTOF-based platforms (Lanucara, Holman, Gray, & Eyers, 2014; Uetrecht, Rose, van Duijn, Lorenzen, & Heck, 2010). IM-MS measures the time it takes for an ion to travel through a tube densely filled with an inert gas (Fig. 2F). The ion's transit time is dependent not only on mass and charge but also on the overall shape of the analyzed protein, so that an assembly with a large surface area will experience more collisions with the gas, and consequently drift more slowly than an assembly with a more compact structure (Fig. 2F). The recorded drift times can be converted into collision cross sections (CCS) which, in turn, can be related to the conformational and topological arrangements of the assembly (Ben-Nissan & Sharon, 2018).

Collision-induced unfolding (CIU) experiments, which couple collisional structural perturbations with ion mobility measurements, enable characterization of the conformation, dynamics, and stability of protein complexes (Niu, Rabuck, & Ruotolo, 2013; Zhong, Han, & Ruotolo, 2014). In this type of experiment, the energy in the collision cell is elevated in a stepwise manner, causing protein activation that may consequently induce conformational changes. The collision voltage at which transitions between conformations occur, the mode of the transition, and the size of intermediates, generate a protein's characteristic unfolding trajectory.

In general, a protein's movement through the mobility cell is highly dependent on the applied wave height, wave velocity, and IMS gas flow rate (Michaelevski, Kirshenbaum, & Sharon, 2010). Manipulating each one of these factors will influence the width of the arrival time distribution (ATD) and shift the drift times. In IM-MS experiments of the 20S proteasome, we apply the aforementioned settings for MS^1 measurements, and typically set the IMS wave velocity to 300 m/s, and the IMS gas flow rate to 60 mL/min. If the mobility plot is dispersed and not symmetrical, we adjust different instrumental settings, as previously described (Michaelevski et al., 2010). We then measure the mobility of the 20S proteasome at three different wave heights, 18, 19, and 20 V. A typical MS spectrum and the corresponding mobilogram of the rat 20S proteasome, measured at a wave height of 20 V, are shown in Fig. 10A and B, respectively. The measurement indicates that the time taken for the 20S proteasome to drift through the mobility cell is around 20–40 ms. In order to extract the CCS value of the particle, we generate a calibration curve by measuring the drift times of known proteins with defined CCS values, and plotting them against their cross sections as previously described (Bush et al., 2010). Note that in order to extract the CCS value of a protein of interest, it is necessary to employ calibrants that are in the same mass range as the protein itself. Furthermore, all measurements of the protein and the calibrant proteins must be performed on the same day, using the exact same instrument conditions.

In this example, we generated a calibration curve based on the CCS values of Alcohol Dehydrogenase (Sigma A3263), Concanavalin A (Sigma C2010) and GroEL (a gift from the lab of Amnon Schwartz, Weizmann Institute of Science, Israel), and calculated the CCS value according to Bush et al. (2010). The calculated CCS of 20,403 ± 66 Å is in good agreement with theoretical CCS values of different eukaryotic 20S proteasomes (*Homo sapiens* PDB # 5LE5, 19,172 Å, *S. cerevisiae*, PDB #19,045 Å,

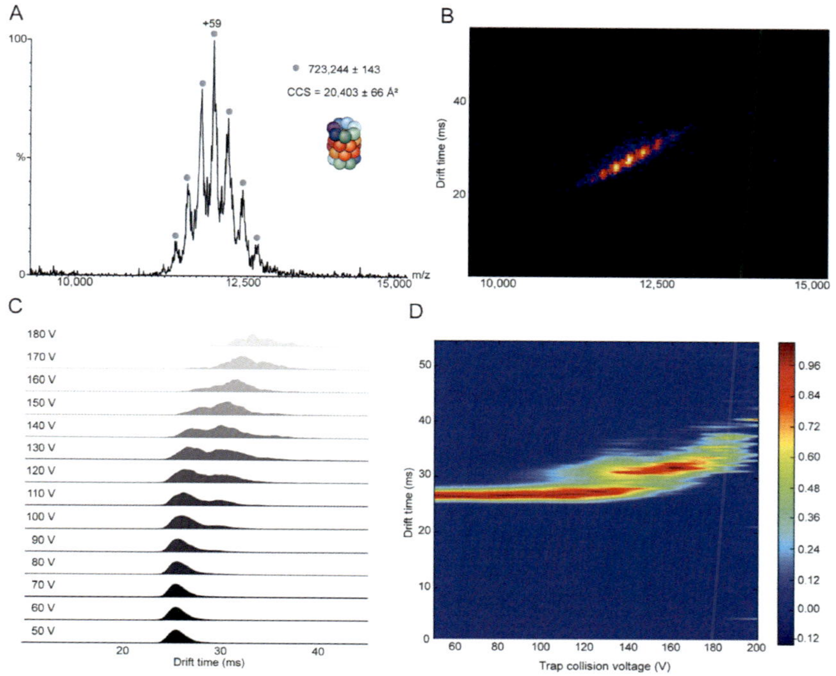

Fig. 10 IM-MS analysis of the 20S proteasome reveals information about the overall shape and stability of the complex. The rat 20S proteasome was measured in a Synapt G2 instrument, at settings optimized for MS^1 analysis. The two-dimensional *m/z* spectrum is shown in (A), and the extracted mobilogram is shown in (B). The calculated CCS is indicated in (A). (C) For CIU analysis, the 20S proteasome was measured over a range of collision energies, gradually elevated in a stepwise manner. The ATDs of the +59 charge state for each collision energy are shown. The gradual activation led to a conformational shift into a more extended structure, starting at 100 V. (D) The CIU contour plot of the rat 20S proteasome indicates a two-step unfolding mechanism. *Colors* denote signal intensity, as indicated.

Bos Taurus, PDB # 1IRU, 19,098 Å), given that the PA algorithm is known to underestimate CCS values by ~13% (D'Atri, Porrini, Rosu, & Gabelica, 2015).

3.6 CIU analysis of the 20S proteasome

IMS measurements can also be harnessed to study the conformational dynamics and stability of protein complexes by tracking their unfolding trajectories at a gradient of trap collision energies (Niu et al., 2013; Zhong et al., 2014). In Fig. 10C, we show the series of ATDs for the +59 charge state of the rat 20S proteasome, which was progressively activated in steps of 5 V in the trap cell, at collision energies between 50 and 180 V. These values were then compiled, using the open-source CIUSuite software (Eschweiler, Rabuck-Gibbons, Tian, & Ruotolo, 2015), to generate a CIU fingerprint (Fig. 10D). Our data indicate that at low collision voltages, the rat 20S proteasome retained its overall compact structure. This configuration persists until reaching a collision voltage of ~155 V; however, already at 100 V a transition occurs to a more extended conformation, giving rise to higher drift-time values. At 160 V, the majority of the particles have taken on the new partially unfolded conformation. At higher collision energies, this conformation gradually increases until the signal decays. Taken together, the results indicate that the 20S proteasome is a highly sturdy particle, which retains its overall structure even at high collision energies, and undergoes two conformational shifts to more extended, but still intact structures during collisional-induced unfolding.

3.7 Interaction of the 20S proteasome with its substrates

Here, we demonstrate how native MS can be utilized to study the relationship between the 20S proteasome and its substrates (Sharon et al., 2006). By itself, the 20S proteasome can degrade unfolded proteins, typically at a high rate (Ben-Nissan & Sharon, 2014; Moscovitz et al., 2015, 2012). In order to measure such a transient interaction between the 20S proteasome and its substrate, the proteasome must first be preincubated with a specific inhibitor. In the following example (Fig. 11), we incubated the *T. acidophilum* 20S proteasome with the clasto-lactacystin-β-lactone active-site inhibitor, prior to mixing it with an unfolded substrate (in this case, the chemically unfolded green fluorescent protein, GFP). To overcome the inherent overlap of charge states between the apo- and substrate-bound proteasomes, we took advantage of the tandem MS approach. As described earlier, the asymmetric dissociation pattern that is a characteristic of CID, leads to the formation of highly charged monomer ions and relatively low charged, stripped complexes. Consequently, the peaks of the stripped complexes display greater separation between the charge states than the parent ions, thus enabling more accurate mass measurements to be performed (Aquilina et al., 2003). Fig. 11 shows a typical tandem MS experiment, in which one charge state of a substrate-occupied 20S proteasome was isolated and subjected to CID at 130 V. Mass measurements indicate that initially, the proteasome is occupied by three GFP molecules. During the first dissociation step, the

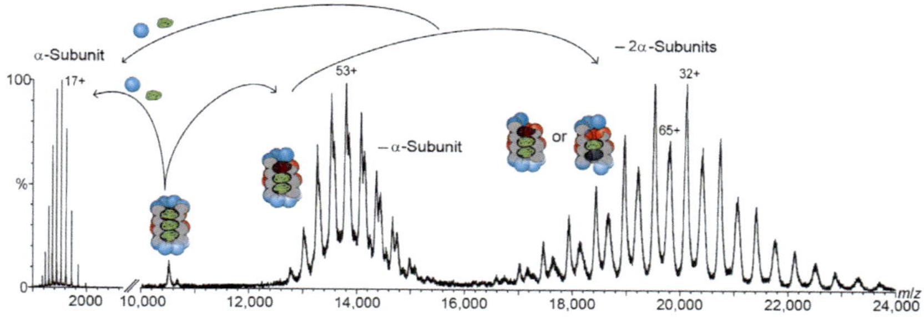

Fig. 11 The 20S proteasome displays the potential to store its substrates for continual degradation. The 20S proteasome from *T. acidophilum* was inhibited with clasto-lactacystin-β-lactone, and reacted with denatured GFP. Following incubation, the complexed 20S proteasome—substrate was measured in a high mass QStar XL instrument and subjected to tandem MS analysis, at a collision energy of 130V. The results show that the 20S proteasome initially binds three GFP molecules. During the first dissociation step, the stripping of one α-subunit exposes the substrate molecule in the antechamber, and triggers its dissociation. In the second dissociation step, an additional α-subunit is lost, either from the same ring or from the opposite one. In the latter case, another substrate molecule is exposed and released. The region between 12,000 and 15,000 *m/z* was magnified threefold. *The figure was adapted with permission from Sharon, M., Witt, S., Felderer, K., Rockel, B., Baumeister, W., & Robinson, C. V. (2006). 20S proteasomes have the potential to keep substrates in store for continual degradation.* The Journal of Biological Chemistry, 281*(14), 9569–9575. doi:10.1074/jbc.M511951200.*

stripping of one α-subunit liberated a caged substrate, presumably from one of the antechambers. In the second dissociation step, another α-subunit was lost; in case this second α-subunit dissociated from the opposite ring, it could expose another substrate molecule and liberate it, too.

These experiments demonstrate that the archaeal 20S proteasome can engage up to three GFP substrates simultaneously, one residing in the central proteolytic chamber, while the other two occupy the two antechambers. Collectively, these findings imply that the 20S proteasome can store substrates prior to their degradation, specifically in cases where translocation rates are slower than proteolysis (Sharon et al., 2006).

3.8 Interaction of the 20S proteasome with its regulators

Degradation of unstructured proteins by the 20S proteasome is a passive process, which does not require energy, and does not rely on specific tagging of the substrate prior to degradation. As such, the 20S proteasome poses significant danger to the cellular protein milieu, which contains a large number of partially and/or fully disordered proteins (Dyson & Wright, 2005; van der

Lee et al., 2014). In recent years, however, we realized that this process does not occur at random, and that the cell contains several proteins with moonlighting activities such as the Parkinson's-associated protein DJ-1 and the oxidoreductase NQO1, which in addition to their previously determined catalytic activities, also inhibit the 20S proteasome (Asher, Tsvetkov, Kahana, & Shaul, 2005; Moscovitz et al., 2015, 2012). The mechanism underlying this inhibitory function of these two proteins remains to be clarified; yet our findings indicate that it involves a physical interaction between the regulator and the 20S proteasome.

Fig. 12 displays the results of tandem MS experiments, performed to examine the interaction between the 20S proteasome and one of its regulators, DJ-1 (Ben-Nissan, Chotiner, Tarnavsky, & Sharon, 2016). In a control reaction, we isolated the charge state series of the 20S proteasome within a window of 2000 m/z units, centered around 12,000 m/z, and elevated the HCD energy to dissociate the complex into its constituent subunits. As a result, we detected several dissociated α-subunits in the lower range of the m/z scale (Fig. 12A). When we performed the same experiment on the 20S proteasome which had been preincubated with DJ-1 (Fig. 12B), in addition to the dissociated α-subunits, we identified another charge state series which corresponds in mass to the DJ-1 monomer. This finding clearly demonstrates that prior to activation in the HCD cell, DJ-1 was physically bound to the 20S proteasome. Similar results were also found for NQO1 (Moscovitz et al., 2012), suggesting for a common inhibitory mechanism for both proteins.

To further investigate the inhibitory effect of the regulators on 20S proteasome activity, we performed degradation assays inside the capillary, and quantitatively tracked the levels of a 20S proteasome substrate over time, in the presence or absence of a 20S proteasome regulator. An example is shown in Fig. 13, where we monitored, in the Q Exactive Plus Orbitrap instrument, the levels of α-synuclein, a known substrate of the 20S proteasome (Moscovitz et al., 2015, 2012). Since the 20S proteasome is a very robust enzyme, relatively insensitive to buffer conditions, we could perform all the degradation assays in 150 mM ammonium acetate. We therefore chose to exchange the buffers of the proteasome, the substrate and the regulator to ammonium acetate, immediately prior to mixing them.

Initially, we preincubated the 20S proteasome with a regulator (CBR3, in this example), at a rate of 1–50 μM, on ice, for 30 min–2 h. We then acquired the spectrum of 10 μM α-synuclein alone, over a period of 10 min each and repeated the measurement three times. Next, we undertook degradation

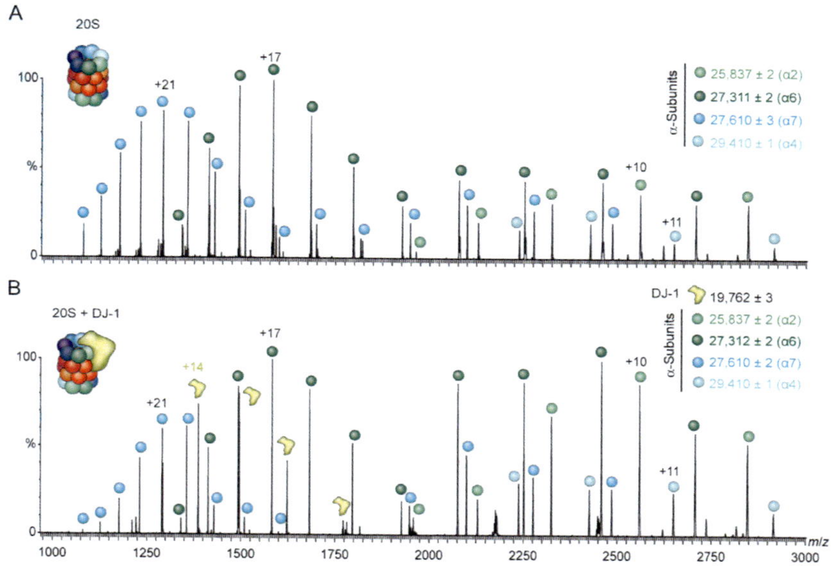

Fig. 12 Tandem MS analysis demonstrates that the 20S proteasome regulator, DJ-1, physically binds the 20S proteasome. The rat 20S proteasome was incubated alone (A), or with DJ-1 (B) overnight at 4°C, and then subjected to native MS analysis in the Q Exactive Plus Orbitrap. The charge states of the 20S proteasome at 12,000 ± 2000 *m/z* were isolated, and subjected to HCD activation at 140 V. The spectra focus on the lower range of the *m/z* spectrum, where the dissociated 20S proteasome subunits appear. The spectrum in (A) shows the dissociation of four α-subunits, whereas the spectrum recorded for the 20S proteasome that was preincubated with DJ-1 (B), shows an additional charge state series, corresponding in mass to the monomeric form of the inhibitor. By extrapolation, we conclude that prior to tandem MS analysis, DJ-1 was physically bound to the 20S proteasome. *The figure was adapted with permission from Ben-Nissan, G., Chotiner, A., Tarnavsky, M., & Sharon, M. (2016). Structural characterization of missense mutations using high resolution mass spectrometry: A case study of the Parkinson's-related protein, DJ-1.* Journal of the American Society for Mass Spectrometry, *27(6), 1062–1070. doi:10.1007/s13361-016-1379-z.*

assays in which we mixed the 20S proteasome with α-synuclein, at final concentrations of 1 and 10 μ*M*, respectively. Since the 20S proteasome is very active, we mix the reactions and immediately spray and record the spectra. Finally, we performed the same experiments using the 20S proteasome, which was preincubated with the regulator.

Since in this case, we focused only on the α-synuclein spectrum, we changed the instrument settings for the measurement of small proteins as follows: we reduced the trapping gas pressure to 1.5 (resulting in an HV pressure of 4.5×10^{-5} mbar, and a UHV pressure of 3.4×10^{-10} mbar), and changed

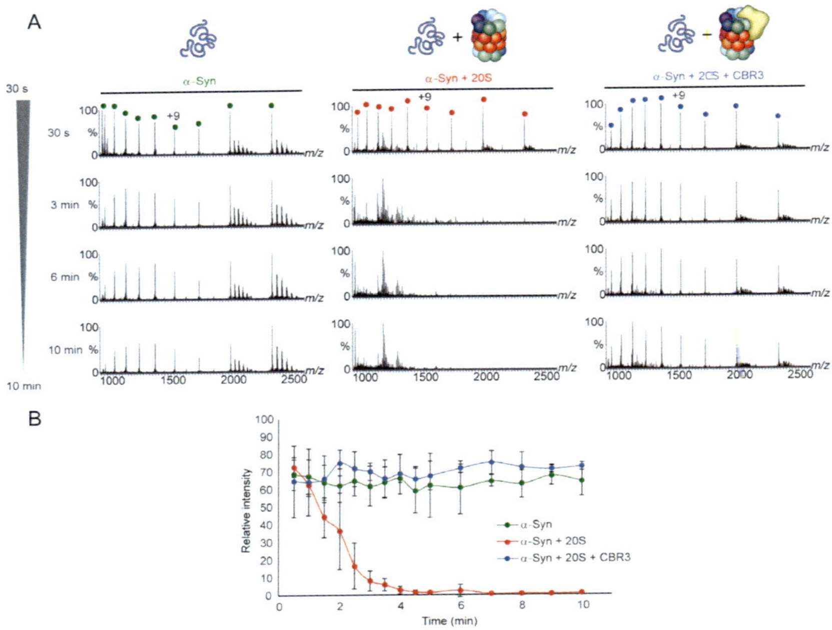

Fig. 13 Native MS can be used to monitor 20S proteasome activity over time. (A) The levels of α-synuclein, a known 20S proteasome substrate (*green circles*), were monitored over a period of 10 min. When mixed with the 20S proteasome, α-synuclein was rapidly degraded (*red circles*), whereas if the 20S proteasome was preincubated with the inhibitor CBR3, its catalytic activity was inhibited, and the levels of α-synuclein remained stable (*blue circles*). Each reaction was mixed and immediately sprayed into the instrument, for continuous measurement over a period of 10 min. Representative spectra are shown. (B) Plots of the different degradation reactions were generated by calculating the intensity of the charge state series of α-synuclein relative to the intensity of the highest peak in each spectrum, at each time point. Each reaction was performed 3 times. Error bars represent standard deviation.

the bent flatapole DC bias and gradient to 1.8 and 15 V, respectively, without applying any HCD energy.

Examples of spectra from the three different reactions, acquired at different time points, are shown in Fig. 13A. On the left column are spectra measured for α-synuclein alone. Since α-synuclein is an unfolded protein, it generates a broad distribution of highly charged peaks, in the lower region of the *m*/*z* spectrum. As can be seen, the overall signal of α-synuclein was stable over the entire 10-min period, and did not decay. Spectra acquired over time for the degradation assay of α-synuclein mixed with the 20S proteasome are shown in the middle panel. At 30 s, the α-synuclein signal was clearly detected and dominated the spectrum, but at 3 min, the relative signal of α-synuclein was essentially at baseline levels. In comparison, when we performed the same experiment using the 20S proteasome which had been preincubated with CBR3 (right panel), the α-synuclein signal intensity remained stable throughout the entire experiment. Such analyses can also be quantified by calculating the intensity of the α-synuclein charge state series relative to the intensity of the highest peak in each spectrum, at each time point, and plotting these values over time to generate corresponding degradation curves (Fig. 13B).

4. Summary and perspective

Our intent in this chapter was to demonstrate how native MS could be used for structural investigations into the 20S proteasome complex. We discussed the currently available native MS methods, and provided guidance as to how to avoid pitfalls in all operations, from initial 20S proteasome purification and sample preparation, to spectra acquisition and analysis. We emphasized how use of different instrument platforms and application of distinct native MS methodologies could expose various layers of information hidden within the proteasome. We anticipate that future improvements in the resolution and sensitivity of native MS instruments will extend the range of potential applications of this approach. For example, it would be interesting to uncover the biological impact of proteasome population heterogeneity, and the dynamic regulation of different proteasome species across different tissues, cellular compartments, cell cycle phases, or in response to various stimuli.

Although we chose to focus on the application of native MS to analysis of the 20S proteasome complex, an array of additional MS approaches creates opportunities for in-depth structural examination of the proteasome system (Chorev, Ben-Nissan, & Sharon, 2015; Olshina et al., 2018). These include approaches such as top-down and quantitative proteomics, which enable identification of different protein isoforms and PTM characteristics (Cox & Mann, 2011; Jensen, 2006; Maiolica, Junger, Ezkurdia, & Aebersold, 2012), hydrogen exchange and radical footprinting, which reveal secondary structures via solvent accessibility (Englander, 2006; Kiselar, Mahaffy, Pollard, Almo, & Chance, 2007; Konermann, Pan, & Liu, 2011; Konermann, Pan, & Stocks, 2011; Wei et al., 2014), and cross-linking experiments which yield information on the protein fold and intersubunit connectivity (Sinz, 2018; Walzthoeni, Leitner, Stengel, & Aebersold, 2013). Ultimately, by integrating data extracted from different MS-derived structural methods, a comprehensive view of the composition, architecture, and dynamics of the proteasome system, including its associated substrates, suppressors, and enhancers, will be attained.

Acknowledgments

M.S. is grateful for the support of a Starting Grant from the European Research Council (ERC) (Horizon 2020)/ERC Grant Agreement No. 636752 and for an Israel Science Foundation (ISF) grant 300/17. M.S. is an incumbent of the Aharon and Ephraim Katzir Memorial Professorial Chair.

References

Aquilina, J. A., Benesch, J. L., Bateman, O. A., Slingsby, C., & Robinson, C. V. (2003). Polydispersity of a mammalian chaperone: Mass spectrometry reveals the population of oligomers in alphaB-crystallin. *Proceedings of the National Academy of Sciences of the United States of America*, *100*(19), 10611–10616. https://doi.org/10.1073/pnas.1932958100.

Asher, G., Tsvetkov, P., Kahana, C., & Shaul, Y. (2005). A mechanism of ubiquitin-independent proteasomal degradation of the tumor suppressors p53 and p73. *Genes & Development*, *19*(3), 316–321. https://doi.org/10.1101/gad.319905.

Banerjee, S., & Mazumdar, S. (2012). Electrospray ionization mass spectrometry: A technique to access the information beyond the molecular weight of the analyte. *International Journal of Analytical Chemistry*, *2012*, 282574. https://doi.org/10.1155/2012/282574.

Belov, M. E., Damoc, E., Denisov, E., Compton, P. D., Horning, S., Makarov, A. A., et al. (2013). From protein complexes to subunit backbone fragments: A multi-stage approach to native mass spectrometry. *Analytical Chemistry*, *85*(23), 11163–11173. https://doi.org/10.1021/ac4029328.

Benesch, J. L. (2009). Collisional activation of protein complexes: Picking up the pieces. *Journal of the American Society for Mass Spectrometry*, *20*(3), 341–348. https://doi.org/10.1016/j.jasms.2008.11.014.

Ben-Nissan, G., Belov, M. E., Morgenstern, D., Levin, Y., Dym, O., Arkind, G., et al. (2017). Triple-stage mass spectrometry unravels the heterogeneity of an endogenous protein complex. *Analytical Chemistry*, *89*(8), 4708–4715. https://doi.org/10.1021/acs.analchem.7b00518.

Ben-Nissan, G., Chotiner, A., Tarnavsky, M., & Sharon, M. (2016). Structural characterization of missense mutations using high resolution mass spectrometry: A case study of the Parkinson's-related protein, DJ-1. *Journal of the American Society for Mass Spectrometry*, *27*(6), 1062–1070. https://doi.org/10.1007/s13361-016-1379-z.

Ben-Nissan, G., & Sharon, M. (2011). Capturing protein structural kinetics by mass spectrometry. *Chemical Society Reviews*, *40*(7), 3627–3637. https://doi.org/10.1039/c1cs15052a.

Ben-Nissan, G., & Sharon, M. (2014). Regulating the 20S proteasome ubiquitin-independent degradation pathway. *Biomolecules*, *4*(3), 862–884. https://doi.org/10.3390/biom4030862.

Ben-Nissan, G., & Sharon, M. (2018). The application of ion-mobility mass spectrometry for structure/function investigation of protein complexes. *Current Opinion in Chemical Biology*, *42*, 25–33. https://doi.org/10.1016/j.cbpa.2017.10.026.

Bose, S., Brooks, P., Mason, G. G., & Rivett, A. J. (2001). Gamma-interferon decreases the level of 26S proteasomes and changes the pattern of phosphorylation. *The Biochemical Journal*, *353*(Pt. 2), 291–297.

Brooks, P., Fuertes, G., Murray, R. Z., Bose, S., Knecht, E., Rechsteiner, M. C., et al. (2000). Subcellular localization of proteasomes and their regulatory complexes in mammalian cells. *The Biochemical Journal*, *346*(Pt. 1), 155–161.

Bush, M. F., Hall, Z., Giles, K., Hoyes, J., Robinson, C. V., & Ruotolo, B. T. (2010). Collision cross sections of proteins and their complexes: A calibration framework and database for gas-phase structural biology. *Analytical Chemistry*, *82*(22), 9557–9565. https://doi.org/10.1021/ac1022953.

Cai, W., Guner, H., Gregorich, Z. R., Chen, A. J., Ayaz-Guner, S., Peng, Y., et al. (2016). MASH suite Pro: A comprehensive software tool for top-down proteomics. *Molecular & Cellular Proteomics*, *15*(2), 703–714. https://doi.org/10.1074/mcp.O115.054387.

Chen, F., Gulbakan, B., Weidmann, S., Fagerer, S. R., Ibanez, A. J., & Zenobi, R. (2016). Applying mass spectrometry to study non-covalent biomolecule complexes. *Mass Spectrometry Reviews*, *35*(1), 48–70. https://doi.org/10.1002/mas.21462.

Chernushevich, I. V., & Thomson, B. A. (2004). Collisional cooling of large ions in electrospray mass spectrometry. *Analytical Chemistry*, *76*(6), 1754–1760. https://doi.org/10.1021/ac035406j.

Chorev, D. S., Ben-Nissan, G., & Sharon, M. (2015). Exposing the subunit diversity and modularity of protein complexes by structural mass spectrometry approaches. *Proteomics*, *15*(16), 2777–2791. https://doi.org/10.1002/pmic.201400517.

Cox, J., & Mann, M. (2011). Quantitative, high-resolution proteomics for data-driven systems biology. *Annual Review of Biochemistry*, *80*, 273–299. https://doi.org/10.1146/annurev-biochem-061308-093216.

D'Atri, V., Porrini, M., Rosu, F., & Gabelica, V. (2015). Linking molecular models with ion mobility experiments. Illustration with a rigid nucleic acid structure. *Journal of Mass Spectrometry*, *50*(5), 711–726. https://doi.org/10.1002/jms.3590.

DeMartino, G. N., & Slaughter, C. A. (1993). Regulatory proteins of the proteasome. *Enzyme & Protein*, *47*(4–6), 314–324.

Dyson, H. J., & Wright, P. E. (2005). Intrinsically unstructured proteins and their functions. *Nature Reviews. Molecular Cell Biology*, *6*(3), 197–208. https://doi.org/10.1038/nrm1589.

Eisele, M. R., Reed, R. G., Rudack, T., Schweitzer, A., Beck, F., Nagy, I., et al. (2018). Expanded coverage of the 26S proteasome conformational landscape reveals mechanisms of peptidase gating. *Cell Reports*, *24*(5), 1301–1315. e1305. https://doi.org/10.1016/j.celrep.2018.07.004.

Englander, S. W. (2006). Hydrogen exchange and mass spectrometry: A historical perspective. *Journal of the American Society for Mass Spectrometry*, *17*(11), 1481–1489. https://doi.org/10.1016/j.jasms.2006.06.006.

Eschweiler, J. D., Rabuck-Gibbons, J. N., Tian, Y., & Ruotolo, B. T. (2015). CIUSuite: A quantitative analysis package for collision induced unfolding measurements of gas-phase protein ions. *Analytical Chemistry*, *87*(22), 11516–11522. https://doi.org/10.1021/acs.analchem.5b03292.

Fabre, B., Lambour, T., Delobel, J., Amalric, F., Monsarrat, B., Burlet-Schiltz, O., et al. (2013). Subcellular distribution and dynamics of active proteasome complexes unraveled by a workflow combining in vivo complex cross-linking and quantitative proteomics. *Molecular & Cellular Proteomics*, *12*(3), 687–699. https://doi.org/10.1074/mcp.M112.023317.

Fabre, B., Lambour, T., Garrigues, L., Ducoux-Petit, M., Amalric, F., Monsarrat, B., et al. (2014). Label-free quantitative proteomics reveals the dynamics of proteasome complexes composition and stoichiometry in a wide range of human cell lines. *Journal of Proteome Research*, *13*(6), 3027–3037. https://doi.org/10.1021/pr500193k.

Fort, K. L., van de Waterbeemd, M., Boll, D., Reinhardt-Szyba, M., Belov, M. E., Sasaki, E., et al. (2017). Expanding the structural analysis capabilities on an Orbitrap-based mass spectrometer for large macromolecular complexes. *Analyst*, *143*(1), 100–105. https://doi.org/10.1039/c7an01629h.

Glickman, M. H., & Ciechanover, A. (2002). The ubiquitin-proteasome proteolytic pathway: Destruction for the sake of construction. *Physiological Reviews*, *82*(2), 373–428. https://doi.org/10.1152/physrev.00027.2001.

Goldberg, A. L. (2003). Protein degradation and protection against misfolded or damaged proteins. *Nature*, *426*(6968), 895–899. https://doi.org/10.1038/nature02263.

Groettrup, M., Kirk, C. J., & Basler, M. (2010). Proteasomes in immune cells: More than peptide producers? *Nature Reviews. Immunology*, *10*(1), 73–78. https://doi.org/10.1038/nri2687.

Groll, M., Ditzel, L., Löwe, J., Stock, D., Bochtler, M., Bartunik, H. D., et al. (1997). Structure of 20S proteasome from yeast at 2.4Å resolution. *Nature*, *386*, 463–471.

Harshbarger, W., Miller, C., Diedrich, C., & Sacchettini, J. (2015). Crystal structure of the human 20S proteasome in complex with carfilzomib. *Structure*, *23*(2), 418–424. https://doi.org/10.1016/j.str.2014.11.017.

Huang, X., Luan, B., Wu, J., & Shi, Y. (2016). An atomic structure of the human 26S proteasome. *Nature Structural & Molecular Biology*, *23*(9), 778–785. https://doi.org/10.1038/nsmb.3273.

Huang, R., Perez, F., & Kay, L. E. (2017). Probing the cooperativity of *Thermoplasma acidophilum* proteasome core particle gating by NMR spectroscopy. *Proceedings of the National Academy of Sciences of the United States of America*, *114*(46), E9846–E9854. https://doi.org/10.1073/pnas.1712297114.

Huber, E. M., Basler, M., Schwab, R., Heinemeyer, W., Kirk, C. J., Groettrup, M., et al. (2012). Immuno- and constitutive proteasome crystal structures reveal differences in substrate and inhibitor specificity. *Cell*, *148*(4), 727–738. https://doi.org/10.1016/j.cell.2011.12.030.

Humbard, M. A., & Maupin-Furlow, J. A. (2013). Prokaryotic proteasomes: Nanocompartments of degradation. *Journal of Molecular Microbiology and Biotechnology*, *23*(4–5), 321–334. https://doi.org/10.1159/000351348.

Jensen, O. N. (2006). Interpreting the protein language using proteomics. *Nature Reviews. Molecular Cell Biology*, *7*(6), 391–403. https://doi.org/10.1038/nrm1939.

Kirshenbaum, N., Michaelevski, I., & Sharon, M. (2010). Analyzing large protein complexes by structural mass spectrometry. *Journal of Visualized Experiments*, (40), e1954. https://doi.org/10.3791/1954.

Kiselar, J. G., Mahaffy, R., Pollard, T. D., Almo, S. C., & Chance, M. R. (2007). Visualizing Arp2/3 complex activation mediated by binding of ATP and WASp using structural mass spectrometry. *Proceedings of the National Academy of Sciences of the United States of America*, *104*(5), 1552–1557. https://doi.org/10.1073/pnas.0605380104.

Kniepert, A., & Groettrup, M. (2014). The unique functions of tissue-specific proteasomes. *Trends in Biochemical Sciences*, *39*(1), 17–24. https://doi.org/10.1016/j.tibs.2013.10.004.

Konermann, L. (2017). Addressing a common misconception: Ammonium acetate as neutral pH "buffer" for native electrospray mass spectrometry. *Journal of the American Society for Mass Spectrometry*, *28*(9), 1827–1835. https://doi.org/10.1007/s13361-017-1739-3.

Konermann, L., Pan, J., & Liu, Y. H. (2011). Hydrogen exchange mass spectrometry for studying protein structure and dynamics. *Chemical Society Reviews*, *40*(3), 1224–1234. https://doi.org/10.1039/c0cs00113a.

Konermann, L., Pan, Y., & Stocks, B. B. (2011). Protein folding mechanisms studied by pulsed oxidative labeling and mass spectrometry. *Current Opinion in Structural Biology*, *21*(5), 634–640. https://doi.org/10.1016/j.sbi.2011.05.004.

Konijnenberg, A., & Sobott, F. (2015). Bouncing off the walls: Excited protein complexes tell their story. *Chemistry & Biology*, *22*(5), 563–564. https://doi.org/10.1016/j.chembiol.2015.05.004.

Krusemark, C. J., Frey, B. L., Belshaw, P. J., & Smith, L. M. (2009). Modifying the charge state distribution of proteins in electrospray ionization mass spectrometry by chemical derivatization. *Journal of the American Society for Mass Spectrometry, 20*(9), 1617–1625. https://doi.org/10.1016/j.jasms.2009.04.017.
Kunjappu, M. J., & Hochstrasser, M. (2014). Assembly of the 20S proteasome. *Biochimica et Biophysica Acta, 1843*(1), 2–12. https://doi.org/10.1016/j.bbamcr.2013.03.008.
Lanucara, F., Holman, S. W., Gray, C. J., & Eyers, C. E. (2014). The power of ion mobility-mass spectrometry for structural characterization and the study of conformational dynamics. *Nature Chemistry, 6*(4), 281–294. https://doi.org/10.1038/nchem.1889.
Latham, M. P., Sekhar, A., & Kay, L. E. (2014). Understanding the mechanism of proteasome 20S core particle gating. *Proceedings of the National Academy of Sciences of the United States of America, 111*(15), 5532–5537. https://doi.org/10.1073/pnas.1322079111.
Li, Y., Tomko, R. J., Jr., & Hochstrasser, M. (2015). Proteasomes: Isolation and activity assays. *Current Protocols in Cell Biology, 67*, 3.43.1–3.43.20. https://doi.org/10.1002/0471143030.cb0343s67.
Liko, I., Allison, T. M., Hopper, J. T., & Robinson, C. V. (2016). Mass spectrometry guided structural biology. *Current Opinion in Structural Biology, 40*, 136–144. https://doi.org/10.1016/j.sbi.2016.09.008.
Lossl, P., van de Waterbeemd, M., & Heck, A. J. (2016). The diverse and expanding role of mass spectrometry in structural and molecular biology. *The EMBO Journal, 35*(24), 2634–2657. https://doi.org/10.15252/embj.201694818.
Ma, X., Loo, J. A., & Wysocki, V. H. (2015). Surface induced dissociation yields substructure of *Methanosarcina thermophila* 20S proteasome complexes. *International Journal of Mass Spectrometry, 377*, 201–204.
Ma, C. P., Slaughter, C. A., & DeMartino, G. N. (1992). Identification, purification, and characterization of a protein activator (PA28) of the 20 S proteasome (macropain). *The Journal of Biological Chemistry, 267*(15), 10515–10523.
Maiolica, A., Junger, M. A., Ezkurdia, I., & Aebersold, R. (2012). Targeted proteome investigation via selected reaction monitoring mass spectrometry. *Journal of Proteomics, 75*(12), 3495–3513. https://doi.org/10.1016/j.jprot.2012.04.048.
Michaelevski, I., Kirshenbaum, N., & Sharon, M. (2010). T-wave ion mobility-mass spectrometry: Basic experimental procedures for protein complex analysis. *Journal of Visualized Experiments,* (41), e1985. https://doi.org/10.3791/1985.
Moscovitz, O., Ben-Nissan, G., Fainer, I., Pollack, D., Mizrachi, L., & Sharon, M. (2015). The Parkinson's-associated protein DJ-1 regulates the 20S proteasome. *Nature Communications, 6*, 6609. https://doi.org/10.1038/ncomms7609.
Moscovitz, O., Tsvetkov, P., Hazan, N., Michaelevski, I., Keisar, H., Ben-Nissan, G., et al. (2012). A mutually inhibitory feedback loop between the 20S proteasome and its regulator, NQO1. *Molecular Cell, 47*(1), 76–86. https://doi.org/10.1016/j.molcel.2012.05.049.
Murata, S., Yashiroda, H., & Tanaka, K. (2009). Molecular mechanisms of proteasome assembly. *Nature Reviews. Molecular Cell Biology, 10*(2), 104–115. https://doi.org/10.1038/nrm2630.
Niu, S., Rabuck, J. N., & Ruotolo, B. T. (2013). Ion mobility-mass spectrometry of intact protein–ligand complexes for pharmaceutical drug discovery and development. *Current Opinion in Chemical Biology, 17*(5), 809–817. https://doi.org/10.1016/j.cbpa.2013.06.019.
Olsen, J. V., Macek, B., Lange, O., Makarov, A., Horning, S., & Mann, M. (2007). Higher-energy C-trap dissociation for peptide modification analysis. *Nature Methods, 4*(9), 709–712. https://doi.org/10.1038/nmeth1060.
Olshina, M. A., Ben-Nissan, G., & Sharon, M. (2018). Functional regulation of proteins by 20S proteasome proteolytic processing. *Cell Cycle, 17*(4), 393–394. https://doi.org/10.1080/15384101.2017.1414682.

Rose, R. J., Damoc, E., Denisov, E., Makarov, A., & Heck, A. J. (2012). High-sensitivity Orbitrap mass analysis of intact macromolecular assemblies. *Nature Methods*, *9*(11), 1084–1086. https://doi.org/10.1038/nmeth.2208.

Rozen, S., Tieri, A., Ridner, G., Stark, A. K., Schmaler, T., Ben-Nissan, G. et al. (2013). Exposing the subunit diversity within protein complexes: A mass spectrometry approach. *Methods*, *59*(3), 270–277. https://doi.org/10.1016/j.ymeth.2012.12.013.

Ruschak, A. M., & Kay, L. E. (2012). Proteasome allostery as a population shift between interchanging conformers. *Proceedings of the National Academy of Sciences of the United States of America*, *109*(50), E3454–E3462. https://doi.org/10.1073/pnas.1213640109.

Ruschak, A. M., Religa, T. L., Breuer, S., Witt, S., & Kay, L. E. (2010). The proteasome antechamber maintains substrates in an unfolded state. *Nature*, *467*(7317), 868–871. https://doi.org/10.1038/nature09444.

Sadre-Bazzaz, K., Whitby, F. G., Robinson, H., Formosa, T., & Hill, C. P. (2010). Structure of a Blm10 complex reveals common mechanisms for proteasome binding and gate opening. *Molecular Cell*, *37*(5), 728–735. https://doi.org/10.1016/j.molcel.2010.02.002.

Sahasrabuddhe, A., Hsia, Y., Busch, F., Sheffler, W., King, N. P., Baker, D., et al. (2018). Confirmation of intersubunit connectivity and topology of designed protein complexes by native MS. *Proceedings of the National Academy of Sciences of the United States of America*, *115*(6), 1268–1273. https://doi.org/10.1073/pnas.1713646115.

Schmidt, M., & Finley, D. (2014). Regulation of proteasome activity in health and disease. *Biochimica et Biophysica Acta*, *1843*(1), 13–25. https://doi.org/10.1016/j.bbamcr.2013.08.012.

Schweitzer, A., Aufderheide, A., Rudack, T., Beck, F., Pfeifer, G., Plitzko, J. M., et al. (2016). Structure of the human 26S proteasome at a resolution of 3.9 Å. *Proceedings of the National Academy of Sciences of the United States of America*, *113*(28), 7816–7821. https://doi.org/10.1073/pnas.1608050113.

Sharon, M. (2013). Biochemistry. Structural MS pulls its weight. *Science*, *340*(6136), 1059–1060. https://doi.org/10.1126/science.1236303.

Sharon, M., Witt, S., Felderer, K., Rockel, B., Baumeister, W., & Robinson, C. V. (2006). 20S proteasomes have the potential to keep substrates in store for continual degradation. *The Journal of Biological Chemistry*, *281*(14), 9569–9575. https://doi.org/10.1074/jbc.M511951200.

Sharon, M., Witt, S., Glasmacher, E., Baumeister, W., & Robinson, C. V. (2007). Mass spectrometry reveals the missing links in the assembly pathway of the bacterial 20S proteasome. *The Journal of Biological Chemistry*, *282*(25), 18448–18457. https://doi.org/10.1074/jbc.M701534200.

Sinz, A. (2018). Cross-linking/mass spectrometry for studying protein structures and protein-protein interactions: Where are we now and where should we go from here? *Angewandte Chemie (International Ed. in English)*, *57*(22), 6390–6396. https://doi.org/10.1002/anie.201709559.

Sobott, F., & Robinson, C. V. (2002). Protein complexes gain momentum. *Current Opinion in Structural Biology*, *12*(6), 729–734.

Solomon, H., Brauning, B., Fainer, I., Ben-Nissan, G., Rabani, S., Goldfinger, N., et al. (2017). Post-translational regulation of p53 function through 20S proteasome-mediated cleavage. *Cell Death and Differentiation*, *24*(12), 2187–2198. https://doi.org/10.1038/cdd.2017.139.

Tsvetkov, P., Myers, N., Moscovitz, O., Sharon, M., Prilusky, J., & Shaul, Y. (2012). Thermo-resistant intrinsically disordered proteins are efficient 20S proteasome substrates. *Molecular BioSystems*, *8*(1), 368–373. https://doi.org/10.1039/c1mb05283g.

Uetrecht, C., Rose, R. J., van Duijn, E., Lorenzen, K., & Heck, A. J. (2010). Ion mobility mass spectrometry of proteins and protein assemblies. *Chemical Society Reviews*, *39*(5), 1633–1655. https://doi.org/10.1039/b914002f.

Unno, M., Mizushima, T., Morimoto, Y., Tomisugi, Y., Tanaka, K., Yasuoka, N., et al. (2002). The structure of the mammalian 20S proteasome at 2.75 Å resolution. *Structure*, *10*(5), 609–618.

van der Lee, R., Buljan, M., Lang, B., Weatheritt, R. J., Daughdrill, G. W., Dunker, A. K., et al. (2014). Classification of intrinsically disordered regions and proteins. *Chemical Reviews*, *114*(13), 6589–6631. https://doi.org/10.1021/cr400525m.

Verma, R., Chen, S., Feldman, R., Schieltz, D., Yates, J., Dohmen, J., et al. (2000). Proteasomal proteomics: Identification of nucleotide-sensitive proteasome-interacting proteins by mass spectrometric analysis of affinity-purified proteasomes. *Molecular Biology of the Cell*, *11*(10), 3425–3439. https://doi.org/10.1091/mbc.11.10.3425.

Voges, D., Zwickl, P., & Baumeister, W. (1999). The 26S proteasome: A molecular machine designed for controlled proteolysis. *Annual Review of Biochemistry*, *68*, 1015–1068. https://doi.org/10.1146/annurev.biochem.68.1.1015.

Walzthoeni, T., Leitner, A., Stengel, F., & Aebersold, R. (2013). Mass spectrometry supported determination of protein complex structure. *Current Opinion in Structural Biology*, *23*(2), 252–260. https://doi.org/10.1016/j.sbi.2013.02.008.

Wei, H., Mo, J., Tao, L., Russell, R. J., Tymiak, A. A., Chen, G., et al. (2014). Hydrogen/deuterium exchange mass spectrometry for probing higher order structure of protein therapeutics: Methodology and applications. *Drug Discovery Today*, *19*(1), 95–102. https://doi.org/10.1016/j.drudis.2013.07.019.

Wilm, M., & Mann, M. (1996). Analytical properties of the nanoelectrospray ion source. *Analytical Chemistry*, *68*(1), 1–8.

Witt, S., Kwon, Y. D., Sharon, M., Felderer, K., Beuttler, M., Robinson, C. V., et al. (2006). Proteasome assembly triggers a switch required for active-site maturation. *Structure*, *14*(7), 1179–1188. https://doi.org/10.1016/j.str.2006.05.019.

Wysocki, V. H., Joyce, K. E., Jones, C. M., & Beardsley, R. L. (2008). Surface-induced dissociation of small molecules, peptides, and non-covalent protein complexes. *Journal of the American Society for Mass Spectrometry*, *19*(2), 190–208. https://doi.org/10.1016/j.jasms.2007.11.005.

Zhong, Y., Han, L., & Ruotolo, B. T. (2014). Collisional and Coulombic unfolding of gas-phase proteins: High correlation to their domain structures in solution. *Angewandte Chemie (International Ed. in English)*, *53*(35), 9209–9212. https://doi.org/10.1002/anie.201403784.

Zhou, M., Huang, C., & Wysocki, V. H. (2012). Surface-induced dissociation of ion mobility-separated noncovalent complexes in a quadrupole/time-of-flight mass spectrometer. *Analytical Chemistry*, *84*(14), 6016–6023. https://doi.org/10.1021/ac300810u.

Zhou, M., & Wysocki, V. H. (2014). Surface induced dissociation: Dissecting noncovalent protein complexes in the gas phase. *Accounts of Chemical Research*, *47*(4), 1010–1018. https://doi.org/10.1021/ar400223t.

Zwickl, P., Lottspeich, F., & Baumeister, W. (1992). Expression of functional *Thermoplasma acidophilum* proteasomes in *Escherichia coli*. *FEBS Letters*, *312*(2–3), 157–160.

CHAPTER TEN

Single-molecule methods for measuring ubiquitination and protein stability

Jason Hon, Ying Lu*
Department of Systems Biology, Harvard Medical School, Boston, MA, United States
*Corresponding author: e-mail address: ying_lu@hms.harvard.edu

Contents

Abstract

The ubiquitin-proteasome system (UPS) contributes to changes in cell state and homeostatic maintenance in humans by modulating the stability of about a third of human proteins. For example, cell-cycle regulation requires a central ubiquitin ligase, the anaphase-promoting complex/cyclosome (APC/C), which starts a ubiquitination cascade leading to the degradation of multiple targets. This targeted degradation is mediated by the 26S proteasome, a 2.5-MDa protein complex, which recognizes and

Methods in Enzymology, Volume 619
ISSN 0076-6879
https://doi.org/10.1016/bs.mie.2018.12.031
© 2019 Elsevier Inc. All rights reserved.

degrades ubiquitinated proteins at rates partially controlled by the variations in ubiquitin chain topology. Substrate selectivity of ubiquitin ligases such as the APC/C and of the 26S proteasome from pools of near-identical targets reflects highly regulated kinetic mechanisms. Single-molecule techniques are powerful tools that allow distinction between differential substrate affinities and identification of reaction intermediates in complex mixtures. Here we describe fluorescence-based single-molecule imaging of in vitro ubiquitination reactions catalyzed by the APC/C and ubiquitin-dependent degradation reactions catalyzed by the 26S proteasome.

1. Introduction

The ubiquitin-proteasome system (UPS) regulates the stability of about a third of human proteins. Understanding how the UPS selectively targets the right substrates for degradation is important for studying the cellular processes controlled through protein degradation and diseases due to misregulation of protein stability. Central to the functionality of the UPS is its unique reaction network, which involves poorly characterized reaction kinetics and intermediates. Here, we describe a single-molecule fluorescence method for investigating the mechanism of protein ubiquitination and proteasomal degradation.

Target proteins are processed by the UPS first by direct ligation of the ubiquitin protein to primary amines on the target surface. In order to achieve this, the C-terminal carboxylate of ubiquitin (Ub) is first activated by the E1 enzyme, which transfers the activated Ub onto the active-site cysteine of an E2 enzyme, forming an E2–Ub thioester bond. About 600 E3 enzymes in human cells determine the substrate specificity and mediate the transfer of Ub from E2s to their targets (Hochstrasser, 1996; Pickart, 2001) (Fig. 1A). Ub molecules often form chains, which can have a variety of Ub–Ub linkages; multiple Ub chains can be conjugated on the same substrate molecule, resulting in complex configurations, sometimes called the "ubiquitin code" (Komander & Rape, 2012). A large number of protein factors containing Ub-binding domains serve as receptors that decode the information in these complex Ub configurations and direct substrates to their appropriate fates. As an example in cell-cycle regulation, a central E3 enzyme, the anaphase-promoting complex/cyclosome (APC/C), promotes a wave of ubiquitination and degradation of about 50 targets (Peters, 2006). The timing of degradation for each substrate is highly choreographed, and this is critical for the proper sequential execution of

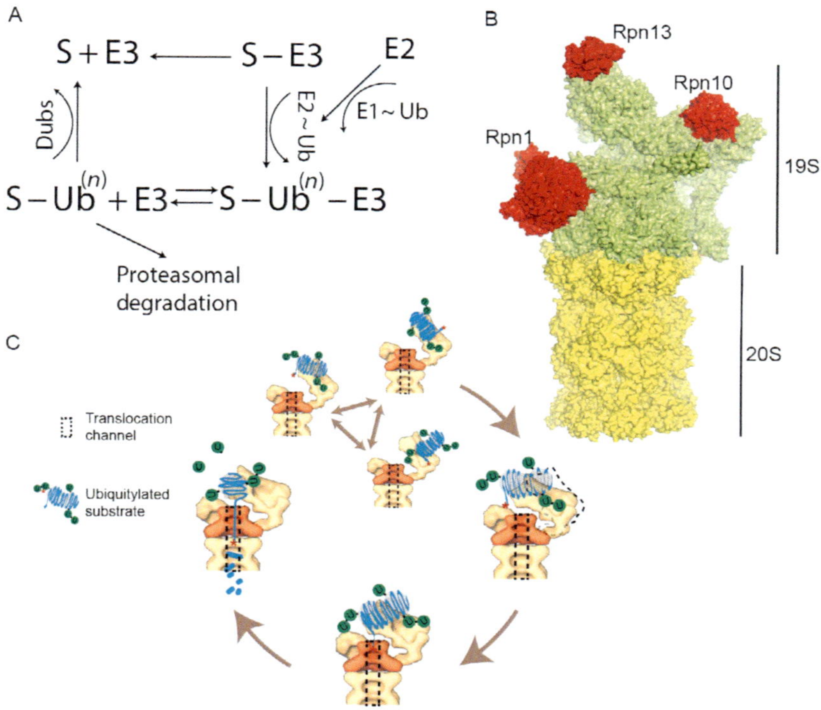

Fig. 1 (A) Schematic of the reactions in the ubiquitin-proteasome systems. *S*, substrate. (B) The 26S proteasome consists of a 20S proteolytic core particle (*yellow*) as well as a 19S regulatory particle which recognizes and unfolds substrates. The ubiquitin receptors Rpn1, Rpn10, and Rpn13 on the regulatory particle are shown in *red*. (C) Schema of the cycle whereby ubiquitinated substrates are recognized, unfolded, and degraded by the 26S proteasome. *Reproduced from Lu, Y., Lee, B. H., King, R. W., Finley, D., & Kirschner, M. W. (2015). Substrate degradation by the proteasome: A single-molecule kinetic analysis.* Science, *348(6231), 1250834. doi:10.1126/science.1250834 and from Lu, Y., Wang, W., & Kirschner, M. W. (2015). Specificity of the anaphase-promoting complex: A single-molecule study.* Science, *348(6231), 1248737. doi:10.1126/science.1248737. Reprinted with permission from AAAS.*

events during mitosis. The relative timing stems from the differential ubiquitination and degradation kinetics of each substrate (Rape, Reddy, & Kirschner, 2006). Short recognition motifs, such as D and KEN boxes, are necessary, but not sufficient, for accurate control of ubiquitination by APC/C, which also depends on specific kinetic mechanisms in the reaction network (Lu, Wang, & Kirschner, 2015).

The 26S proteasome, a 2.5-MDa protein complex, recognizes ubiquitinated proteins and degrades them at rates reflecting their Ub configurations, playing an important role in determining the overall specificity of protein degradation (Finley, 2009; Saeki, 2017) (Fig. 1B). The proteasome has twofold symmetry, with the core particle (CP, also called the 20S proteasome) capped on its two ends by regulatory particles (RP, also called the 19S regulator) (Chen et al., 2016). The CP harbors six proteolytic sites in a central chamber to which access is controlled by two gated orifices at the ends of the CP cylinder. After binding to the Ub receptors in the RP, which include Rpn10, Rpn1, Rpn13 (red in Fig. 1B), and Sem1 (not shown), the Ub-substrate conjugates are captured through a poorly understood mechanism by the translocation apparatus within a heterohexameric ATPase ring of the RP. Due to geometrical constraints, substrates must completely unfold before processive translocation through the ATPase ring and into the CP core (Chen et al., 2016; Lu et al., 2017). This unfolding process is likely powered by collective conformational transitions of the ATPases (Zhu et al., 2018). Rpn11, an essential deubiquitinase at the translocation entrance, removes Ub chains en bloc from a translocating peptide to ensure unobstructed translocation (Yao & Cohen, 2002). Conformational transitions of the ATPases also control movements of the lid subcomplex of the RP within the 26S proteasome, activity of Rpn11, and gating of the CP, thereby coordinating multiple events in the degradation process (schematized in Fig. 1C) (Zhu et al., 2018). We refer interested readers to excellent review papers for additional information on the biology of the UPS (Bard et al., 2018; Collins & Goldberg, 2017; Hochstrasser, 1996; Tomko & Hochstrasser, 2013; Yau & Rape, 2016).

The methods described here combine fluorescence-based single-molecule imaging on a chemically modified glass surface with in vitro ubiquitination and degradation reactions (Lu, Lee, King, Finley, & Kirschner, 2015; Lu, Wang, et al., 2015). Single-molecule assays can precisely measure weak interactions and isolate transient reaction intermediates as well as differentiate multiple binding modes. The reactions described often require relatively high concentrations of reagents, and it is therefore key to optimize surface-coating (passivation) chemistry to minimize nonspecific binding of reaction components, in conjunction with intensity calibration to achieve quantitative measurements. While we principally focus on ubiquitination of substrates by the APC/C and degradation of substrates by the 26S proteasome in both purified reaction systems and functional cell lysates, the methods described here can be readily modified to cater to specific studies.

2. A single-molecule fluorescence assay for detecting substrate ubiquitination by the APC/C and substrate degradation by the 26S proteasome

Nonspecific interactions of molecules with specialized surfaces can introduce experimental artifacts through sample loss and denaturation. Furthermore, nonspecific interactions of labeled molecules with these surfaces can introduce statistical artifacts into the subsequent analysis. Here we describe a polyethylene glycol (PEG)-based method used to passivate the imaging surface (i.e., reduce the probability of either nonspecific binding or side reactions of proteins and small molecules with the surface). This method allows us to use biotinylated PEG to make streptavidin-interacting surfaces to immobilize the substrates and protein complexes for imaging by total internal reflection fluorescence (TIRF) microscopy (Fig. 2A).

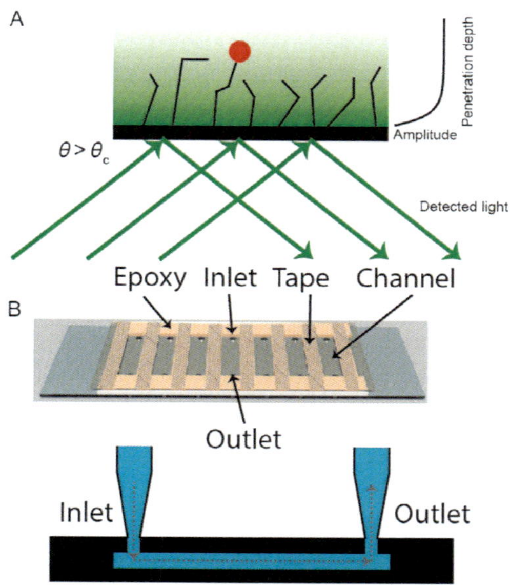

Fig. 2 General imaging and encapsulation scheme. (A) A schema of the imaging principle. In TIRF microscopy, the excitation wave penetrates weakly past the surface with exponentially decaying amplitude (*green gradient*), and excites a surface-tethered fluorophore, whose emission is detected by a camera. (B) Flow cell schema, showing the components described in Section 2.1.4.

2.1 Coverslip preparation

2.1.1 Initial cleaning

Standard 25 × 50 mm No. 1 glass coverslips (VistaVision, VWR) are first cleaned by removing residual oil from manufacture and lightly etched in KOH to expose hydroxyl groups on the glass surface. Coverslips should be functionalized after cleaning without prolonged storage.

Procedure:

1. Load coverslips on a metal slide rack and cleaned by sonication (Jakan PS-40A single-setting sonicator) for 30 min in warm detergent (7x™ solution from MP Bio, ~50°C) then rinse thoroughly in Milli-Q water.
2. Sonicate coverslips 30 min in 100% ethanol at room temperature and then rinse in Milli-Q water.
3. Sonicate coverslips 50 min in 1 *M* KOH at room temperature and then rinse in Milli-Q water.
4. Sonicate coverslips 30 min in Milli-Q water at room temperature.
5. Dry coverslips with a steady flow of compressed air.

2.1.2 Silanization

Primary amine functional groups are introduced onto the glass coverslip surfaces using a bifunctional silanization reagent.

Procedure:

1. Flame the dried coverslip for 3 s on both sides and allow it to cool in air.
2. Immerse cleaned coverslips in a solution of 5% acetic acid and 1.5% *N*-(2-aminoethyl)-3-aminopropyltrimethoxysilane (Unitedchem) in pure methanol for 20 min on an orbital shaker. Sonicate the coverslips for 1 min in the middle of the reaction to remove attached bubbles.
3. Sonicate amine-functionalized coverslips in Milli-Q water for 60 min to reduce fluorescent contaminations introduced by commercial methanol.
4. Dry coverslips in nitrogen gas to minimize oxidation and store in vacuum-sealed bags at −20°C for up to 12 months without loss of quality.

2.1.3 Polyethylene glycol coating

Amine-functionalized coverslips are reacted with a mixture of polyethylene glycol-succinimidyl valerate (PEG-SVA) with and without a biotin functional group. The *N*-hydroxysuccinimide (NHS) functional group within SVA is amine reactive and therefore interacts with the amine-functionalized slides (Section 2.1.2) to immobilize the PEG reagents via amide bonds.

Procedure:

1. Thoroughly mix 18% (w/v, final concentration) 5 kDa mPEG-SVA (LaysonBio), 0.45% (w/v, final concentration) 5 kDa biotin-PEG-SVA (LaysonBio) in freshly prepared 100 m*M* $NaHCO_3$ (pH 8.3), 500 m*M* K_2SO_4. The ratio of biotinylated to nonbiotinylated PEG may be adjusted to accommodate specific applications.
2. Centrifuge at 8000 × *g* for 1 min. Transfer the supernatant to a fresh 1.5-mL Eppendorf tube; the solution will become cloudy after remixing.
3. Add Milli-Q water drop-wise to clarify the solution, to reach approximately the "clouding point." The NHS functional group has a ~20 min lifetime at pH 8.3, and therefore these steps must be completed promptly.
4. Spread 120 μL of the PEG mixture between two coverslips in a humid chamber and incubate for 4 h, omitting light.
5. Separate surfaces and rinse in Milli-Q water, dry under nitrogen gas, and store in vacuum-sealed bags at −20°C. The coated side has a different wetting property, and thus can be easily distinguished from the uncoated by flushing with water. It is practical to mark the treated side of the surface with a tweezer.

2.1.4 Flow cell assembly

Flow cells are constructed to rapidly replace buffer over the imaging surface and to encapsulate the reaction mixture during imaging. The final assembly is shown in Fig. 2B.

Procedure:

1. Drill opposing holes ~0.5 mm in diameter on glass slides using Dremel tools with a small diamond drill bit (Kingsley).
2. Create a series of channels connecting each pair of holes by applying strips of ~0.1 mm thick double-stick tape in between neighboring pairs.
3. Adhere PEG-passivated coverslips to the tape to form arrays of identical enclosures so that the PEG-passivated surface faces the channel interior.
4. Apply 5 min epoxy glue to opposite ends of the enclosures to seal the flow cells.
5. Establish flow by inserting tubing or pipette tips into one of the drilled holes and introducing negative pressure to the opposing hole, drawing fluid into the flow cell (assembly shown in Fig. 2B).
6. If necessary, the surface can be passivated again using a short PEG to further reduce nonspecific binding. Prepare 15% 1 kDa mPEG-SC

(LaysonBio) in freshly prepared 100 mM $NaHCO_3$ (pH 8.3) 500 mM K_2SO_4, and incubate for 30 min at room temperature. Rinse the channels with Milli-Q water thoroughly.

7. Store flow cells in vacuum-sealed bags at −20°C until use.

2.1.5 Imaging surface quality assessment

Coverslips and flow cells are prepared for assays to measure substrate ubiquitination by the APC/C or substrate degradation by the 26S proteasome through monitoring of the fluorescence emission intensity of labeled ubiquitin and labeled APC/C antibody. To assess suitability of the imaging surfaces for these assays, two control experiments must be performed: first, an experiment to determine the ratio of specific to nonspecific binding; and second, an experiment involving photobleaching of the fluorophores, which allows calibration of the fluorescence intensity to ubiquitin stoichiometry.

2.1.5.1 Determination of the level of nonspecific binding of fluorescent ubiquitin

Procedure:

1. Rinse the channels in a flow cell with Milli-Q water.
2. Inject 100 mg/mL bovine serum albumin (BSA) in PBS into the channel (see Section 2.1.4, step 5) and incubate for 30 min at room temperature.
3. Rinse 2 × with Milli-Q water.
4. Incubate the channels with 2 mg/mL streptavidin (Thermo Fisher Scientific) and 100 mg/mL BSA in PBS for 1 min at room temperature.
5. Rinse 3 × with PBS, and quickly empty the channel.
6. Seal the injection ports of the flow cell with ~0.5 μL mineral oil to prevent loss of passivation due to drying. This will not affect subsequent injection of reaction mix.
7. Incubate surface with a known concentration (~200 nM) of either Dylight550-Ub or Alexa647-Ub in imaging buffer or HeLa cell extract and image immediately as described in Section 2.2 to determine the quality of the surface. The level of nonspecific binding can also be determined using other fluorescently labeled proteins, such as fluorescent antibodies, which tend to show substantially weaker nonspecific binding than ubiquitin.
8. Count the number of fluorescent molecules in the field of view. For a surface of good quality, we normally identify <20 molecules of Ub or <5 molecules of antibody in a field of view (80 × 80 μm).

2.1.5.2 Calibration of fluorescence intensity to determine ubiquitin stoichiometry

Procedure:

1. Perform steps 1–6 in Section 2.1.5.1.
2. Incubate surface in ~3 n*M* preformed biotin- and Dylight550-labeled ubiquitin chains (see Section 3.5.)
3. Image the surface (see Section 2.2) and identify fluorescence vs time trajectories. Analyze trajectories to identify photobleaching events. The intensity loss at each photobleaching step is uniform for Dylight550-Ub and Alexa647-Ub chains, which is taken as the unit intensity of each Ub. The fluorescence intensity at each level may be used to assign the intensity for each ubiquitin stoichiometry (Fig. 3B). Typically we find that the uncertainty in measuring this stoichiometry is approximately 30%, as determined from comparison of the normal distributions associated with each level. This is likely due to uneven laser illumination over the field of view.

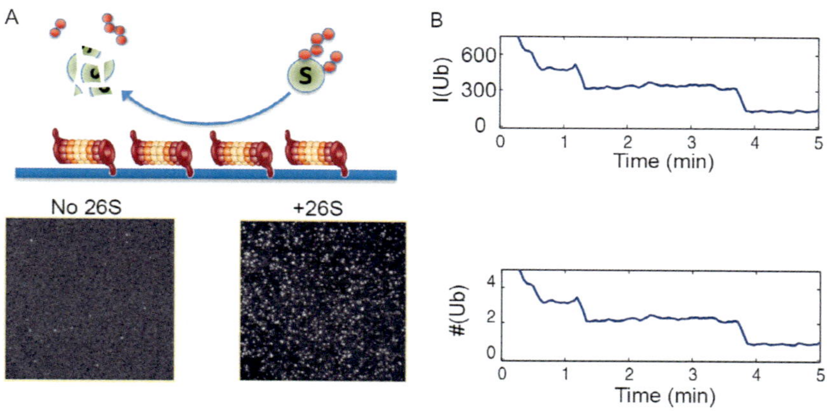

Fig. 3 (A) Experimental setup as described in Section 2.2.1. Surface fluorescence due to nonspecific and specific binding of fluorescent ubiquitin is shown later, as described in Section 2.1.5.1. (B) Intensity calibration as described in Section 2.1.5.2, showing that progressive photobleaching of fluorescent ubiquitin chains can be used to assess the number of ubiquitin monomers per chain. *Reproduced from Lu, Y., Lee, B. H., King, R. W., Finley, D., & Kirschner, M. W. (2015). Substrate degradation by the proteasome: A single-molecule kinetic analysis.* Science, 348*(6231), 1250834. doi:10.1126/science.1250834 and from Lu, Y., Wang, W., & Kirschner, M. W. (2015). Specificity of the anaphase-promoting complex: A single-molecule study.* Science, 348*(6231), 1248737. doi:10.1126/science.1248737. Reprinted with permission from AAAS.*

2.2 Imaging

Labeled APC/C, 26S proteasome substrates, or ubiquitin molecules are imaged using TIRF imaging. The optical configurations have been described in great detail elsewhere (Roy, Hohng, & Ha, 2008); briefly, TIRF imaging allows fluorescence emission vs time trajectories from many, on the order of 10^3, molecules to be simultaneously recorded at a millisecond–second timescale. Later, we describe methods for immobilizing labeled ubiquitin or APC/C molecules onto passivated coverslips (see Section 2.2), which may be used to record trajectories containing information on the dynamics of substrate degradation by the proteasome, or substrate ubiquitination by the APC/C.

In TIRF imaging experiments (schematized in Fig. 2A), molecules are generally tethered to a solid support so that many events can be recorded. While the density of surface-tethered molecules must be maintained high enough to collect required statistical strength for the experimental design, the surface-tethered molecule density must be low enough to distinguish contributions from individual molecules. This tethering is achieved by mixing biotinylated PEG into the passivation layer (see Section 2.2) and adding the tetrameric streptavidin, which is immobilized to the surface by binding to the biotinylated PEG molecules and in turn provides binding sites for biotinylated target molecules. With these rough guidelines in mind, in this section we will describe TIRF-based single-molecule assays for imaging substrate ubiquitination by the APC/C and substrate degradation by the 26S proteasome.

2.2.1 *Single-molecule imaging of substrate degradation by the proteasome*

This section describes a method for immobilizing 26S proteasomes to passivated surfaces using a biotin-labeled anti-20S antibody, then subsequently imaging binding and dissociation of substrates conjugated with fluorescent ubiquitin on the 26S proteasome. The 26S proteasome purified according to Section 3.6 contains essentially no Usp14. This allows identifying substrate translocation/degradation through a stepwise and processive deubiquitination driven by Rpn11 (Fig. 4A and B). When the above condition is not met, fluorescently labeled substrates can be used to signal peptide proteolysis (Fig. 4C and D).

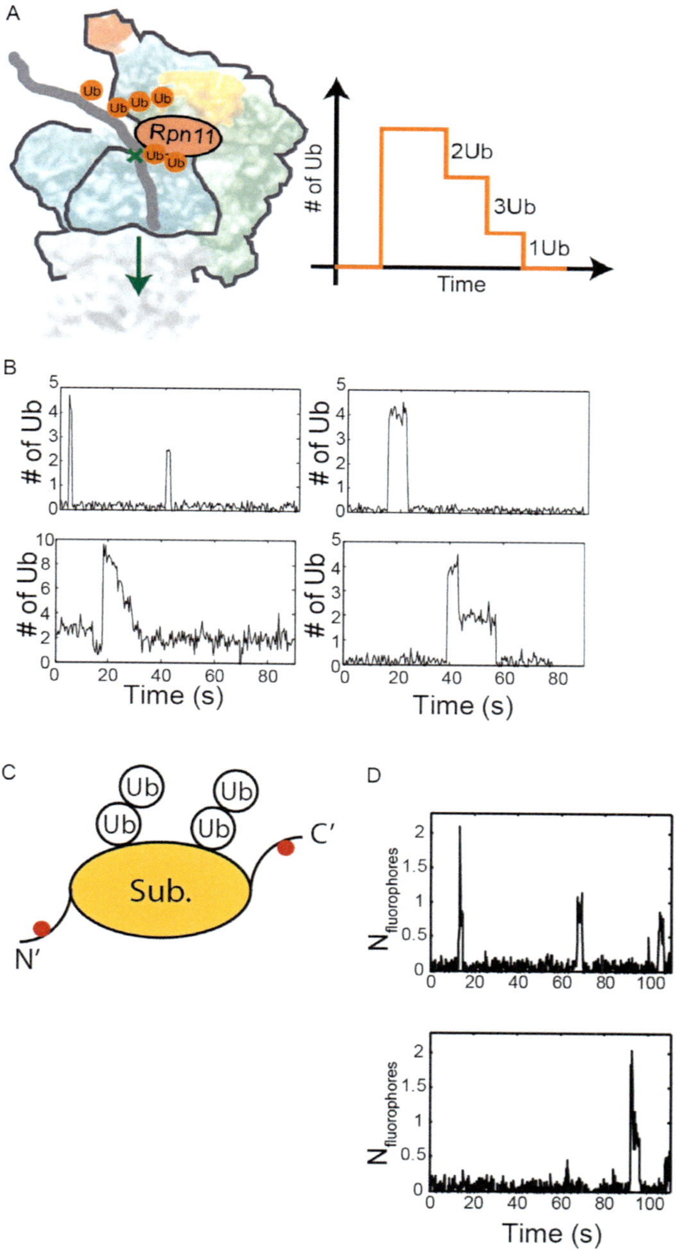

Fig. 4 (A) A schema showing that ubiquitin can be removed from substrates en bloc by the deubiquitinase Rpn11 on the 26S proteasome during translocation, which leads to the "stepped" traces (B) featuring a processive reduction in fluorescent ubiquitin. *(Continued)*

2.2.1.1 Single-molecule imaging of interactions of ubiquitinated substrates with the 26S proteasome

Procedure:

1. Perform steps 1–6 in Section 2.1.5.1.
2. Prepare 100 μL of 15 n*M* purified 26S proteasome and biotinylated 15 n*M* MCP21 anti-20S proteasome antibody in imaging buffer (26SIB), and incubate the mix for 15 min at room temperature.
3. Load the flow cell channel with the solution from step 2, and incubate for 3 min.
4. Flush the channel with about 100 μL imaging buffer containing at least 5 n*M* Dylight550-ubiquitin-labeled substrate.
5. Image the surface immediately if no inhibitors are added, using a Nikon Ti TIRF microscope equipped with a 561 nm laser at 30 mW with a 200 ms frame-rate. For experiments containing inhibitors, such as (1,10)-phenanthroline, allow the mixture to equilibrate for at least 20 min at room temperature, then begin imaging. See Section 2.3 for details on data analysis.

2.2.1.2 Single-molecule imaging substrate degradation by the proteasome

Procedure:

1. Perform steps 1–3 in Section 2.2.1.1.
2. Flush the channel with about 100 μL imaging buffer containing at least 5 n*M* ubiquitinated substrate labeled at the N- and C-termini with Dylight550.
3. Image the surface immediately if no inhibitors are added, using a Nikon Ti TIRF microscope equipped with a 561 nm laser at 30 mW with a 200 ms frame-rate. For experiments containing inhibitors, such as

Fig. 4—Cont'd (B) Following intensity calibration, the copy number of ubiquitins on a substrate molecule can be measured in real time after engaging with proteasome. Two types of traces can be observed from this experiment: (1) "Structureless" traces suggesting simple binding and disassociation (*top*); (2) "Stepped" traces suggesting Rpn11-dependent deubiquitination events. (C) A schema showing a substrate (cyclinB N-terminus) with fluorophore labels at the substrate N- and C-termini. (D) Two-stepped events are detected in fluorescence vs time trajectories (here, converted into units of number of fluorophores, $N_{fluorophores}$ vs time, using identical methods as in Section 2.1.5.2), which indicate processing of the substrate. *Reproduced from Lu, Y., Lee, B. H., King, R. W., Finley, D., & Kirschner, M. W. (2015). Substrate degradation by the proteasome: A single-molecule kinetic analysis.* Science, *348(6231), 1250834. doi:10.1126/science.1250834. Reprinted with permission from AAAS.*

(1,10)-phenanthroline, allow the mixture to equilibrate for at least 20 min at room temperature, then begin imaging. See Section 2.3 for details on data analysis.

2.2.2 Single-molecule imaging of substrate ubiquitination by the APC/C

Carefully prepared, functional cell extracts can mimic in situ cellular environments, allowing the study of a number of important biological activities under near-physiological conditions. Single-molecule methods have proved to be highly valuable in complex chemical environments in extracting quantitative information about dynamic reactions such as protein ubiquitination and deubiquitination (Lu, Wang, et al., 2015). In this assay, we label ubiquitin, APC/C, and substrates using organic fluorophores, and carry out the imaging in prepared HeLa cell extracts in order to study the kinetics of ubiquitin conjugation and deconjugation in a native environment. The basic setup of the flow chamber and microscope is identical to that of the proteasome assay. Example results are shown in Fig. 5.

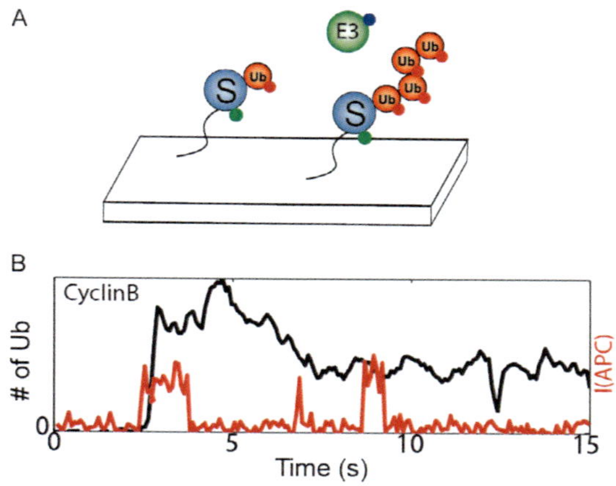

Fig. 5 (A) Experimental setup. Fluorescently labeled substrate (S), ubiquitin, and APC/C are imaged simultaneously by TIRF microscopy, as discussed in Section 2.3.2.2. (B) An example trace showing ubiquitination and APC/C binding on a cyclinB molecule in G1 HeLa cell extract. *Black*, Alexa647-ubiquitin fluorescence vs time trajectory, and *red*, APC/C fluorescence vs time trajectory. APC/C binding events coincide with productive extension of the ubiquitin chain on the same substrate molecule. *Reproduced from Lu, Y., Wang, W., & Kirschner, M. W. (2015). Specificity of the anaphase-promoting complex: A single-molecule study.* Science, *348(6231), 1248737. doi:10.1126/science.1248737. Reprinted with permission from AAAS.*

Procedure:

1. Perform steps 1–6 in Section 2.1.5.1.
2. Thaw 100 μL HeLa cell extract, and spin at 20,000 × *g* for 10 min at 4°C. Take the supernatant and supplement with 1 × Energy Mix (see Table 1) and 50 n*M* E1.
3. Deplete endogenous free ubiquitin by absorbing to E2. Incubate the extract with 20 μ*M* His-UbcH10 for 20 min at room temperature. Simultaneously label APC/C by adding 500 n*M* Dylight550 anti-APC4 antibody to the extract. Deplete extract of 20 μ*M* His-UbcH10

Table 1 Full buffer compositions abbreviated in the main text

Buffer	Composition
PBS	10 m*M* Na_2HPO_4 and 1.8 m*M* KH_2PO_4 pH 7.4, 137 m*M* NaCl, 2.7 m*M* KCl
26SIB	25 m*M* Tris–HCl pH 7.5, 50 m*M* NaCl, 10 m*M* $MgCl_2$, 3 m*M* ATP, 20 m*M* imidazole, 5 mg/mL BSA, 0.5 m*M* DTT
UBAB	25 m*M* Tris–HCl pH 7.5, 50 m*M* NaCl, 10 m*M* $MgCl_2$
HB	50 m*M* Tris–HCl 7.5, 150 m*M* NaCl, 5 m*M* $MgCl_2$, 0.5 m*M* DTT, 1 m*M* PMSF, 0.1% Tween-20, 5% glycerol
MB	25 m*M* Bis-Tris pH 6.5, 100 m*M* NaCl
LiB	25 m*M* Tris–HCl, pH 8.0, 10 m*M* $MgCl_2$, 3 m*M* ATP, 0.5 m*M* DTT
hLB	50 m*M* NaH_2PO_4 pH 7.5, 100 m*M* NaCl, 10% glycerol, 5 m*M* $MgCl_2$, 0.5% NP-40, 5 m*M* ATP, and 1 m*M* DTT
hWB	50 m*M* Tris–HCl pH 7.5, 1 m*M* $MgCl_2$, and 1 m*M* ATP
TBS	25 m*M* Tris–HCl pH 7.5, 150 m*M* NaCl
SB	20 m*M* Tris–HCl pH 7.5, 5 m*M* KCl, 1.5 m*M* $MgCl_2$, 1 m*M* DTT, protease-inhibitor tablet (Roche)
ATP-regenerating system (10 ×) (Energy mix)	10 × UBAB, 200 m*M* phosphocreatine, 1 mg/mL creatine kinase, 10 m*M* ATP

by incubation with washed Ni-agarose beads. Verify completeness of ubiquitin depletion by an antiubiquitin western blot.
4. Supplement extract with at least 1 μ*M* fluorescently labeled ubiquitin (e.g., Alexa647-Ub or the lysine-less Alexa647-Ub0K.)
5. Supplement extract with about 20 n*M* Alexa488-labeled substrate which is also biotinylated at the C-terminus (see later). 5 m*M* Trolox, 2.5 m*M* protocatechuic acid (PCA), and 0.5 mg/mL protocatechuate-3,4-dioxygenase (PCD) are added as the oxygen scavenging system and triplet state quencher to protect dye molecules, especially Alexa488 and Alexa647, from photobleaching.
6. Load the prepared extract in the flow cell, and incubate for 3 min. Image the surface immediately using a Nikon Ti TIRF microscope with three laser lines: 491 nm (50 mW, 500 ms exposure time), 561 nm (50 mW, 750 ms exposure time), and 640 nm (125 mW, 3 s exposure time). See Section 2.3 for details on data analysis.

2.3 Analysis of single-molecule images

The goal of the analysis of single-molecule time-series data is to determine and quantify the kinetic mechanism that was most likely sampled by the measurement. For information on the experimental preparations, see Section 2.2. Specialized software for identifying single molecules and analyzing intensity trajectories may be obtained from various sources, some of which are listed here (Bronson, Fei, Hofman, Gonzalez, & Wiggins, 2009; Cox et al., 2012; Juette et al., 2016; Persson, Lindén, Unoson, & Elf, 2013; Qin, Auerbach, & Sachs, 2000; Van De Meent, Bronson, Wiggins, & Gonzalez, 2014).

2.3.1 Identification and preparation of fluorescence vs time trajectories

TIRF movies containing successive images of fluorescence amplitudes of surface-tethered molecules contain a spatio-temporally varying signal roughly given by $S(t, x, y) = I(t, x, y) + N(x, y) + B(t, x, y)$ where t denotes time, x and y are the pixel coordinates, S is a random variable denoting the observed signal, I is a random variable denoting the fluorescence intensity, N is a random variable denoting normally distributed noise from various sources, and B is a random variable denoting the Poisson distributed noise due to freely diffusing fluorophores in the background. Following others (Smith, Joseph, Rieger, & Lidke, 2010), we assume that for a tethered fluorophore imaged at position (x_c, y_c), the intensity profile of a single fluorophore is given by a point-spread function approximated by a 2D

Gaussian function, whose observed values are Poisson distributed: $I(t,x,y) = \text{Poiss}\left(\frac{1}{2\pi\sigma}\text{Exp}\left(-\frac{1}{2\sigma}\left((x-x_c)^2 + (y-y_c)^2\right)\right)\right)$.

Procedure:

1. Estimate $B(t,x,y)$ inhomogeneous background fluorescence by a mean-filtering routine, and the inhomogeneous background variance of $N(x,y)$ from the 15 dimmest pixels in a 5×5 region surrounding (x,y). The optimal setting may need to be adapted to the specific imaging system and camera.
2. Locate pixels that are local maxima in a single 2D image, (x_{ci}, y_{ci}) for each i maxima, which are candidate pixels to contain intensity samples from the light emissions from surface-tethered fluorescent molecules.
3. Register successive images to calculate image drift using local maxima pixels above a signal-to-noise threshold of ~5 by iterative rotations and translations.
4. Estimate $I(t, x_{ci}, y_{ci})$ by fitting with a 2D Gaussian function with background $B(t,x,y)$, reestimated from step 2.

2.3.2 Analysis of single-molecule trajectories

Trajectories recorded from the single-molecule TIRF experiments described here are typically modeled as signal vs time trajectories containing information describing transitions between a discrete number of distinct molecular states. These states are parameterized by the fluorescence intensity of the molecule while in those states as well as the rates of transition from state to state. For such a model, the probability of a molecule to be in a given state, $\rho_i(t)$, obeys: $\frac{d\rho_i(t)}{dt} = -\sum_i k_{ij}\rho_i(t)$ where k_{ij} is the rate of transition from state i to state j and is subject to the condition $k_{ii} = -\sum_{j\neq i} k_{ij}$. We here consider only two types of simplified model: a first in which $k_{ij}=0$, $i \neq j-1$ or $i \neq j$, which models the ubiquitin stoichiometry experiment described in Section 2.1.5.2, and a second, which assumes that there are only two states and allows us to solve for the time-normalized dwell time distribution $\tau(t, 2|1) = k_{12}Exp(-k_{12}t)$. A similar calculation gives $\tau(t, 1|2) = k_{21}Exp(-k_{21}t)$. The main strategy of the analysis in both models is to divide trajectories into n subtrajectories that are synchronized so that $\rho_{in}(0) = 0$ and which terminate when a transition has occurred. For the first model, we are interested in how many distinct classes are separated by solely downward transitions, while in the second model we are interested in calculation of the rate constants k_{12} and k_{21}.

2.3.2.1 Quantification of ubiquitin stoichiometry

Procedure:

1. Obtain fluorescence vs time trajectories (see Section 2.3.1).
2. Identify "stepped" events within each trajectory (Fig. 4B), for which the fluorescence signal drops from one mean value to a new one that is significantly lower within 2 frames.
3. Work backward from the background state for which $I_0 \sim Norm(0, \sigma_{bg})$, count up in stepped events, mark time points consistent with the current event as I_n, until reaching the beginning of the trajectory. The intensity of a chain with n labeled ubiquitin molecules, I_n, is assumed distributed as $I_n \sim Norm(\mu_n, \sigma_n)$ where μ_n is the mean of I_n over all trajectories and time points, and σ_n the corresponding standard deviation.

2.3.2.2 Quantification of ubiquitinated substrate binding and dissociation from 26S proteasomes

Procedure:

1. Obtain fluorescence vs time trajectories (see Section 2.3.1).
2. Calibrate the fluorescence amplitudes to ubiquitin stoichiometry (see Sections 2.1.5.2 and 2.3.2.1).
3. Logically divide the time points of fluorescence vs time trajectories into either a most likely bound state (Number of Ub > 0.7) or a most likely dissociated state (Number of Ub < 0.7).
4. Divide the trajectories into n subtrajectories which start in either state and terminate in the other state and measure the lengths of these subtrajectories ("dwell times," $n\tau(i|j)$).
5. Estimate the rate constants k_{ij} by direct fitting of the histogram-binned representation of $n\tau(t,\ i|j)$ to an exponential distribution.

2.3.2.3 Quantification of fluorescence trajectories of colocalized APC/C, biotinylated APC/C substrate, and ubiquitin

Procedure:

1. Obtain fluorescence vs time trajectories (see Section 2.3.1).
2. Calibrate the fluorescence amplitudes to ubiquitin stoichiometry (see Sections 2.1.5.2 and 2.3.2.1).
3. Synchronize trajectories collected at distinct frame-rates.
4. Identify APC/C binding events by sharp increases in intensity in the APC/C fluorescence vs time trajectories (Fig. 5B).
5. Measure the dwell time of each APC/C binding event, and associate the dwell time with the state of ubiquitination.

3. Purification and sample preparation

This section describes the purification of substrates and protein complexes that form the basis of the assays described in the previous sections. A list of the relevant buffers is provided in Table 1.

3.1 Purification of ubiquitin and substrates

Procedure:

1. Poly-histidine tags are genetically inserted into genes encoding ubiquitin and substrates which are subsequently expressed from a T7 promoter in *Escherichia coli* (BL21(DE3), NEB) at 37°C with 0.5 m*M* IPTG for 3h. Substrates requiring C-terminal biotinylation are purified using the IMPACT™ kit (Thermo Fisher Scientific).
2. Cells are lysed in a standard purification buffer (HB) by sonication and purified protein is isolated using Ni-NTA resin (Bornhurst & Falke, 2000), following manufacturer's instructions.
3. Proteins eluted from the Ni-NTA resin are further purified with size-exclusion and ion-exchange chromatography.

3.2 Labeling of antibodies

Procedure:

1. Antibody biotinylation and fluorescent labeling are performed using the APEX antibody labeling kit from Thermo Fisher Scientific according the manual. We adjust the amount of reactive dye or biotin to limit the degree of labeling to less than three biotins or fluorophores per antibody.

3.3 Fluorescent labeling of 26S proteasome substrates

Procedure:

1. Two cysteines are genetically inserted into the sequence encoding the substrate. In Fig. 4C and D, the substrate is the N-terminal portion of cyclinB with D6C and E96C mutations.
2. Purified substrate is mixed with at least 2× molar excess of cysteine-reactive fluorescent dye (Dylight550-maleimide) in MB for 4h at room temperature. The reaction is quenched by addition of 2m*M* DTT.

3. Labeled substrate is purified and separated from unlabeled substrate by anion exchange chromatography using a HiTrap Q HP column at pH 8.5.
4. Eluted proteins are concentrated to 200 μ*M* and snap frozen with 5% glycerol.

3.4 Fluorescent labeling of ubiquitin

Procedure:

1. Human ubiquitin with an N-terminal tag MGSSHHHHHHSSGLVPR| GSH**C**GGG is expressed and purified from *E. coli* as described in Section 3.1.
2. Purified ubiquitin is mixed with at least 2× molar excess of cysteine-reactive fluorescent dye (Alexa647-maleimide or Dylight550-maleimide) in MB for 4 h at room temperature. The reaction is quenched by addition of 2 m*M* DTT.
3. Poly-His tag is removed (between the LVPR and GS of the tag in step 1) by addition of thrombin protease (Thermo Fisher).
4. Labeled ubiquitin is purified and separated from unlabeled ubiquitin by anion exchange chromatography using a HiTrap Q HP column at pH 8.5.
5. Eluted proteins are concentrated to 200 μ*M* and snap frozen with 5% glycerol.

3.5 Formation of ubiquitin chains of specific lysine linkages

Procedure:

1. Fluorescently labeled ubiquitin is buffer exchanged into LiB with an ATP-regeneration system using a 7 kDa Zeba desalting column.
2. A reaction mixture of 200 μ*M* fluorescently labeled ubiquitin, 100 n*M* E1 and 20 μ*M* E2 (for K48 chains, E2-25k; for K11 chains, Ube2SΔC). E1 and E2 proteins are purchased from BostonBiochem. The reaction is incubated at 37°C for 16 h.

3.6 Purification of human 26S proteasomes

Procedure:

1. A detailed purification procedure can be found in Wang et al. (2007). Briefly, begin by overexpressing hexahistidine-TEV cleavage-in vivo biotinylation-hexahistidine (HTBH)-tagged hRPN11 using a retrovirus in HEK293 cells.

2. Harvest about 40 15-cm plates of cells. Dounce-lyse cells in hLB, and immediately incubate cleared lysate with NeutrAvidin agarose resin (ThermoScientific) overnight at 4°C.
3. Rinse resin with excess hLB and hWB.
4. Elute by cleavage with TEV protease (Invitrogen). Note that human proteasome purified under above condition contains essentially no Usp14. If required, recombinant Usp14 can be coupled to proteasome in hWB (Lee et al., 2016).

3.7 Preparation of cell-cycle synchronized HeLa cell extract

Procedure:

1. Synchronize two 1 L HeLa S3 spinner cultures at prometaphase using a thymidine/nocodazole double-block protocol.
2. Release spinner cultures for ~3 h at 37°C into G1.
3. Harvest cells and subject to N_2 cavitation in 75% volume SB at 500 psi.
4. Clear extract by centrifugation at 35,000 × *g* for 30 min at 4°C and snap freeze immediately in liquid nitrogen in 100 μL aliquots.

3.8 Affinity purification of the APC/C

Procedure:

1. Incubate protein-A agarose with anti-Cdc27(AF3.1, Abcam) antibody, which immobilizes the antibody on the agarose resin. Cdc27 is an APC/C subunit.
2. Incubate G1 HeLa cell extract with 10% (v/v) AF3.1 agarose for 4 h at 4° C on a rotor.
3. Wash the beads with buffer SB, and elute APC/C with a competitive peptide (sequence: QLHAAESDEFQLHAAESDEF) for 1.5 h in the presence of 150 m*M* NaCl. Eluted APC/C is snap-frozen with 10% glycerol in liquid nitrogen.

3.9 Conjugation of model substrates with biotin-labeled peptides

Procedure:

1. Recombinant substrates, such as securin, geminin, cyclinA, and N-terminal cyclinB, are purified using the IMPACT™ kit (NEB).
2. Eluted substrates are diluted to about 100 μ*M* in a buffer containing 25 m*M* Tris–HCl pH 7.5, 150 m*M* NaCl, 0.05% Tween-20, 10 m*M* 2-mercaptoethanesulfonic acid sodium salt (MESNA), and 500 μ*M*

biotin-containing peptide (sequence: CGGG[biotin]). The sample is incubated at room temperature for 5 h to allow coupling of the peptide to the C-terminal thioester of the substrate to form a stable peptide bond.
3. Remove unconjugated peptides by gel filtration, for instance a Superdex 75 column (GE Life Sciences)

3.10 Ubiquitination of substrates by the APC/C

Procedure:
1. Incubate 50 μ*M* ubiquitin, 10 n*M* APC/C, 100 n*M* E1, 2 μ*M* UbcH10, 2 mg/mL BSA, 500 n*M* substrate, the ATP-regenerating system in UBAB buffer at 30°C for 2 h.
2. Verify substrate ubiquitination by antiubiquitin Western blot. Substrates can be used directly in degradation assays without further purification.

3.11 Proteasomal degradation assay

Procedure:
1. Ubiquitinate the substrate by the APC/C as described in Section 3.10
2. Mix reaction products 1:4 with UBAB supplemented with 10 n*M* 26S proteasome, 2 mg/mL BSA, and 3 m*M* ATP, and incubate at 37°C.
3. Sample the reaction at 0, 3, 6, 15 min and determine the extent of substrate degradation by antiubiquitin Western blot.

4. Summary

The complexity of ubiquitin chain variants recognized by the 26S proteasome and the diversity of substrates modified by the APC/C suggest that cells employ intricate kinetic mechanisms to control the timing and specificity of protein degradation. Single-molecule techniques allow distinction between differential substrate affinities and kinetics in complex mixtures. The methods described here allow determination of ubiquitin stoichiometry and can capture transient substrate–enzyme interactions. These tools should be adaptable to studies of other complex reactions.

Acknowledgments

We thank Dr. Marc Kirschner for help in developing the method. This work is supported by an Edward Mallinckrodt, Jr. foundation award to Dr. Lu.

References

Bard, J. A. M., Goodall, E. A., Greene, E. R., Jonsson, E., Dong, K. C., & Martin, A. (2018). Structure and function of the 26S proteasome. *Annual Review of Biochemistry*, *87*, 697–724. https://doi.org/10.1146/annurev-biochem-062917-011931.

Bornhurst, J. A., & Falke, J. J. (2000). Purification of proteins using polyhistidine affinity tags. *Methods in Enzymology*, *326*, 245–254. https://doi.org/10.1016/S0076-6879(00)26058-8.

Bronson, J. E., Fei, J., Hofman, J. M., Gonzalez, R. L., & Wiggins, C. H. (2009). Learning rates and states from biophysical time series: A Bayesian approach to model selection and single-molecule FRET data. *Biophysical Journal*, *97*(12), 3196–3205. https://doi.org/10.1016/j.bpj.2009.09.031.

Chen, S., Wu, J., Lu, Y., Ma, Y. B., Lee, B. H., Yu, Z., et al. (2016). Structural basis for dynamic regulation of the human 26S proteasome. *Proceedings of the National Academy of Sciences of the United States of America*, *113*(46), 12991–12996. https://doi.org/10.1073/pnas.1614614113.

Collins, G. A., & Goldberg, A. L. (2017). The logic of the 26S proteasome. *Cell*, *169*(5), 792–806. https://doi.org/10.1016/j.cell.2017.04.023.

Cox, S., Rosten, E., Monypenny, J., Jovanovic-Talisman, T., Burnette, D. T., Lippincott-Schwartz, J., et al. (2012). Bayesian localization microscopy reveals nanoscale podosome dynamics. *Nature Methods*, *9*(2), 195–200. https://doi.org/10.1038/nmeth.1812.

Finley, D. (2009). Recognition and processing of ubiquitin-protein conjugates by the proteasome. *Annual Review of Biochemistry*, *78*, 477–513. https://doi.org/10.1146/annurev.biochem.78.081507.101607.

Hochstrasser, M. (1996). Ubiquitin-dependent protein degradation. *Annual Review of Genetics*, *30*, 405–439. https://doi.org/10.1146/annurev.genet.30.1.405.

Juette, M. F., Terry, D. S., Wasserman, M. R., Altman, R. B., Zhou, Z., Zhao, H., et al. (2016). Single-molecule imaging of non-equilibrium molecular ensembles on the millisecond timescale. *Nature Methods*, *13*, 341–344. https://doi.org/10.1038/nmeth.3769.

Komander, D., & Rape, M. (2012). The ubiquitin code. *Annual Review of Biochemistry*, *81*, 203–229. https://doi.org/10.1146/annurev-biochem-060310-170328.

Lee, B. H., Lu, Y., Prado, M. A., Shi, Y., Tian, G., Sun, S., et al. (2016). USP14 deubiquitinates proteasome-bound substrates that are ubiquitinated at multiple sites. *Nature*, *532*(7599), 398–401. https://doi.org/10.1038/nature17433.

Lu, Y., Lee, B. H., King, R. W., Finley, D., & Kirschner, M. W. (2015). Substrate degradation by the proteasome: A single-molecule kinetic analysis. *Science*, *348*(6231), 1250834. https://doi.org/10.1126/science.1250834.

Lu, Y., Wang, W., & Kirschner, M. W. (2015). Specificity of the anaphase-promoting complex: A single-molecule study. *Science*, *348*(6231), 1248737. https://doi.org/10.1126/science.1248737.

Lu, Y., Wu, J., Dong, Y., Chen, S., Sun, S., Ma, Y. B., et al. (2017). Conformational landscape of the p28-bound human proteasome regulatory particle. *Molecular Cell*, *67*(2), 322–333.e6. https://doi.org/10.1016/j.molcel.2017.06.007.

Persson, F., Lindén, M., Unoson, C., & Elf, J. (2013). Extracting intracellular diffusive states and transition rates from single-molecule tracking data. *Nature Methods*, *10*(3), 265–269. https://doi.org/10.1038/nmeth.2367.

Peters, J. M. (2006). The anaphase promoting complex/cyclosome: A machine designed to destroy. *Nature Reviews. Molecular Cell Biology*, *7*(9), 644–656. https://doi.org/10.1038/nrm1988.

Pickart, C. M. (2001). Mechanisms underlying ubiquitination. *Annual Review of Biochemistry*, *70*, 503–533. https://doi.org/10.1146/annurev.biochem.70.1.503.

Qin, F., Auerbach, A., & Sachs, F. (2000). A direct optimization approach to hidden Markov modeling for single channel kinetics. *Biophysical Journal, 79*(4), 1915–1927. https://doi.org/10.1016/S0006-3495(00)76441-1.

Rape, M., Reddy, S. K., & Kirschner, M. W. (2006). The processivity of multiubiquitination by the APC determines the order of substrate degradation. *Cell, 124*(1), 89–103. https://doi.org/10.1016/j.cell.2005.10.032.

Roy, R., Hohng, S., & Ha, T. (2008). A practical guide to single-molecule FRET. *Nature Methods, 5*(6), 507–516. https://doi.org/10.1038/nmeth.1208.

Saeki, Y. (2017). Ubiquitin recognition by the proteasome. *Journal of Biochemistry, 161*(2), 113–124. https://doi.org/10.1093/jb/mvw091.

Smith, C. S., Joseph, N., Rieger, B., & Lidke, K. A. (2010). Fast, single-molecule localization that achieves theoretically minimum uncertainty. *Nature Methods, 7*(5), 373–375. https://doi.org/10.1038/nmeth.1449.

Tomko, R. J., Jr., & Hochstrasser, M. (2013). Molecular architecture and assembly of the eukaryotic proteasome. *Annual Review of Biochemistry, 82*, 415–445. https://doi.org/10.1146/annurev-biochem-060410-150257.

Van De Meent, J. W., Bronson, J. E., Wiggins, C. H., & Gonzalez, R. L. (2014). Empirical Bayes methods enable advanced population-level analyses of single-molecule FRET experiments. *Biophysical Journal, 106*(6), 1327–1337. https://doi.org/10.1016/j.bpj.2013.12.055.

Wang, X., Chen, C. F., Baker, P. R., Chen, P. L., Kaiser, P., & Huang, L. (2007). Mass spectrometric characterization of the affinity-purified human 26S proteasome complex. *Biochemistry, 46*(11), 3553–3565. https://doi.org/10.1021/bi061994u.

Yao, T., & Cohen, R. E. (2002). A cryptic protease couples deubiquitination and degradation by the proteasome. *Nature, 419*(6905), 403–407. https://doi.org/10.1038/nature01071.

Yau, R., & Rape, M. (2016). The increasing complexity of the ubiquitin code. *Nature Cell Biology, 18*(6), 579–586. https://doi.org/10.1038/ncb3358.

Zhu, Y., Wang, W. L., Yu, D., Ouyang, Q., Lu, Y., & Mao, Y. (2018). Structural mechanism for nucleotide-driven remodeling of the AAA-ATPase unfoldase in the activated human 26S proteasome. *Nature Communications, 9*(1), 1360. https://doi.org/10.1038/s41467-018-03785-w.

CHAPTER ELEVEN

In vitro analysis of proteasome-associated USP14 activity for substrate degradation and deubiquitylation

Srinivasan Muniyappan, Byung-Hoon Lee*
Department of New Biology, DGIST, Daegu, Republic of Korea
*Corresponding author: e-mail address: byung-hoon_lee@dgist.ac.kr

Contents

Abstract

The ubiquitin-proteasome pathway plays an essential role in maintaining protein homeostasis and regulates almost every aspect of cellular processes in eukaryotes. Emerging evidence indicates that the proteasome does not work as a simple unidirectional molecular machinery for substrate proteolysis. In fact, proteasome activity should be tightly regulated, and the proteasome itself can be dynamically engaged in the degradation cycle.

Methods in Enzymology, Volume 619
ISSN 0076-6879
https://doi.org/10.1016/bs.mie.2018.12.028
© 2019 Elsevier Inc.
All rights reserved.

Proteasome-mediated degradation can occur through multistep mechanisms such as ubiquitin-dependent substrate recognition, deubiquitination, and ATP-driven unfolding and translocation of the substrate into 20S chamber for proteolytic cleavage. Deubiquitination is particularly interesting because this reaction may impose a critical checkpoint for substrate turnover on the proteasome. Notably, there are three major deubiquitinating enzymes (DUBs) on human proteasomes: USP14, UCH37 and RPN11. USP14 can spare the substrate from degradation prior to the proteasome's commitment step, suggesting that USP14 inhibition may stimulate proteasomal degradation of undesirable proteins under certain proteotoxic conditions. Furthermore, USP14 deubiquitinates multichain conjugates, the first among ~100 DUBs found to have this striking specificity. In this chapter, we describe in vitro methods to test proteasome-associated USP14 activity for substrate degradation and deubiquitylation.

1. Introduction

Proteins mediate a broad range of essential functions in cellular physiology and thus constitute the primary molecular workforce within the cells. Protein turnover is critical for maintaining cellular protein homeostasis, and dysregulation of this process often causes devastating human diseases, such as proteopathies (Balch, Morimoto, Dillin, & Kelly, 2008). In eukaryotic cells, protein substrates are degraded by two major pathways: the ubiquitin-proteasome system (UPS) and lysosomal degradation pathway (Ciechanover, 2005). The UPS degrades cytosolic and nuclear proteins selectively by recognizing ubiquitin as a destruction signal (Hochstrasser, 1996). In eukaryotes, the great majority (at least 80%) of intracellular proteins can be degraded by 26S proteasome, which is a ~2.5-MDa multiprotein protease complex consisting of the 19S regulatory particle (RP) and 20S core particle (CP) (Finley, 2009; Schrader, Harstad, & Matouschek, 2009). Proteasome-mediated degradation is a highly regulated and complex process that can be governed by a number of proteasome-associated components. For example, ubiquitin receptors (for ubiquitin chain recognition), an E3 ubiquitin ligase (for ubiquitin chain elongation), and deubiquitinating enzymes (DUBs; for ubiquitin chain cleavage) are located on the RP, where they mediate elaborate ubiquitin conjugate processing and editing (Finley, 2009).

DUBs are also emerging as promising drug targets since they reverse ubiquitination and may antagonize the destruction of harmful proteins (Harrigan, Jacq, Martin, & Jackson, 2018). Of note, the human proteasome contains three major and distinctive classes of DUBs-USP14 (in yeast, Ubp6), UCH37/UCHL5, and RPN11/POH1 (Finley, 2009). USP14 and UCH37 may act before the ubiquitinated substrate is irreversibly engaged in the degradation cycle and rescue the substrate prior to the proteasomal commitment step, whereas RPN11 is coupled to degradation in an ATP-dependent manner (de Poot, Tian, & Finley, 2017). Our recent studies have identified highly selective USP14 inhibitors, IU1 and its derivatives, by high-throughput screening and subsequent structure–activity relationship studies, and their specificity was also confirmed by recent cocrystal structures of inhibitor-bound USP14 (Wang et al., 2018). USP14 inhibitor treatment accelerates the degradation of a variety of proteasome substrates within cells, including proteotoxic proteins in neurodegenerative disease models (Boselli et al., 2017; Lee et al., 2010; Moon & Lee, 2018), most notably the prion protein PrP (Homma et al., 2015; McKinnon et al., 2016). IU1 is also cytoprotective under specific stress conditions (Chen, Zhu, Johns, & Yang, 2018; Lee et al., 2010; VerPlank, Lokireddy, Feltri, Goldberg, & Wrabetz, 2018). Remarkably, IU1 was found to provide substantial phenotypic rescue of *parkin* and *pink1* null mutants in *D. melanogaster*, apparently as a result of its promotion of basal mitophagy (Chakraborty et al., 2018).

We found that USP14 prefers substrates that are ubiquitinated at multiple sites (e.g., supernumerary chains) and can cut the base of the chains by a proximal cleavage mechanism (see Figs. 1 and 4) (Lee et al., 2016). Therefore, USP14 on the proteasome displays unprecedented and surprisingly sophisticated specificity, discriminating very sensitively among substrates on the basis of diverse configurations of the ubiquitin conjugates. As a result of this specificity, USP14 inhibitors are expected to modulate the degradation of a limited subset of proteasome substrates.

For the study of USP14, we have established robust in vitro assay systems for examining proteasome-associated USP14 activities (Lee, Finley, & King, 2012; Lee et al., 2010, 2016). In this chapter, we describe essential materials and procedures to perform USP14 DUB assays. The schematic workflow is outlined in Fig. 1.

2. Purification of human proteasome

Human proteasomes can be purified from HEK293 cells stably expressing RPN11 tagged with C-terminal biotinylation and TEV cleavage sites (a kind gift from L. Huang, UC Irvine) as previously described (Lee et al., 2012, 2016). Detailed procedures are given below.

Step 1. By standard mammalian cell culture in DMEM containing 10% FBS, 1% penicillin/streptomycin (Pen/Strep), and 1% Gln, grow tagged

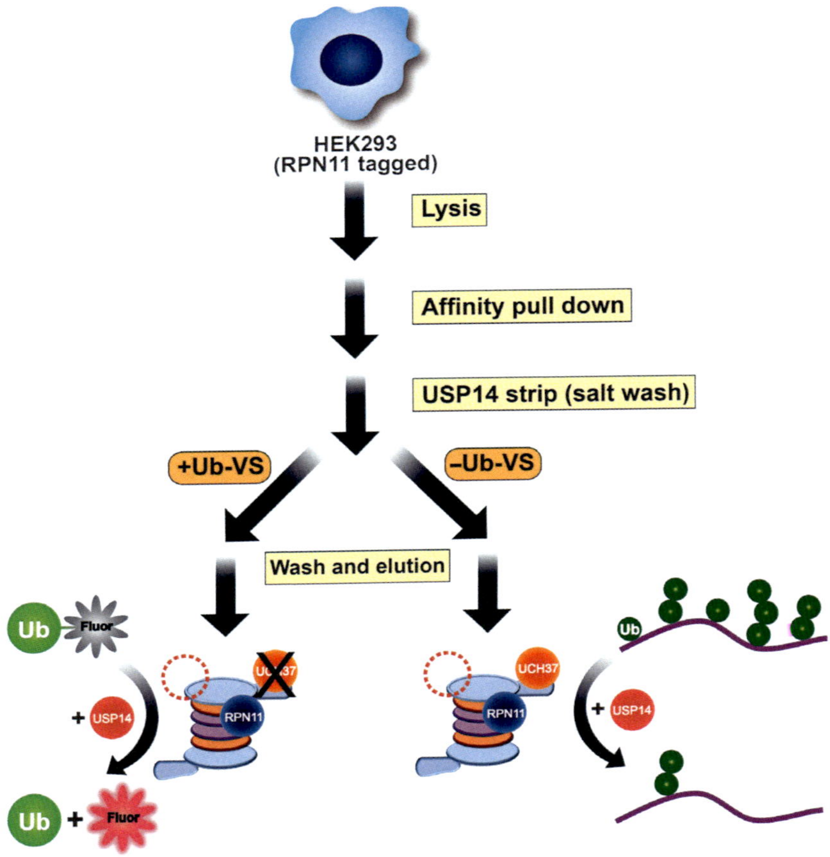

Fig. 1 The schematic workflow for the preparation of human proteasomes and in vitro reconstitution assays with proteasome, USP14, and substrates (e.g., fluorogenic ubiquitin adducts or ubiquitin-protein conjugates). *Dotted circle* indicates that endogenous USP14 is stripped off by salt wash during proteasome purification. Ub-VS treatment inactivates UCH37 on the proteasome (*X mark* on the UCH37).

RPN11-expressing HEK293 cells in 1 L of monolayer culture to full confluence.

Step 2. The cells are washed once in ice-cold PBS (about 15 mL of packed volume) and resuspended in at least 2 bed volumes of chilled lysis buffer A (50 m*M* NaH_2PO_4 [pH 7.5], 100 m*M* NaCl, 10% glycerol, 5 m*M* $MgCl_2$, 0.5% NP-40, 5 m*M* ATP, and 1 m*M* DTT) containing protease inhibitors (complete cocktail tablet [Roche or equivalent source] and 1 m*M* phenylmethylsulfonyl fluoride [PMSF]).

Step 3. Lyse the resuspended cells with a tight-fitting type B pestle of 40 mL Dounce homogenizer (Pyrex or Wheaton). After 15 strokes, clear the cell lysate by centrifuging at 20,000 × *g*, at 4°C, for 30 min.
Step 4. Incubate the cleared lysate with a buffer A-prewashed 2 mL slurry of NeutrAvidin agarose resin (Thermo Fisher) for at least 2 h to overnight at 4°C. Pack the beads-containing lysate into a chromatography column by gravity flow and wash with excess buffer A containing protease inhibitors (at least 20 bed volumes) followed by 10 bed volumes of buffer B (50 m*M* Tris–HCl [pH 7.5]) containing 10% glycerol and 1 m*M* ATP-$MgCl_2$ in the cold room.
Step 5. Further wash the proteasome-bound beads with 5 bed volumes of buffer B containing 1 m*M* ATP-$MgCl_2$ at room temperature. The resin is then treated with 0.5–1 bed volume of buffer B containing 1 m*M* ATP-$MgCl_2$ and at least 2 μ*M* of ubiquitin-vinyl sulfone (Ub-VS; R&D Systems) (1.5–2 mL total volume, accounting for beads).
Step 6. Incubate the Ub-VS-containing slurry for at least 3 h at 30°C with occasional agitation by gently flicking the column. For Ub-VS-untreated control, perform the procedure in parallel with the same total volume of corresponding buffer without Ub-VS.
Step 7. Remove unreacted residual Ub-VS by washing the beads with 30 bed volumes of buffer B containing 1 m*M* ATP-$MgCl_2$. The beads are then treated with 2 bed volumes of buffer B containing 1 m*M* ATP-$MgCl_2$ and 6 μL of TEV protease (10 U/μL; Thermo Fisher) and incubated for at least 1 h at 30°C with occasional agitation.
Step 8. Elute Ub-VS-treated and untreated human proteasomes from the column by gravity and further collect each 0.5 mL of 2–3 fractions by applying additional buffer B containing 1 m*M* ATP-$MgCl_2$ onto the beads. Measure the protein concentration, pool the desired fractions, add glycerol to a final concentration of 10% (v/v), prepare multiple aliquots, and snap freeze them in liquid nitrogen for storage at −80°C. Analyze the final product of purified proteasomes by SDS–PAGE/Coomassie Brilliant Blue (CBB) staining and in-gel peptidase assay to check their integrity and quality (Fig. 2A). (Optional) To deplete ATP from the proteasome prep, purified proteasomes (at the step before adding glycerol) are incubated with reaction buffer C (50 m*M* Tris–HCl [pH 7.5] and 5 m*M* $MgCl_2$) containing 20 m*M* glucose and 0.2 U/μL hexokinase (Sigma) for 1 h at 30°C, and then add glycerol to a final concentration of 10% for storage at −80°C after snap freezing.

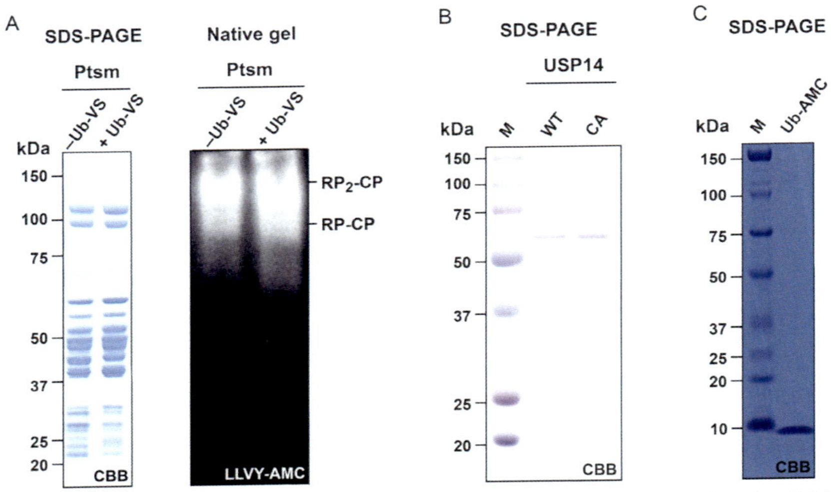

Fig. 2 Preparation of essential reagents for proteasome-associated USP14 deubiquitination assays. (A) SDS–PAGE and Coomassie Brilliant Blue (CBB) staining of purified proteasomes (Ptsm) treated or untreated with Ub-VS (*left panel*, 6.5 μg each). Native gel and in-gel LLVY-AMC staining of each proteasome (*right panel*, 6.5 μg each). (B) SDS–PAGE and CBB staining of purified USP14 WT and C114A (each 0.5 μg). (C) SDS–PAGE and CBB staining of purified Ub-AMC (1 μg).

2.1 Comments on human proteasome purification

1. Endogenous USP14 is stripped off from the proteasome by washing with lysis buffer A containing 100 m*M* NaCl (Step 4). This step eliminates any interference from proteasome-bound endogenous USP14 in the later DUB assays, for which various recombinant USP14 proteins can be reconstituted with proteasomes (Figs. 1–4).
2. Ub-VS, a thiol protease DUB-specific suicide inhibitor, is used to inactivate the basal DUB activity from UCH37 on the proteasome. Note that UCH37 is highly active for fluorogenic Ub adducts such as ubiquitin-7-amino-4-methylcoumarin (Ub-AMC) or ubiquitin-rhodamine 110 (Ub-Rho110), while this UCH type of enzyme shows very poor DUB activity for isopeptide-linked (i.e., Ub C-terminus linked to a lysine side chain) Ub conjugates (see Sections 6 and 7; Figs. 3 and 4). By contrast, the metalloprotease RPN11 is expected to have only modest activity for fluorogenic Ub adducts under the conditions of our DUB assays.
3. As an alternative to salt washing or Ub-VS treatment, one may generate and employ gRNA/CRISPR knockout human cell lines devoid of

USP14 and/or UCH37 for proteasome purification, but the integrity of 26S proteasome from such cell lines should be validated.

4. For DUB assays with "ADP-proteasome" which was used to suppress substrate degradation, ATP must be acutely depleted from purified proteasomes to preserve their integrity because proteasome purified in the presence of ADP may be dissociated during the course of the purification procedures described above.

3. Purification of recombinant USP14 protein

Human USP14 protein of high yield and purity can be purified from *E. coli*. GST-, intein-chitin binding domain (CBD)-, or His-tagged USP14 have all been successfully purified with high activity and little proteolysis. In this section, we describe a method for purifying N-terminal His-tagged human USP14 (Fig. 2B).

Step 1. Transform the expression vector pET15b-USP14 (kindly provided by G. Tian, D. Finley lab, Harvard Medical School) into *E. coli* Rosetta 2 (Novagen) strain.

Step 2. After inoculating an overnight liquid culture with a single colony, grow 1 L of the *E. coli* cells in LB medium containing 50 μg/mL of ampicillin and 34 μg/mL of chloramphenicol at 37°C in a shaking incubator until OD_{600} reaches 0.5–0.6. Induce protein expression by adding IPTG to a final concentration of 0.5 m*M*, and grow the cells at 16°C overnight.

Step 3. Harvest the cells by centrifuging 25 min at 3300 × *g*, 4°C. Discard the supernatant and resuspend the cell pellets with 20 mL of ice-cold lysis buffer (50 m*M* NaH_2PO_4 [pH 8.0], 100 m*M* NaCl, 10% glycerol, and 25 m*M* imidazole) containing 1 m*M* β-mercaptoethanol (β-ME) and protease inhibitors.

Step 4. Lyse the cells by using a French press (at 1000–1200 psi) or by sonication (according to the manufacturer's instructions), and clear the cell lysate by centrifuging 30 min at 20,000 × *g*, 4°C. Meanwhile, prepare a 2 mL slurry of Ni-NTA agarose resin (Qiagen) by washing with 10–20 bed volumes of lysis buffer.

Step 5. Combine the cleared lysate with the washed resin and incubate with rotation for 1 h in the cold room. Pack the resin-containing lysate into an Econo-Column chromatography column (Bio-Rad) by gravity flow in the cold room. Wash the beads 4 times with 20 mL of wash buffer (lysis buffer containing 300 m*M* NaCl) per wash.

Step 6. Elute 5–7 equal fractions using 2 bed volumes of elution buffer (lysis buffer containing 250 m*M* of imidazole), measure the protein concentration, and check each fraction by SDS–PAGE gel followed by CBB staining. Pool the desired fractions for chromatography using AKTA Fast Protein Liquid Chromatography (FPLC) system (GE Healthcare).

Step 7. FPLC is performed according to the manufacturer's instructions. Briefly, an anion exchange HiTrap Q HP column (GE Healthcare) is equilibrated with HiTrap Q buffer (20 m*M* Tris–HCl [pH 7.5] and 1 m*M* DTT), loaded with the sample, and then eluted with a gradient of 0%–100% of 1 *M* NaCl at a flow rate of 1 mL/min. Combine and concentrate the desired fractions, and subsequently load the sample onto the preequilibrated HiLoad Superdex 200 PG column (16/600; GE Healthcare) for running ~1.25 bed volume of gel filtration buffer (20 m*M* Tris–HCl [pH 7.5], 100 m*M* NaCl, and 1 m*M* DTT) with a flow rate of 0.8 mL/min. Pool the desired fractions, concentrate to a final concentration of at least 1 mg/mL by using an Amicon ultracentrifugation filter (30 kDa MW cut-off; Millipore), add glycerol to a final concentration of 10%, check by SDS–PAGE/CBB staining (Fig. 2B), and snap freeze multiple aliquots for storage at −80°C.

3.1 Comments on USP14 protein purification

1. We routinely perform FPLC for preparing extremely high purity of recombinant USP14 protein, but even after Ni-NTA beads elution (at Step 6), it is expected that the USP14 protein is generally good in purity as judged by SDS–PAGE/CBB staining.

4. Generation of fluorogenic ubiquitin adducts as model substrates for USP14 assays

Various model substrates for DUB assays, such as Ub adducts with small leaving groups, reporter-based Ub substrates, and Förster Resonance Energy Transfer (FRET)- or Fluorescence Polarization (FP)-based Ub substrates are commercially available from suppliers (e.g., R&D Systems, LifeSensors, UbiQ, Ubiquigent, Sigma, and Enzo Life Sciences). Among them, Ub-AMC and Ub-Rho110 are most routinely used as fluorogenic DUB substrates. The fluorescence moieties of AMC and Rho110 are quenched when appended to Ub but they become fluorescent after liberation by DUB catalytic activity. FP- or FRET-based fluorophore-labeled diubiquitin substrates have advantages over fluorogenic Ub adducts especially for linkage-specific DUBs. However, these substrates work poorly for proteasome-associated USP14 since this enzyme prefers to cut the base of the chains from the substrate. Ub-AMC and Ub-Rho110 can be purchased from commercial suppliers for most of in vitro applications. Previously, for high-throughput USP14 inhibitor screening, we also successfully synthesized and purified a large amount of Ub-AMC to high purity by intein chemistry (Lee et al., 2012) (Fig. 2C). In this section, we briefly outline the methods for obtaining custom-made Ub-AMC.

4.1 Generation of ubiquitin thioesters

Step 1. Ub thioester preparation is mostly performed according to the IMPACT-CN manual (New England Biolabs [NEB]). By standard methods, transform the expression vector pTYB2-Ub_{75}-intein-CBD (custom-made with pTYB2 parental vector from NEB) into strain BL21 (Novagen), plating on media containing 50 μg/mL of ampicillin. Grow 1 L of the *E. coli* culture at 37°C until the OD_{600} reaches ~0.6. Add IPTG to a final concentration of 0.5 m*M* to induce protein expression, and grow the cells for 3 h at 30°C.
Step 2. Harvest and lyse the cells in Ub-intein lysis buffer (50 m*M* HEPES [pH 6.5] and 100 m*M* NaOAc) containing protease inhibitors. The cleared cell lysate is loaded onto a 15 mL slurry of chitin-binding resin (NEB), preequilibrated in Ub-intein lysis buffer, in an Econo-Column with a flow rate of 0.5–0.8 mL/min. Wash the beads with at least 10 bed volumes of Ub-intein lysis buffer in the cold followed by quick flush with 30–50 mL of the lysis buffer containing 50 m*M* 2-mercaptoethanesulfonic acid, sodium salt (MESNa; Sigma) at room temperature. Apply 1 bed volume of the lysis buffer containing 50 m*M* MESNa to the resin for on-column cleavage, 37°C, overnight.
Step 3. Elute the Ub_{75}-MESNa thioester by gravity and apply 2 bed volumes of lysis buffer without containing MESNa for further elution. Combine and concentrate the eluate to ~3 mg/mL by using an Amicon ultracentrifugation filter (5 kDa MW cut-off; Millipore), and proceed to Ub-AMC synthesis (below) or snap freeze the Ub thioesters for low temperature storage at −80°C until use.

4.2 Ub-AMC synthesis and purification

Step 4. To synthesize Ub-AMC, first dissolve ~54 mg of Gly-AMC salt (Glycosynth or Gold Biotechnology) in 4 mL of 70% dimethyl formamide (DMF) (Sigma) through vigorous vortexing. Add 200 μL of 2 *M* *N*-hydroxysuccinimide (Sigma) in DMF and then slowly add 1 mL of Ub-MESNa thioester (~3 mg) to the Gly-AMC solution while vortexing, and mix well. Initiate the reaction by adding 250 μL of 2 *M* HEPES base (pH reaches approximately 10) and incubate overnight at 37°C.

Step 5. Stop the reaction by adding 250 μL of 5% HCl while vortexing and solubilize any precipitates. Remove insoluble materials by centrifuging 30 min at 20,000 × *g*, at 4°C, and recover the supernatant. Dialyze out excess unreacted Gly-AMC in dialysis buffer (50 m*M* NaOAc [pH 4.5]) by four changes of 4 L of buffer, each ~2 h, in the cold room.

Step 6. Concentrate up to 5–10 mL by using an Amicon ultracentrifugation filter (5 kDa MW cut-off; Millipore) and centrifuge the final mixture at 20,000 × *g*, at 4°C, for 20 min before performing FPLC. Load the cleared sample onto a cation exchange HiTrap SP HP column (GE Healthcare) that is preequilibrated with dialysis buffer and then elute with a gradient of 0%–70% of 1 *M* NaCl at a flow rate of 1 mL/min. Analyze the fractions by SDS–PAGE and CBB staining, collect the desired fractions, and concentrate by ultracentrifugation, using an Amicon filter (5 kDa cut-off). Check the end product by SDS–PAGE/CBB staining (Fig. 2C), and snap freeze multiple aliquots for storage at −80°C.

4.3 Comments on Ub-AMC synthesis

1. Precipitation generally occurs during the reaction at Step 4. If there is massive precipitation, try to vary the reaction conditions, such as reducing the amount of Gly-AMC, increasing the amount of DMF for solubilization, or decreasing the amount of Ub thioester.
2. During the FPLC run, the residual Gly-AMC will be removed, and self-hydrolyzed Ub_{75}, which is a major contaminant in the reaction, will elute earlier than Ub-AMC due to its C-terminal carboxyl group (e.g., hydrolyzed Ub_{75} at ~20% buffer B and Ub-AMC at ~30% buffer B, respectively).
3. By mass spectrometry, molecular weights of purified Ub-AMC and hydrolyzed Ub_{75} were 8725 and 8508 Da, respectively (data not shown).
4. Likewise, Ub-Rho110 can be synthesized as previously reported (Hassiepen et al., 2007).

5. Generation of ubiquitin conjugates as substrates for USP14 assays

Although fluorogenic Ub adducts should provide convenient and quantitative readouts as DUB substrates, they are not suitable to monitor the proteasomal degradation that can be regulated by USP14 DUB activity. Moreover, DUB substrates with small leaving groups may not reflect physiological DUB specificity compared to the isopeptide-linked Ub-protein conjugates. As noted above, UCH37 on the proteasome predominates over USP14 for cleavage of fluorescent model substrates (e.g., Ub-AMC or Ub-Rho110), but its DUB activity is typically weak for more physiological test substrates such as Ub-protein conjugates. By contrast, USP14 specifically deubiquitinates proteasome-bound conjugates that are ubiquitinated at multiple sites and by en bloc chain cleavage (Lee et al., 2016). Thus, USP14 activity can discriminate among diverse chain architectures of Ub-protein conjugates in the context of the proteasome. We routinely prepare various Ub conjugates by employing immunopurified anaphase-promoting complex/cyclosome (APC/C) from *Xenopus* egg extracts as the E3 Ub ligase and an N-terminal fragment of cyclin B1 as substrate (Dimova et al., 2012; Lee et al., 2016). Essential procedures are briefly described below.

Step 1. Purify the disordered N-terminal 88-residue segment of cyclin B1 (NCB1; wild type or K to R mutants) (kindly provided by N. Dimova, R. King lab, Harvard Medical School) containing an HA-tag and a 6xHis-tag at its N-terminus and C-terminus, respectively, as previously described (Dimova et al., 2012).

Step 2. Interphase egg extracts from *Xenopus laevis* can be prepared and induced to enter mitosis by adding nondegradable cyclin B1 (MBP-Δ 90) as described previously (Dimova et al., 2012; Murray, 1991). Immunopurify APC/C from ~1 mL of mitotic egg extract with 18 μg of anti-Cdc27 antibodies (0.2 mg/mL; Santa Cruz Biotechnologies) bound to 40 μL (settled volume) of protein A-Sepharose 4B fast flow beads (Sigma) (Dimova et al., 2012; Lee et al., 2016).

Step 3. Ubiquitination reactions are carried out by incubating 2.5 μ*M* MBP-E1, 3.8 μ*M* UbcH10, 133 μ*M* ubiquitin (wild type or mutant; various commercial sources), and 2.3 μ*M* NCB1 on the APC-bound beads in the reaction buffer (20 m*M* Tris–HCl [pH 7.5], 100 m*M* KCl, 2 m*M* ATP, and 2.5 m*M* $MgCl_2$) for at least 90 min while vortexing on a rotator at room temperature. Where appropriate, 3.8 μ*M* of Ubc4 (for making multi-Ub chain conjugate with different topology), 2 μ*M* of Ube2S (along with 0.7 μ*M* of UbcH10, for K11-linked chain elongation), or other type of substrates can be used. E1 and E2 enzymes were purified as previously reported (Dimova et al., 2012).

Step 4. Collect the reaction mixture following a brief centrifugation to pellet the APC/C beads. Add 1 bed volume of reaction buffer lacking salt to the APC/C beads and wash the beads by gently flicking the tube. Retrieve the supernatant and combine with the reaction mixtures.

5.1 Comments on ubiquitin conjugate synthesis

1. Depending on the E2s, whether Ub is wild type or mutant, and whether NCB1 is wild type or mutant, various types of conjugates will be produced. Typically, the UbcH10 with wild-type Ub can generate highly heterogeneous multichain conjugates with short chains and different linkages (Kirkpatrick et al., 2006). In this type of conjugate, with more than one chain present, the "extra" chains are referred to as supernumerary chains.
2. For conjugation with preformed Ub chains, 18.4 μ*M* (K11-, K48-, or K63-linked Ub_4) or 10.5 μ*M* (K63-linked Ub_8) (LifeSensors or R&D systems) of the free chain is incubated with 0.7–1 μ*M* NCB1 and UbcH10 as an E2.
3. As an alternative to immuno-pulldown from the *Xenopus* mitotic egg extract, APC/C complexes can be purified by employing a recently developed biGBac expression system (Weissmann et al., 2016).
4. $Sic1^{PY}$-Ub conjugates generated with the Rsp5 E3 ligase and Ubc4 E2 are easy to prepare and have also been successfully used for USP14 DUB assays (Lee et al., 2010; Saeki, Isono, & Toh, 2005).

6. Proteasome-associated USP14 deubiquitination assays with fluorogenic ubiquitin adducts

As described above, UCH37 shows a far stronger DUB activity than USP14 for fluorogenic Ub adducts, which had been a major hurdle to study the role of USP14 on the proteasome (Lee et al., 2010) (Fig. 3). Recently, we established a method to specifically isolate USP14 activity on the proteasome by inactivating UCH37 with Ub-VS, a DUB-specific irreversible thiol protease inhibitor during the proteasome purification

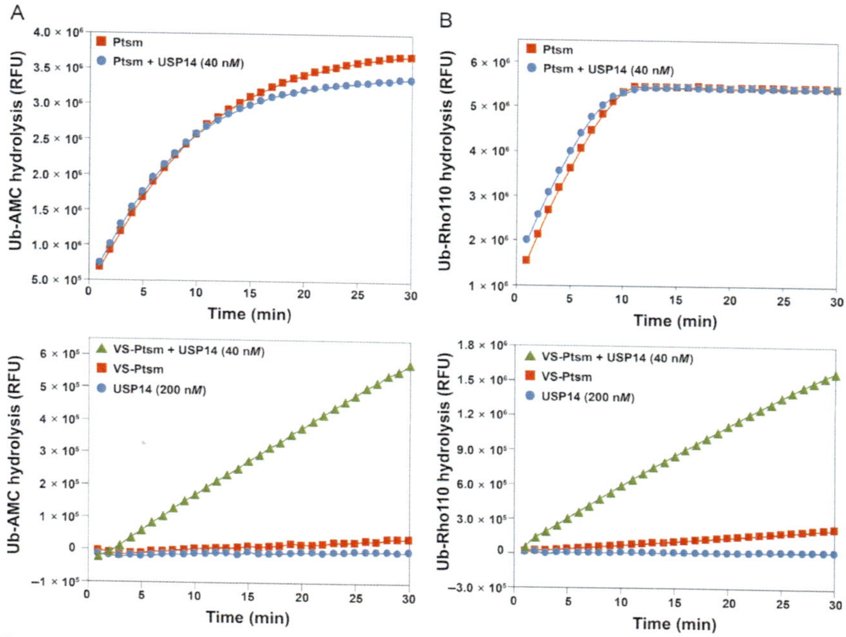

Fig. 3 In vitro analysis of proteasome-associated USP14 activities with the fluorogenic ubiquitin adducts. (A) Ub-AMC hydrolysis assays of USP14 activities with Ptsm (*top panel*) or VS-Ptsm (*bottom panel*). (B) Ub-Rho110 hydrolysis assays of USP14 activities with Ptsm (*top panel*) and VS-Ptsm (*bottom panel*). Ptsm or VS-Ptsm: 1 n*M*, Ub-AMC or Ub-Rho110: 1 μ*M*, RFU: relative fluorescence units. Note that for the *top panel* in (B), the fluorescence intensity is saturated at ~10 min due to the detection limit of the microplate reader.

(Figs. 1 and 2A) (Lee et al., 2010). When such Ub-VS-treated human proteasomes (hereafter, VS-Ptsm) were reconstituted with recombinant USP14, USP14 activity can be increased up to ~1000-fold over free USP14 (i.e., USP14 in the absence of proteasome), an extreme degree of activation that has never been observed with other DUBs (Lee et al., 2010). These data also strongly suggest that USP14 activity will be tightly regulated only in the context of proteasome association. In this section, we describe in vitro assays for measuring proteasome-associated USP14 DUB activities with fluorogenic Ub adducts as substrates.

Step 1. Prepare the Ub-AMC assay buffer (50 m*M* Tris–HCl [pH 7.5], 1 m*M* EDTA, and 1 mg/mL ovalbumin [Sigma]) containing freshly added 1 m*M* DTT and 1 m*M* ATP-$MgCl_2$. Set up a 384-well microplate (low volume, nonbinding, flat-bottom; Corning).

Step 2. Thaw appropriate amounts of USP14, VS-Ptsm, Ptsm (Ub-VS-untreated proteasome, as a control), and Ub-AMC (or Ub-Rho110) for at least triplicate measurements per condition. For standard assays, plan the final concentration of USP14, VS-Ptsm (or Ptsm), and Ub-AMC (or Ub-Rho110) to be 4–400 n*M*, 1 n*M*, and ~1 μ*M*, respectively. The total assay volume per well is 20 μL. Prepare a titration series of recombinant USP14 by diluting the stock protein in the assay buffer. Where appropriate, preincubate 2 n*M* of VS-Ptsm or Ptsm with USP14 for at least 10 min. Dispense 10 μL of USP14 solution in the presence or absence of proteasomes into each well of the 384-well plate.
Step 3. Prepare master mixtures containing 2 μ*M* of Ub-AMC or Ub-Rho110 in assay buffer. Dispense 10 μL of the mixture into each well containing USP14 or USP14/proteasome to initiate the reaction.
Step 4. Monitor the Ub-AMC (or Ub-Rho110) hydrolysis in real time at room temperature by measuring fluorescence every minute for about 30 min at Ex365/Em460 (or Ex485/Em535 for Ub-Rho110) with an Envision plate reader (Perkin Elmer) equipped with a mirror of appropriate band-pass cut-off (e.g., LANCE/DELFIA, 400 nm for Ub-AMC and FITC, 505 nm for Ub-Rho110) (Fig. 3).
Step 5. Determine the DUB activity of USP14-mediated Ub adduct cleavage by subtracting the signal of Ub-AMC (or Ub-Rho110) alone and/or VS-Ptsm alone, and then calculate the slopes in a given linear time span to calculate fold-activation of USP14 by proteasome.

6.1 Comments on USP14 deubiquitination assays with fluorogenic Ub adducts

1. With Ub-VS-untreated proteasome (Ptsm), USP14 activation by the proteasome is difficult to assess due to high basal activity from UCH37 (Fig. 3, top panels).
2. VS-Ptsm may have some residual DUB activity depending on the proteasome preparation (Fig. 3, bottom panels).
3. Free form of USP14 even at 200 n*M* shows very low activity, while as little as tens of n*M* USP14 can be highly activated in association with VS-Ptsm (Fig. 3, bottom panels).
4. If necessary, a catalytically inactive mutant of USP14 (e.g., USP14 C114A) or specific small-molecule inhibitors of USP14 (e.g., IU1 and its derivatives) can be employed to test the catalytic effects of USP14.
5. Ovalbumin and nonbinding microplate reduce the nonspecific binding and yield more consistent readouts for these assays.

7. Proteasome-associated USP14 deubiquitination assays with ubiquitin-protein conjugates

By high-throughput chemical screening and structure–activity relationship studies, we identified potent inhibitors for specifically targeting the catalytic activity of proteasome-bound USP14 (Boselli et al., 2017; Lee et al., 2010). In cells and in vitro studies, USP14 inhibitors showed selective effects on the turnover of a subset of proteasome substrates, suggesting that USP14 activity may discriminate among different proteasome-bound substrates (Boselli et al., 2017; Chen et al., 2018; Lee et al., 2010, 2016). Based on this finding, we also discovered that USP14 has a marked preference for multichain ubiquitin conjugates and an en bloc cleavage mechanism (Lee et al., 2016). We successfully established in vitro methods to investigate proteasome-associated USP14 activities with various ubiquitin conjugates as substrates (Lee et al., 2016), and in this section, detailed procedures will be described.

Step 1-1. Prepare the degradation assay buffer (50 m*M* Tris–HCl [pH 7.5]) containing freshly added 5 m*M* ATP-$MgCl_2$. Thaw appropriate amounts of purified human proteasomes (not treated with Ub-VS), recombinant USP14, and the desired Ub conjugates, and put the reagents on ice. Ub conjugates may include Ub_n-NCB1 (e.g., typical multishort chain conjugate), K0Ub_n-NCB1 (e.g., multimono Ub conjugate generated with Lys-less Ub), Ub_n-1KNCB1 (e.g., single chain conjugate generated on single Lys NCB1), and more defined Ub conjugates generated with preformed free chains. The final concentration of proteasome, USP14, and Ub conjugate is 3–5, 80, and ~110 n*M*, respectively.

Step 1-2. For USP14 deubiquitination assays uncoupled from degradation, ATP-depleted "ADP-proteasome" (e.g., ADP-Ptsm) can be prepared as described above. ADP-Ptsm is used for assays in a buffer of 50 m*M* Tris–HCl (pH 7.5) supplemented with 3–5 m*M* "ATP-free" ADP-$MgCl_2$. For removing residual ATP contamination, commercial ADP (Calbiochem) was first dissolved in water at 0.5–1 *M* at pH 7.0, and then treated with hexokinase and glucose in the reaction buffer as described above for ATP depletion.

Step 2. Purified human proteasomes are incubated with each Ub conjugate at room temperature, and where indicated, USP14 is preincubated with proteasome for at least 5 min before adding the substrates.

Step 3. Take a 12μL aliquot from the reaction mixture at each time point (e.g., 0, 5, and 10min), terminate the reaction by immediately mixing with predispensed 3.3μL of 5× SDS–PAGE sample buffer, and heat the sample for 10min at 70°C. For real time 0min without any time lag, enzyme mixture and substrate mixture are separately quenched into 5× SDS–PAGE sample buffer prior to incubation.

Step 4. The samples are subsequently subjected to SDS–PAGE in a 4%–12% Bis–Tris gradient gel in MES or MOPS buffer (Thermo Fisher) and anti-HA (NCB1) immunoblot analysis (Fig. 4).

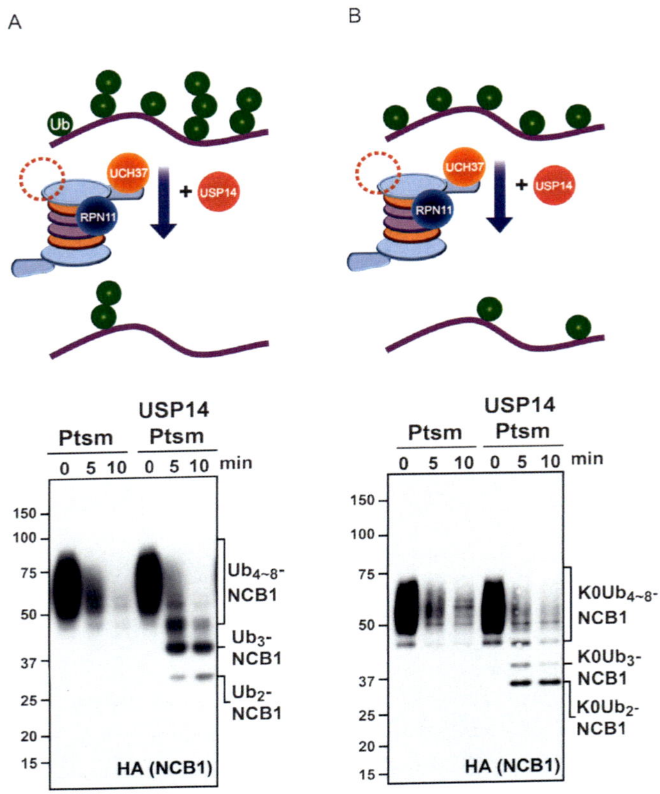

Fig. 4 In vitro degradation and deubiquitination assays of USP14-reconstituted proteasomes with ubiquitin conjugates. Approximately 110n*M* of polyubiquitinated NCB1 (Ub_n-NCB1) (A), multiply monoubiquitinated NCBI ($K0Ub_n$–NCB1) (B), or single chain conjugate on K64-only NCB1 (Ub_n–1KNCB1)

(Continued)

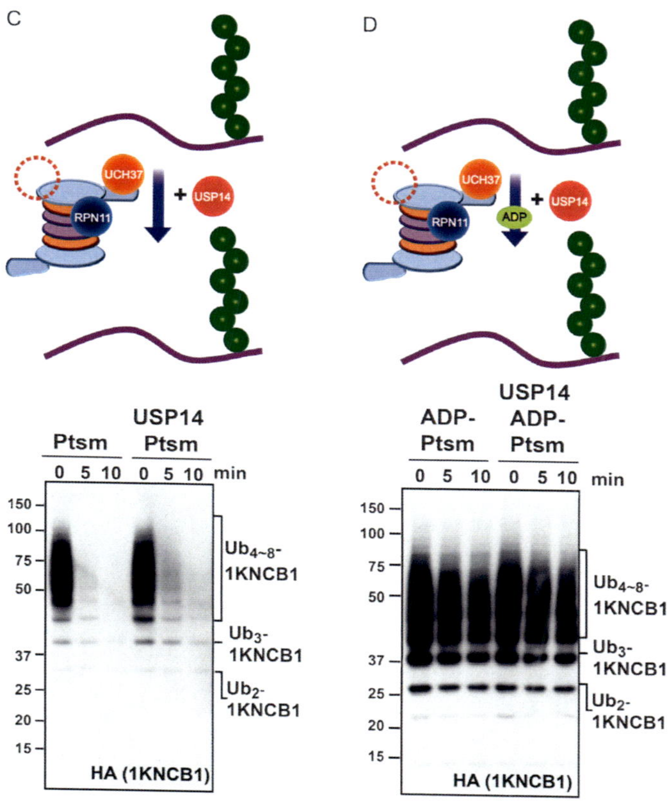

Fig. 4—Cont'd (C) was incubated with purified proteasomes (Ptsm, 4 n*M* each); to test the role of ATP, ~110 n*M* of Ub_n–1KNCB1 was incubated with ADP-proteasome (ADP-Ptsm, 4 n*M*) in the presence of ADP (D). Where indicated, 80 n*M* USP14 was preincubated with proteasomes for 5 min. Samples were analyzed by SDS–PAGE/immunoblotting for HA-tag. *This figure is adapted from Lee, B. H., Lu, Y., Prado, M. A., Shi, Y., Tian, G., Sun, S., et al. (2016). USP14 deubiquitinates proteasome-bound substrates that are ubiquitinated at multiple sites.* Nature, 532, *398–401 with some modification.*

7.1 Comments on USP14 deubiquitination assays with ubiquitin conjugates

1. Proteasome-bound USP14 efficiently cleaves the substrates that are ubiquitinated at multiple sites, but this enzyme scarcely acts on single chain conjugates, regardless of their degradation rate (Fig. 4).
2. In vitro deubiquitination assays with VS-Ptsm vs Ptsm produced very comparable results for Ub conjugates, indicating little contribution of UCH37 to Ub conjugate cleavage by human proteasomes (data not shown).

3. If necessary, USP14 C114A and small-molecule USP14 inhibitors can be employed to test the catalytic effects of USP14.
4. To monitor the released Ub or Ub chains from these reactions, rather than the HA-tagged substrate, it is necessary for the Ub-substrate conjugates to be further purified prior to the reaction to remove excess unreacted free Ub or chains as described previously (Lee et al., 2016).
5. The deubiquitinating activity of USP14 is very fast on favorable substrates such as Ub_n-NCB1, with a timescale as short as millisecond to seconds, depending on conditions. Consequently, it is difficult, when manual sampling is used, to take time points quickly enough to assess early deubiquitination events. We have addressed this important problem by adapting both quench-flow and single-molecule procedures to the study of deubiquitination (Lee et al., 2016). Significant instrumentation is required in both cases, and the detailed methods are described in Lee et al. (2016).

8. Summary

In this chapter, we describe practical information to prepare enzymes and substrates for USP14 DUB assays, and also provide working procedures to examine USP14 activities in association with human proteasomes. USP14 is a critical regulator of proteasome function since it can serve as a potent inhibitory factor on the proteasome. USP14 is highly activated upon binding the proteasome: we observed that free USP14 shows little hydrolysis toward fluorogenic Ub adducts, but surprisingly, its activity can be enhanced up to ~1000-fold in association with the proteasome (Fig. 3) (Lee et al., 2010). By employing Ub conjugates as more physiological substrates, we also found that USP14 has a novel selectivity, very efficiently working on multichain-bearing conjugates, which occurs only upon proteasome binding (Fig. 4) (Lee et al., 2016). The lack of USP14 activity on the single chain substrate suggests that supernumerary chain cleavage on proteasome-bound substrates is the principal role of USP14 in cells. In summary, many of our critical findings have been enabled by in vitro methods that we developed for USP14 and proteasome studies. The protocols here provide detailed guidelines for others to exploit these methods to elucidate additional, unexplored aspects of USP14 biology.

Acknowledgments

We thank D. Finley (Harvard Med School) and S.C. Son (DGIST) for critical comments on the manuscript. We are grateful to H.J. Park for graphical assistance. B.-H.L. gratefully acknowledges the National Research Foundation of Korea (NRF) grant, DGIST Start-up Fund, and HRHR grant funded by the Ministry of Science, ICT, and Future Planning (2017R1A2B4006671, 2018010095, and 2018010146) for funding support.

References

Balch, W. E., Morimoto, R. I., Dillin, A., & Kelly, J. W. (2008). Adapting proteostasis for disease intervention. *Science, 319*, 916–919.

Boselli, M., Lee, B. H., Robert, J., Prado, M. A., Min, S. W., Cheng, C., et al. (2017). An inhibitor of the proteasomal deubiquitinating enzyme USP14 induces tau elimination in cultured neurons. *The Journal of Biological Chemistry, 292*, 19209–19225.

Chakraborty, J., von Stockum, S., Marchesan, E., Caicci, F., Ferrari, V., Rakovic, A., et al. (2018). USP14 inhibition corrects an in vivo model of impaired mitophagy. *EMBO Molecular Medicine, 10*, e9014.

Chen, L., Zhu, G., Johns, E. M., & Yang, X. (2018). TRIM11 activates the proteasome and promotes overall protein degradation by regulating USP14. *Nature Communications, 9*, 1223.

Ciechanover, A. (2005). Proteolysis: From the lysosome to ubiquitin and the proteasome. *Nature Reviews. Molecular Cell Biology, 6*, 79–87.

de Poot, S. A. H., Tian, G., & Finley, D. (2017). Meddling with fate: The proteasomal deubiquitinating enzymes. *Journal of Molecular Biology, 429*, 3525–3545.

Dimova, N. V., Hathaway, N. A., Lee, B. H., Kirkpatrick, D. S., Berkowitz, M. L., Gygi, S. P., et al. (2012). APC/C-mediated multiple monoubiquitylation provides an alternative degradation signal for cyclin B1. *Nature Cell Biology, 14*, 168–176.

Finley, D. (2009). Recognition and processing of ubiquitin-protein conjugates by the proteasome. *Annual Review of Biochemistry, 78*, 477–513.

Harrigan, J. A., Jacq, X., Martin, N. M., & Jackson, S. P. (2018). Deubiquitylating enzymes and drug discovery: Emerging opportunities. *Nature Reviews. Drug Discovery, 17*, 57–78.

Hassiepen, U., Eidhoff, U., Meder, G., Bulber, J. F., Hein, A., Bodendorf, U., et al. (2007). A sensitive fluorescence intensity assay for deubiquitinating proteases using ubiquitin-rhodamine110-glycine as substrate. *Analytical Biochemistry, 371*, 201–207.

Hochstrasser, M. (1996). Ubiquitin-dependent protein degradation. *Annual Review of Genetics, 30*, 405–439.

Homma, T., Ishibashi, D., Nakagaki, T., Fuse, T., Mori, T., Satoh, K., et al. (2015). Ubiquitin-specific protease 14 modulates degradation of cellular prion protein. *Scientific Reports, 5*, 11028.

Kirkpatrick, D. S., Hathaway, N. A., Hanna, J., Elsasser, S., Rush, J., Finley, D., et al. (2006). Quantitative analysis of in vitro ubiquitinated cyclin B1 reveals complex chain topology. *Nature Cell Biology, 8*, 700–710.

Lee, B. H., Finley, D., & King, R. W. (2012). A high-throughput screening method for identification of inhibitors of the deubiquitinating enzyme USP14. *Current Protocols in Chemical Biology, 4*(4), 311–330.

Lee, B. H., Lee, M. J., Park, S., Oh, D. C., Elsasser, S., Chen, P. C., et al. (2010). Enhancement of proteasome activity by a small-molecule inhibitor of USP14. *Nature, 467*, 179–184.

Lee, B. H., Lu, Y., Prado, M. A., Shi, Y., Tian, G., Sun, S., et al. (2016). USP14 deubiquitinates proteasome-bound substrates that are ubiquitinated at multiple sites. *Nature, 532*, 398–401.

McKinnon, C., Goold, R., Andre, R., Devoy, A., Ortega, Z., Moonga, J., et al. (2016). Prion-mediated neurodegeneration is associated with early impairment of the ubiquitin-proteasome system. *Acta Neuropathologica*, *131*, 411–425.

Moon, S., & Lee, B. H. (2018). Chemically induced cellular proteolysis: An emerging therapeutic strategy for undruggable targets. *Molecules and Cells*, *41*, 933–942.

Murray, A. W. (1991). Cell cycle extracts. *Methods in Cell Biology*, *36*, 581–605

Saeki, Y., Isono, E., & Toh, E. A. (2005). Preparation of ubiquitinated substrates by the PY motif-insertion method for monitoring 26S proteasome activity. *Methods in Enzymology*, *399*, 215–227.

Schrader, E. K., Harstad, K. G., & Matouschek, A. (2009). Targeting proteins for degradation. *Nature Chemical Biology*, *5*, 815–822.

VerPlank, J. J. S., Lokireddy, S., Feltri, M. L., Goldberg, A. L., & Wrabetz, L. (2018). Impairment of protein degradation and proteasome function in hereditary neuropathies. *Glia*, *66*, 379–395.

Wang, Y., Jiang, Y., Ding, S., Li, J., Song, N., Ren, Y., et al. (2018). Small molecule inhibitors reveal allosteric regulation of USP14 via steric blockade. *Cell Research*, *28*, 1186–1194.

Weissmann, F., Petzold, G., VanderLinden, R., Huis In't Veld, P. J., Brown, N. G., Lampert, F., et al. (2016). biGBac enables rapid gene assembly for the expression of large multisubunit protein complexes. *Proceedings of the National Academy of Sciences of the United States of America*, *113*, E2564–E2569.

CHAPTER TWELVE

Methods for studying the regulation of membrane traffic by ubiquitin and the ESCRT pathway

Evan L. Guiney[†], Lu Zhu[†], Richa Sardana, Scott D. Emr*, Matthew G. Baile
Weill Institute for Cell and Molecular Biology, Department of Molecular Biology and Genetics, Cornell University, Ithaca, NY, United States
*Corresponding author: e-mail address: sde26@cornell.edu

Contents

Abstract

Covalent modification of proteins with ubiquitin dynamically regulates their function and fate. The ubiquitination of most plasma membrane proteins initiates endocytosis and ESCRT-mediated sorting to the lysosomal lumen for degradation. Powerful genetic approaches in the budding yeast *Saccharomyces cerevisiae* have been particularly

[†] Contributed equally.

Methods in Enzymology, Volume 619
ISSN 0076-6879
https://doi.org/10.1016/bs.mie.2018.12.041
© 2019 Elsevier Inc.
All rights reserved.

instrumental in the discovery and elucidation of these molecular mechanisms, which are conserved in all eukaryotes. Here we provide two detailed protocols and tools for studying ubiquitination-dependent membrane trafficking mechanisms in yeast. The first utilizes fusions between a protein of interest and an auxotrophic marker to screen for mutants that affect ubiquitin-mediated endocytosis. The second method artificially ubiquitinates a protein of interest, allowing downstream trafficking steps to be studied independently from the regulatory signals that initiate endocytosis.

1. Introduction

A hallmark of eukaryotic cells is the presence of multiple functionally distinct membrane-enclosed organelles. The functional identity of these organelles is determined by their lipid compositions and by the unique repertoire of proteins embedded in their membranes (Munro, 2004). The correct delivery and timely removal of the organellar membrane proteins are therefore critical for maintaining cellular functions (MacGurn, Hsu, & Emr, 2012). Advances in our understanding of membrane protein trafficking have elucidated the important role of reversible attachment of ubiquitin to the membrane protein cargos in regulating their fate (Hicke & Dunn, 2003). The chief outcome of extended ubiquitination of a cargo is its sorting for degradation. In the secretory pathway, protein ubiquitination in the ER serves as a signal for delivery of the protein to the proteasome for degradation in the cytoplasm (Ellgaard, Molinari, & Helenius, 1999). However, almost all post-ER, secretory as well as endocytic, ubiquitinated cargos traffic to the lysosome for degradation (MacGurn et al., 2012; Piper & Luzio, 2007). The ubiquitinated transmembrane protein cargos are first sorted into the lumen of the endosome through a process orchestrated by a series of multiprotein complexes (ESCRT 0, I, II, III, and Vps4 AAA ATPase) that generate multivesicular bodies (MVBs) (Henne, Stenmark, & Emr, 2013; Hurley & Emr, 2006). Fusion of MVBs with the lysosomal membrane delivers the cargos to the lysosomal lumen where they are degraded by hydrolytic enzymes. Most of the sorting machinery components are conserved across eukaryotes and were initially uncovered by creative genetic selections in budding yeast, *Saccharomyces cerevisiae* (Jones, 1977; Robinson, Klionsky, Banta, & Emr, 1988; Rothman & Stevens, 1986).

In this chapter, we highlight two key approaches that have been useful in the identification of ubiquitination as well as sorting machinery for cargos destined for degradation in the yeast lysosome-like vacuole. The first approach takes advantage of the fact that sorting into MVBs sequesters the proteins from the cytoplasm. Therefore, fusion of selectable reporters to the cytoplasmic domains of transmembrane proteins allows one to design genetic screens to identify the machinery and the mechanisms that regulate the degradation of the chimeric proteins. A few such genetic selections employed by our lab were highly successful in uncovering components of the vacuolar protein sorting (VPS) pathway (Bankaitis, Johnson, & Emr, 1986; Robinson et al., 1988) and the ESCRT complexes (CPY-Suc2, His3-CPS screens) (Odorizzi, Babst, & Emr, 1998; Odorizzi, Katzmann, Babst, Audhya, & Emr, 2003), for the analysis of the cargo-Rsp5 adaptor interaction surface (Mup1-GFP-His3 screen) (Guiney, Klecker, & Emr, 2016), as well as for the identification of machinery required for the degradation of resident vacuolar membrane proteins (Ypq1-SBP-GFP-His3 screen) (Zhu, Jorgensen, Li, Chuang, & Emr, 2017).

The second approach allows specific investigation of the vacuolar sorting steps downstream of ubiquitination. This method takes advantage of the rapamycin-induced dimerization between the FKBP domain of human FKBP12 and the FRB domain of human mTOR (Belshaw, Ho, Crabtree, & Schreiber, 1996; Haruki, Nishikawa, & Laemmli, 2008; Heitman, Movva, Hiestand, & Hall, 1991; Zhu et al., 2017). Coexpressing fusions of FRB-ubiquitin and FKBP-"membrane protein cargo" in the presence of rapamycin mimics the ubiquitination of the membrane protein, and hence bypasses the ubiquitination step and allows screening for machinery required post-ubiquitination. This approach, termed the RapiDeg assay, was especially useful in elucidating that ESCRT complexes can directly function on the vacuolar membrane in yeast, as well as for studying endocytic steps downstream of ubiquitination (Zhu et al., 2017).

2. Auxotrophic marker fusions for monitoring and screening ubiquitin-dependent membrane trafficking

Auxotrophic markers, wild-type alleles of genes encoding essential nutrient biosynthesis enzymes, are commonly used in yeast to maintain plasmids or select for disrupted genes in strains that contain nonfunctional alleles of the same gene. Here, an auxotrophic marker is fused to a membrane protein, and used to assess its ubiquitin-dependent membrane trafficking. When the PM transporter-auxotrophic marker fusion protein (herein referred to as

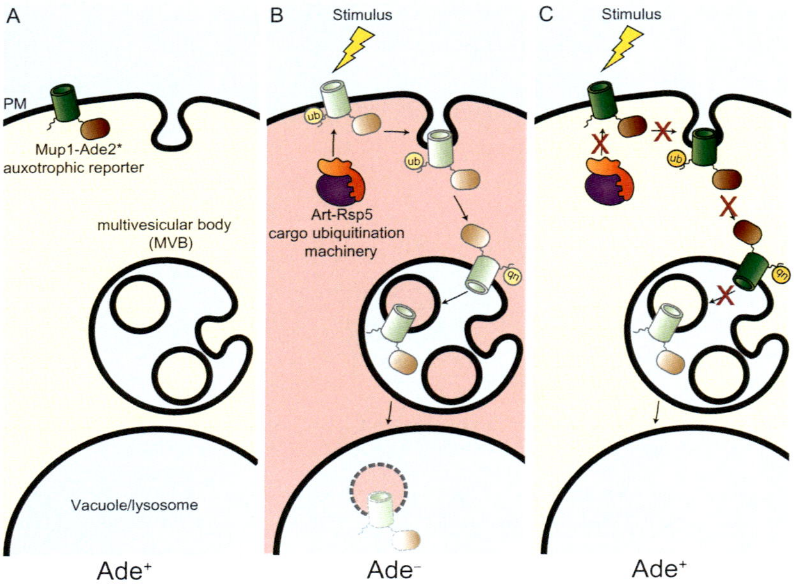

Fig. 1 Trafficking of the auxotrophic reporter. (A) Under nonstimulated conditions, Mup1-Ade2* is localized to the PM, and the fused Ade2* is exposed to the cytosol, resulting in yeast prototrophic for, i.e., able to synthesize, adenine (Ade). (B) After stimulating endocytosis of Mup1-Ade2* (usually with its substrate, methionine), it is ubiquitinated; and subsequently endocytosed, sorted into intraluminal vesicles at the MVB by the ESCRT machinery, and eventually delivered to the vacuole. After MVB sorting, Ade2* is sequestered from the cytoplasm, causing the yeast to be auxotrophic for Ade. (C) Mutants that block any step of ubiquitination or sorting before the ESCRTs result in yeast prototrophic for Ade despite the presence of the endocytic stimulus.

an "auxotrophic reporter") is localized on the PM, the auxotrophic marker is available in the cytoplasm, causing the cell to be prototrophic for that specific nutrient (Fig. 1A); when endocytosis is induced, the auxotrophic reporter is sorted into MVBs by the ESCRT machinery, removing the auxotrophic marker from the cytoplasm and making the cell auxotrophic for (or unable to synthesize) that nutrient (Fig. 1B).

In addition to the auxotrophic marker, it is advantageous to include GFP (or another fluorescent protein) in the auxotrophic reporter. The ability to analyze the auxotrophic reporter by microscopy provides a simple method for comparing its behavior to the native cargo of interest. Including GFP in the auxotrophic reporter also allows for postselection secondary screening.

Here we present a protocol for designing an auxotrophic reporter screen, using Mup1-Ade2*-GFP (Fig. 2A) as an example auxotrophic reporter.

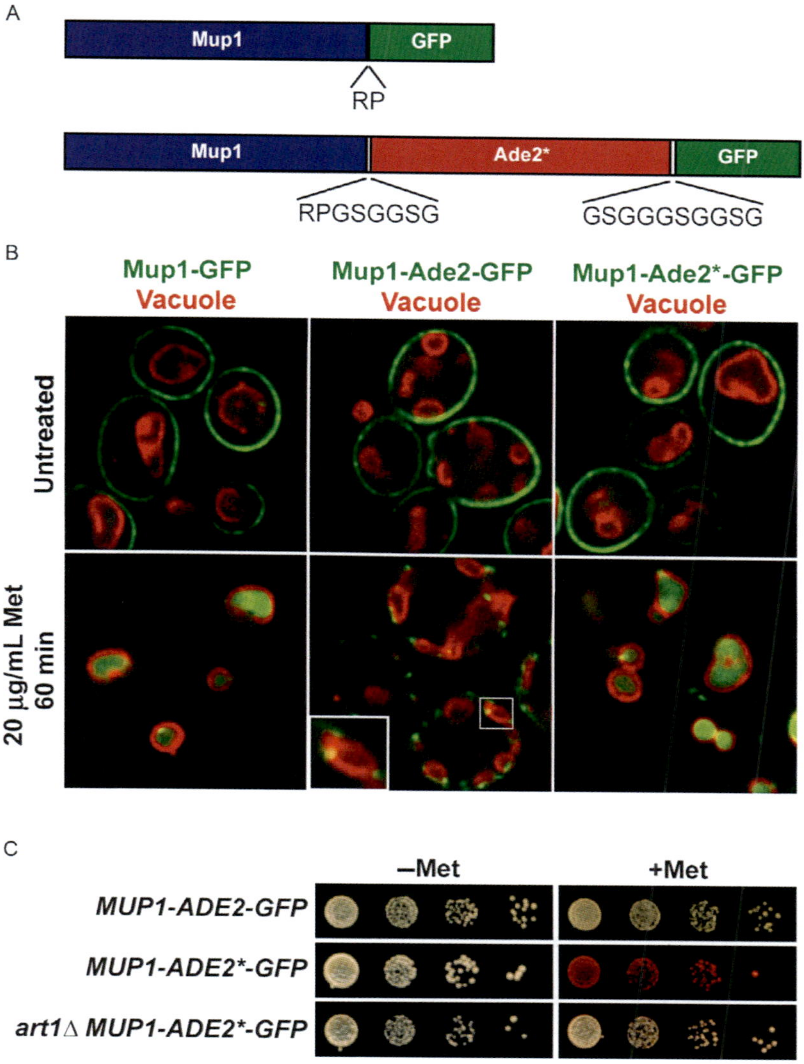

Fig. 2 Mup1-Ade2*-GFP endocytosis. (A) Schematic of the Mup1-GFP and Mup1-Ade2*-GFP fusion proteins. (B) Fluorescence microscopy analysis of Mup1-GFP, Mup1-Ade2-GFP, and Mup1-Ade2*-GFP endocytosis. Upon treatment with 20 µg/mL Met for 60 min, Mup1-GFP was completely endocytosed (*left panels*). Mup1-Ade2-GFP did not reach the vacuole lumen after methionine treatment. Additionally, vacuoles were fragmented and cortical when Mup1-Ade2-GFP was expressed (*middle panels*). Mup1-Ade2*-GFP, which includes a series of selected mutations in Ade2, alleviated the defects in endocytosis and vacuole morphology (*right panels*). (C) 10-fold serial dilutions of cells expressing the indicated *MUP1* fusion were spotted on media containing 10 µg/mL Ade. Cells expressing *MUP1-ADE2-GFP* and *MUP1-ADE2*-GFP* formed white colonies, indicating that Ade2 is cytosolic. On media that additionally contains 100 µg/mL Met, cells expressing *MUP1-ADE2*-GFP* formed red colonies because Mup1-Ade2*-GFP was endocytosed. Due to its endocytosis defects, cells expressing *MUP1-ADE2-GFP* formed, white colonies even in the presence of Met. Mup1-Ade2*-GFP endocytosis, for example, is blocked in an *art1*Δ strain.

This strategy can be modified to identify mutants in any endocytic cargo. Additionally, a nonmutagenized auxotrophic reporter can instead be used to identify and characterize the downstream steps of cargo recognition and trafficking.

2.1 Reagents, buffers, and media

2.1.1 Yeast strains

SEY6211 (*MAT***a** *ura3-52 his3Δ-200 leu2-3,112 trp1-901 ade2-101 suc2Δ9*) (Robinson et al., 1988)
YMB797 (SEY6211; *mup1Δ::TRP1 VPH1::VPH1-mCherry-HIS3MX6*).

2.1.2 Plasmids

pEG162 (pRS415-P_{MUP1}-*MUP1-GFP*) (Guiney et al., 2016)
pEG375 (pRS415- P_{MUP1}-*MUP1-ADE2*-GFP*)
pEG350 (pRS415-P_{MUP1}-*ADE2*-GFP*)

The peptide linker between Mup1 and Ade2 is RPGSGGSG (it introduces an EagI site into the coding DNA sequence), and the Ade2 start codon is omitted. The peptide linker between Ade2 and GFP is GSGGGSGGSG (Fig. 2A).

2.1.3 Buffers

10× Taq buffer:
100 m*M* Tris–HCl pH 8.3
500 m*M* KCl
15 m*M* $MgCl_2$

10× error prone dNTPs:
10 m*M* dCTP
10 m*M* dTTP
2 m*M* dATP
2 m*M* dGTP

2.1.4 Recipe for SCD solid media

Solvent: water
0.17% (w/v) yeast nitrogen base
0.5% (w/v) ammonium sulfate
0.07% (w/v) complete yeast nutrient supplement
2% (w/v) dextrose
2% (w/v) agar

2.1.5 Recipe for complete yeast nutrient supplement (dry mix)

0.5 g adenine
2.0 g histidine
4.0 g leucine
2.0 g lysine
2.0 g methionine
2.0 g tryptophan
2.0 g tyrosine
2.0 g uracil
Nutrients should be thoroughly mixed and stored as a dry powder.
Dropout growth media is made by omitting the desired nutrient.

2.1.6 Recipe for YPDA

Solvent: water
1% (w/v) yeast extract
2% (w/v) peptone
2% (w/v) dextrose
0.02 mg/mL adenine

2.2 Generation and validation of auxotrophic reporters

As with any fusion protein, it is important to validate that each of the proteins comprising the auxotrophic reporter retains its activity after fusion. Notably, many nutrient biosynthetic enzymes are very large, are often part of multiprotein complexes, and play important metabolic roles in cells. Thus, appending an auxotrophic marker to a PM cargo protein has the potential to alter its localization, function, or ability to endocytose. Likewise, tethering an auxotrophic marker to the PM may reduce its function or aberrantly recruit additional complex members to the cargo protein. It is therefore critical to test that the auxotrophic reporter retains the functions of both the cargo protein and the auxotrophic reporter.

2.2.1 Designing the auxotrophic reporter

In principle, any auxotrophic marker will work as part of the auxotrophic reporter. Our lab has successfully employed His3 (Guiney et al., 2016; Odorizzi et al., 2003; Zhu et al., 2017) and Ade2 (Emr Lab, unpublished data) fusions, each of which has its own unique advantages and drawbacks.

For His3 auxotrophic reporters, titrating the competitive inhibitor 3-amino-1,2,4-triazole (3-AT) into the media allows for precise control over the amount of the His3 fusion protein that is tolerated for growth.

Typically, even when endocytosis and degradation of His3 fusion proteins appear complete as assessed by microscopy or immunoblotting, residual amounts of the undegraded auxotrophic reporter are sufficient for growth on SCD-His media without 3-AT. Generally, 1–30 m*M* 3-AT yields satisfactorily stringent growth conditions, but this must be tested empirically. Additionally, using His3 is primarily useful for finding mutants that block degradation of the auxotrophic reporter; finding mutants that promote degradation of a normally stable His3 auxotrophic reporter is substantially more labor intensive, requiring replica plating and screening for colonies that do not survive on SCD-His plates.

Ade2 auxotrophic reporters offer the advantage to be able to identify mutants that either promote or block endocytosis. In addition to their inability to grow on SCD-Ade medium, *ade2* mutants accumulate a red pigment when grown on media with low supplemented adenine (Ade) (Smirnov, Smirnov, Budowsky, Inge-Vechtomov, & Serebrjakov, 1967). Thus, mutants that block endocytosis and degradation of the Ade2 auxotrophic reporter can be selected for growth on SCD-Ade or the appearance of white colonies, while mutants that cause constitutive endocytosis can be identified by the formation of red colonies. However, as with His3 auxotrophic reporters, Ade2 fusions can aggregate; Mup1-Ade2-GFP remains in cortical puncta after 60 min treatment with methionine, conditions that result in complete sorting of Mup1-GFP to the vacuole lumen (Fig. 2B). Notably, vacuoles are fragmented and localized to the cell cortex when Mup1-Ade2-GFP is expressed. Correspondingly, when spotted on solid media containing methionine, cells expressing Mup1-Ade2-GFP do not substantially endocytose Mup1, resulting in white colonies (Fig. 2C).

In an effort to make Mup1-Ade2-GFP better suited for screening, we mutagenized the *ADE2* region (using a pipeline similar to that described later) and screened for mutants that formed red colonies in the presence of methionine, but white colonies in its absence. The mutagenized *ADE2* allele (containing the mutations: N23D, R101G, F188I, N315S, N363S, and I436T), herein referred to as *ADE2**, adequately synthesizes adenine, and as part of the auxotrophic reporter Mup1-Ade2*-GFP, endocytoses with kinetics similar to Mup1-GFP. Importantly, it no longer gives rise to aberrant vacuole morphology (Fig. 2B). Thus, Mup1-Ade2*-GFP will be used as an illustrative example in the following protocol. While we anticipate that the *ADE2** allele will alleviate the problems associated with fusing Ade2 to any cargo protein, there are no mechanistic data explaining how fusing Ade2 to Mup1 delays its endocytosis and causes abnormal vacuole morphology, nor is there any model for how the *ADE2** allele restores Mup1 function. Therefore, additional mutagenesis may be required to generate an *ADE2* allele that is compatible with other cargo proteins.

2.2.2 Validating the auxotrophic reporter

(A) Construct the auxotrophic reporter plasmid

A centromeric plasmid is well suited for a random mutagenesis screen of the cargo portion of the auxotrophic reporter. However, if using the auxotrophic reporter for a genome-wide screen to identify other factors influencing its ubiquitin-dependent endocytosis, genomic integration of the auxotrophic marker, either at the endogenous locus or using an integrating vector, results in more stable expression and is therefore preferable. The Mup1 auxotrophic reporter, pRS415-*MUP1-ADE2*-GFP*, was constructed by inserting *ADE2** between *MUP1* and *GFP* in the plasmid pEG162 (Guiney et al., 2016).

(B) Generate the parental yeast strain

Any laboratory yeast strain will work, provided that it is appropriately auxotrophic to be compatible with the auxotrophic reporter. For example, when using the plasmid pRS415-*MUP1-ADE2*-GFP*, the parent yeast strain must be *leu2* to maintain the transformed plasmid and *ade2* for analysis of the presence of the auxotrophic reporter.

Additional control strains that block endocytosis or delivery of the auxotrophic reporter to the vacuole should also be generated at this step, if possible. For example, methionine-induced endocytosis of Mup1 is known to require the Art1-Rsp5 ubiquitin ligase complex (Lin, MacGurn, Chu, Stefan, & Emr, 2008); thus deleting *ART1* in the parental strain provides a helpful negative control.

(C) Transform the auxotrophic reporter into the parental yeast strains

There are many basic yeast transformation protocols, any of which will work for this purpose (Gietz, 2014; Gietz & Woods, 2001; Ito, Fukuda, Murata, & Kimura, 1983; Rivera, Magana-Ortiz, Gomez-Lim, Fernandez, & Loske, 2014).

(D) Visualize the auxotrophic reporter's localization by microscopy

This step is important to ensure that the auxotrophic reporter properly traffics to the PM, endocytoses, and sorts to the vacuole upon treatment with substrate. Initial characterization of Mup1-Ade2-GFP revealed that endocytosis was aberrant and confirmed that the mutagenized *ADE2* allele, *ADE2**, resulted in normal endocytosis of the auxotrophic reporter (Fig. 2B).

Growth conditions and microscopy techniques will vary depending on the membrane protein being studied. The micrographs shown here were acquired using a DeltaVision RT inverted epi-fluorescence microscope equipped with DV Light SSI 7 Color LED and DV Elite metal-oxide semiconductor camera. Images were deconvolved using sofWoRx 3.5.1. An important consideration regarding Mup1-Ade2*-GFP, and potentially many PM nutrient transporters, is that Mup1 is endocytosed in an Art1-independent manner upon glucose starvation (Lang et al., 2014). Thus, preventing the cells from reaching saturation and replenishing the culture with fresh media will prevent endocytosis and vacuolar sorting of the auxotrophic marker. We typically allow cells to grow for at least 24 h at 30°C, diluting them before they reach $OD_{600} = 1.0$ as many times as necessary.

(E) Test the auxotrophic reporter by growth

Once the auxotrophic reporter is validated to mimic the cargo protein for endocytosis kinetics, it must be tested to ensure that the auxotrophic marker supports growth on the appropriate selection media that will be employed for screening.

This step is an excellent time to define the growth conditions required for the screen. The substances and concentrations required to induce endocytosis, as well as temperature, will vary depending on the protein being studied. Using Mup1-Ade2*-GFP as the auxotrophic reporter, the goal is to find conditions where the nonmutagenized auxotrophic reporter forms white colonies in the absence of methionine (where it is not endocytosed) and forms red but still viable colonies in the presence of methionine. Notably, a low concentration of adenine is essential for adenine auxotrophs to be able to support growth while still developing red coloration. After testing multiple adenine and methionine concentrations, we found that the ideal conditions for the Mup1 auxotrophic reporter Mup1-Ade2*-GFP were SCD − Leu + 100 μg/mL Met + 10 μg/mL Ade to promote endocytosis, yielding red colonies (Fig. 2C).

2.3 Mutagenesis screen to identify residues required for ubiquitin-dependent membrane trafficking and degradation

Once validated, the auxotrophic reporter can be used to isolate mutations in the target protein that affect its ubiquitination and vacuolar sorting. This strategy has been successfully used to identify critical residues required for Mup1 endocytosis (Guiney et al., 2016). Here, we present an updated version of that protocol using the Mup1-Ade2*-GFP reporter. The *MUP1* portion of the auxotrophic reporter is mutagenized by error prone PCR, a mutant library assembled in yeast by homology directed gap-repair, and endocytosis assessed by the presence or absence of Ade2 in the cytosol.

2.3.1 Cloning the gap repair destination vector

Using standard cloning techniques, construct a gap repair destination vector; it will exactly match the vector created in Section 2.2.1, except the *MUP1* ORF will be replaced with a single restriction site for linearization (Bessa, Pereira, Moreira, Johansson, & Queiros, 2012; Weir & Keeney, 2014).

2.3.2 Generating the mutant library

Random mutagenesis is accomplished by error prone PCR. The *MUP1* ORF will be PCR amplified using the nonproofreading Taq DNA polymerase, a biased dNTP ratio, and low concentrations of Mn^{2+}, to generate *MUP1* mutants which will be ligated into the destination vector via gap repair.

Efficient gap repair recombination requires significant homology between the PCR insert and the destination vector. 60 bp of homology on each side of the gap is sufficient for high repair efficiency. 60 bp primers should be designed to anneal immediately 5′ and 3′ of the gap/restriction site designed in Section 2.2.2 (Bessa et al., 2012; Weir & Keeney, 2014).

The following protocol is adapted from Wilson and Keefe (2001) and modified to achieve a lower mutagenic rate (approximately one codon change/1600 bp).

2.3.3 Mutagenic PCR reaction

1 × Taq buffer
1 × Error prone dNTP mix
0.1 m*M* $MnCl_2$
3% (v/v) DMSO
1 μ*M* forward primer
1 μ*M* reverse primer
1 U Taq polymerase
10 ng template DNA

Bring up to a final volume of 50 μL using distilled H_2O

(1) 94°C 2 min 30 s
(2) 94°C 30 s
(3) 54°C 30 s
(4) 72°C 2 min/kb
(5) Go to step 2, 24 more times
(6) 72°C 10 min
(7) 10°C forever

2.3.4 Troubleshooting mutagenesis

In most cases, one mutation per mutagenized region is preferred, as multiple mutations require laborious downstream analysis to separate causative from noncausative mutations. Therefore, it is recommended to isolate and sequence a few clones from the mutagenic PCR library before initiating the full-scale screen. If the mutagenesis rate is higher than desired, reducing the number of PCR amplicon doublings (rather than PCR cycles per se, as error prone PCR efficiency is often less than 100%) has the greatest impact; lowering $MnCl_2$, $MgCl_2$, or reducing the biased dNTP ratio also has an impact. See Wilson and Keefe (2001) for additional troubleshooting recommendations.

2.3.5 Gap repair transformation

The following protocol specifies the amount of each reagent required for a single plate, enough to screen 500–2000 colonies, depending on the transformation and gap repair efficiencies of the parental yeast strain and the destination vector-mutagenized insert pair. Scale the master mix according to the number of plates needed to achieve sufficient coverage of the mutagenized region. For Mup1, at 574 amino acids long, the number of possible single amino acid mutants is $19 \times 574 \approx 10{,}000$; increasing coverage by a factor of 3 gives an expected sampling of 95% of the total mutational diversity. For Mup1 this would entail screening 30,000 colonies, or ~30 plates. Additional coverage may also be required based on the fraction of single mutated molecules in the mutagenized PCR library.

It is also essential to include a nonmutagenized control transformation so that the background growth rates or red/white intensity (for Ade2 fusions) can be assessed; if instead of Ade2 the auxotrophic marker used is selecting for growth (such as His3 or Ura3), then it is also essential to plate a portion of the transformed cells onto nonselective media to assess the total number of gap-repair events. Gap repair is required to recircularize the gapped plasmid, which would otherwise be degraded.

(1) Inoculate a single yeast colony in 5mL YPDA. Incubate overnight with shaking at 26°C.

(2) The next morning, dilute the cells to 0.15 OD_{600}/mL, in a volume large enough to collect 1.5 OD_{600} per plate. Incubate at 26°C, shaking until $OD_{600} = 0.6$ (two doublings).

(3) Harvest 1.5 OD_{600} per plate of cells by centrifugation, 5min at $3000 \times g$.

(4) Wash the cells with sterile water and then spin again.

(5) Resuspend the cells in 50μL (per plate) 100m*M* LiOAc. Incubate at room temperature for 15min.

(6) Prepare a new microfuge tube containing 240 μL 50% (w/v) fresh PEG-4000, 36 μL 1 *M* LiOAc, 35 μL 1 *M* DTT, 2 μL10 mg/mL ssDNA, 250 ng linearized destination vector, and a 10-fold molar excess of mutagenized insert. Add H_2O to bring final volume to 350 μL per tube.
(7) Add 50-μL of cells to each tube. Mix thoroughly, but gently.
(8) Incubate for 30 min at room temperature.
(9) Transfer to a 42°C heat block for 15 min.
(10) Pellet cells by centrifugation, 30 s at 21,000 × *g*.
(11) Aspirate and wash with 1 mL sterile H_2O. Centrifuge cells again.
(12) Aspirate and resuspend each tube in 700 μL YPDA.
(13) Incubate at 30°C with shaking for 3 h.
(14) Pellet cells by centrifugation, 30 s at 21,000 × *g*.
(15) Aspirate and wash with 1 mL sterile H_2O. Centrifuge cells again.
(16) Resuspend cells in 100 μL sterile H_2O and spread on selective media plates as determined in Section 2.2.2, step E. For red/white screening of Mup1-Ade2*-GFP, we used SCD − Leu + 100 μg/mL Met + 10 μg/mL Ade.

2.3.6 Isolation of candidate mutant colonies

The red pigment accumulation takes time to develop, and due to the limited adenine provided by the medium, Ade^- colonies typically require 4–7 days of growth at 30°C before colonies of sufficient size appear. Once red and white colonies can be positively distinguished, pick cells from white (putative endocytosis-defective) colonies and isolate single colonies by streaking onto SCD − Leu + 100 μg/mL Met + 10 μg/mL Ade plates to confirm the white colony-color phenotype.

2.3.7 Plasmid isolation and retesting

Once the mutants are identified, the plasmids bearing the desired mutations will be isolated from yeast using the protocols found in Weir and Keeney (2014), transformed into *E. coli*, and isolated via a standard miniprep protocol. The isolated plasmid is then retransformed into the original yeast strain to verify that the plasmid harboring the mutated Mup1, rather than a genomic mutation, is causing the colony to remain white.

2.3.8 Secondary screening strategies

Consideration of secondary screening steps is important to eliminate mutants that give rise to white colonies via mechanisms other than blocking ubiquitination and endocytosis. Visually screening mutant auxotrophic reporter localization by fluorescence microscopy can identify mutants that fail to localize at the correct organelle, such as mutants that do not leave the endoplasmic reticulum (Guiney et al., 2016). Additionally, because Mup1 endocytosis is triggered by substrate (methionine) transport, transport-deficient Mup1-Ade2* mutants result in white colonies. Testing for sensitivity to the toxic methionine analog ethionine will identify find mutants that retain Mup1 transport activity despite blocked endocytosis (Emr Lab, unpublished data). This strategy may be used for other permeases for which toxic substrate analogs exist.

2.3.9 Sequencing mutants

Once the mutant plasmids are confirmed and isolated, sequence the entire mutagenized region to identify the causative mutation. If multiple mutations are present in a single plasmid, they must be separated to identify the phenotypically relevant mutation(s), and all mutations should be generated in an untagged (or GFP-tagged) version of *MUP1* to confirm the endocytosis phenotype. These experiments are the most labor-intensive part of the screen.

2.3.10 Mutant characterization

After mutants of interest are successfully isolated, further experiments can identify the mechanism by which the mutations block endocytosis. For example, mutants unable to be ubiquitinated can be revealed by immunoprecipitation of Mup1 followed by immunoblotting against ubiquitin (Guiney et al., 2016); or mutations that prevent binding between the permease and ubiquitin ligase adaptor can be elucidated using bi-molecular fluorescence complementation (Hovsepian et al., 2017). Also, suppressor screens using the mutant cargo, where the ubiquitin ligase adaptor is mutagenized using a similar pipeline as above, can be used to identify the potential cargo binding site on the adaptor (Guiney et al., 2016).

3. Rapamycin-induced protein degradation assay for studying membrane trafficking events downstream of ubiquitination

Selective ubiquitin-mediated endocytosis of PM transporters occurs in response to specific stimuli and initiates with the recruitment of the appropriate ART adaptor–Rsp5 complex which modifies the cargo with ubiquitin (Fernandez-Murray, Ngo, & McMaster, 2013; Hein, Springael, Volland, Haguenauer-Tsapis, & Andre, 1995; Hovsepian et al., 2017; Lin et al., 2008; Nikko & Pelham, 2009; Nikko, Sullivan, & Pelham, 2008; Paiva et al., 2009). Moreover, the Rsp5 adaptors themselves are subject to regulation via multiple and complex signaling pathways (Alvaro, Aindow, & Thorner, 2016; Herrador et al., 2015; Hovsepian et al., 2017; Llopis-Torregrosa et al., 2016; MacGurn, Hsu, Smolka, & Emr, 2011; Nikko & Pelham, 2009; O'Donnell, Huang, Thorner, & Cyert, 2013). However, for many PM transporters the stimulus to induce endocytosis, the corresponding adaptor protein, or the signaling mechanisms that regulate the adaptor protein remain unknown. Such considerations make it difficult to study the trafficking steps and machinery acting downstream of ubiquitination for vacuolar sorting of cargo proteins until initiating signaling events are uncovered.

Translational fusions of the cargo protein and ubiquitin can seemingly bypass upstream signaling events, resulting in premature recognition of the newly synthesized cargo for vacuolar sorting before it reaches its normal functional destination (MacDonald, Buchkovich, Stringer, Emr, & Piper, 2012; Stringer & Piper, 2011). To overcome this problem, we developed the *Rap*amycin-*i*nduced *Deg*radation (RapiDeg) assay, which allows temporal control and direct targeting of ubiquitin to cargo proteins to bypass the normal ubiquitination step.

The RapiDeg assay utilizes the rapamycin-induced dimerization between the FRB domain of human mTOR and the FKBP12 protein (Banaszynski, Liu, & Wandless, 2005; Haruki et al., 2008; Zhu et al., 2017). The background strain contains the *tor1-1* mutation, which still functions but is defective for rapamycin binding; and a null mutation for *FPR1*, a major target for rapamycin binding in yeast (Heitman et al., 1991). These mutations allow cell growth in the presence of rapamycin and ensure the exclusive pairing of the FRB domain only with the FKBP motif fused to the protein of interest. The RapiDeg strain coexpresses tandem FKBP12 (2xFKBP) motifs fused to any membrane cargo of interest, and an FRB domain N-terminally fused to a noncleavable chain of three ubiquitins (FRB-3xUb) (Fig. 3A). Addition of rapamycin mimics cargo ubiquitination by recruiting FRB-3xUb to FKBP fused to the cargo (Fig. 3B), resulting in ESCRT-mediated vacuolar sorting of the membrane protein (Zhu et al., 2017). Unlike the proteasome, the ESCRT-0 components Vps27 and Hse1 preferentially bind to K63-linked polyubiquitin chains (Nathan, Kim, Ting, Gygi, & Goldberg, 2013). Since the linear ubiquitin fusions resemble the structure of K63-linked ubiquitin chains (Komander, 2009), the RapiDeg system specifically sorts the cargo to the vacuolar lumen. The addition of a fluorescent tag, such as GFP, on the FKBP-cargo allows visualization of the cargo before and after rapamycin treatment.

In addition to being a useful tool to study membrane trafficking events, RapiDeg also provides an amenable tool for inducible and specific degradation (and therefore inactivation) of membrane proteins in the endomembrane system. Our lab has successfully utilized this strategy for downregulating membrane proteins at the PM, vacuole membrane, Golgi, and endosomes (Zhu et al., 2017). Here, we use this approach to describe a protocol to induce the endocytosis and degradation of the yeast plasma membrane arginine transporter, Can1.

3.1 Reagents, buffers, and media

3.1.1 Plasmids

MLp479: pRS305-P$_{TDH3}$-*FRB-3xUb* (Zhu et al., 2017)
MLp379: pFA6a-*GFP-2xFKBP-TRP1* (Zhu et al., 2017)

3.1.2 Strains

SEY6210.1: *MAT***a** *ura3-52 his3Δ-200 leu2-3,112 trp1-901 lys2-801 suc2Δ9* (Robinson et al., 1988)
LZY273: SEY6210.1; *tor1-1 fpr1Δ::natMX*
LZY562: SEY6210.1; *tor1-1 fpr1Δ::natMX leu2–3,112*::pRS305-P$_{TDH3}$*FRB-3xUb CAN1::CAN1-GFP-2xFKBP-TRP VPH1::VPH1-mCherry-kanMX6* (Zhu et al., 2017)

3.2 Generating and validating the RapiDeg strain

The RapiDeg background strain can be generated using standard yeast genetics techniques. Both the *tor1-1* and *fpr1Δ* mutations are essential to prevent growth inhibition during rapamycin treatment and allow yeast to grow on YPD supplemented with 1 μg/mL rapamycin.

3.2.1 Generating the RapiDeg strain

FRB-3xUb is expressed from a high expression, constitutive *TDH3* promoter (Mumberg, Muller, & Funk, 1995) to ensure that it is not limiting for the rapamycin-induced recruitment to the cargo. Additionally, a few key modifications to the FRB-3xUb construct increase the robustness of the assay. First, mutating the C-terminal Gly-Gly of the first and second Ub to Ala prevents cleavage of FRB-3xUb by deubiquitinating enzymes.

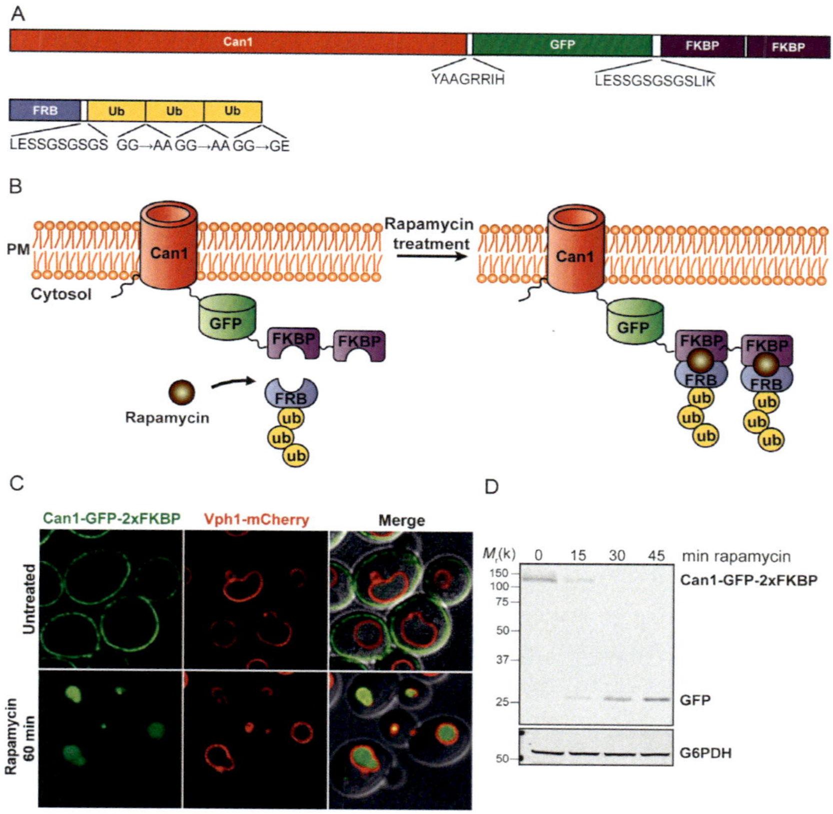

Fig. 3 Can1 endocytosis via RapiDeg. (A) Schematic of the Can1-GFP-2xFKBP and FRB-3xUb fusion proteins. (B) The RapiDeg assay, using Can1-GFP-2xFKBP as an example. Rapamycin addition causes the recruitment of FRB-3xUb to the 2xFKBP domains fused to the C-terminus of Can1-GFP, effectively causing its ubiquitination and thus, endocytosis. (C) Fluorescence microscopy analysis of cells expressing Can1-GFP-2xFKBP and Vph1-mCherry as a vacuole membrane marker was analyzed before (*top*) and after 60 min of rapamycin treatment (1 μg/mL, *bottom*). Can1-GFP-2xFKBP can be completely sorted to the vacuole lumen using RapiDeg. (D) Immunoblot analysis of Can1-GFP-2xFKBP degradation following rapamycin treatment (1 μg/mL) for the indicated times.

Second, mutating the C-terminal Gly of the third Ub to Glu prevents ligation of FRB-3xUb to other Ub chains on endogenous proteins (Fig. 3A). We also recommend genomic integration of FRB-3xUb using an integrative vector (Sikorski & Hieter, 1989) to allow stable expression levels. GFP-2xFKBP is fused to the C-terminus of Can1 at its endogenous genomic locus by amplifying a Can1 C-terminal targeting construct from the plasmid MLp379 (Longtine et al., 1998; Zhu et al., 2017).

3.2.2 Validating the RapiDeg strain

As with all fusion proteins, it is important to verify that the addition of 2xFKBP to the C-terminus does not alter the function of the cargo protein. In the case of Can1-GFP-2xFKBP, Can1 function can be tested by growth on canavanine—inactivation of Can1 makes cells resistant to the analog—or by monitoring its substrate-induced endocytosis by microscopy (Lin et al., 2008). It is also important to test that the degradation of the cargo protein is dependent on the addition of rapamycin and is due to the recruitment of FRB-3xUb to the FKBP domain of the cargo protein. Critical control strains to generate here are: a strain with *CAN1-GFP-2xFKBP* but without *FRB-3xUb*, and a strain with *CAN1-GFP* lacking the 2xFKBP motif but that expresses FRB-3xUb. Testing that the degradation is a direct result of FRB-3xUb recruitment to the FKBP-tagged cargo protein is especially important when analyzing a cargo such as Can1, since its endocytosis is TORC1 dependent (MacGurn et al., 2011).

3.3 RapiDeg assay

(1) Inoculate a single yeast colony in 5 mL YPDA. Incubate at 30°C overnight with shaking.

(2) The next morning, dilute the cells to 0.15 OD_{600}/mL. Incubate at 30°C, with shaking, until $OD_{600} = 0.6$ (two doublings).

(3) Add rapamycin to a final concentration of 1 μ*M* and incubate at 30°C. Collect cells at suitable time points to examine Can1-GFP-2xFKBP degradation using fluorescence microscopy and immunoblot analysis (Fig. 3C and D). The time required for degradation will vary depending on the cargo protein, strain background, and temperature. For many tested proteins, complete degradation can be achieved in as little as 15 min (Zhu et al., 2017), but the kinetics were significantly slower for some cargos, such as Mup1 (Emr Lab, unpublished data).

3.3.1 Troubleshooting the RapiDeg assay

The GFP-2xFKBP fusion is rather large, and as such may inhibit cargo protein biosynthesis, trafficking, or degradation. In such cases, extending the linker region between the cargo and GFP, or between GFP and the 2xFKBP may help. Additionally, we have found that an N-terminal 2xFKBP tag works better for some cargo proteins, especially for proteins such as SNAREs whose C-terminus is essential for its function (Emr Lab, unpublished data). For some proteins, the 2xFKBP domain may be inaccessible and therefore unable to bind rapamycin or FRB-3xUb. An important test would be to add a secondary fluorescent tag to FRB-3xUb (or simply FRB) to ensure that it is recruited to the site of the cargo protein upon rapamycin treatment.

3.4 Applications of the RapiDeg assay

The RapiDeg assay provides the ability to specifically downregulate membrane proteins without prior knowledge of the upstream mechanisms required to initiate their endocytosis or downregulation. The basic RapiDeg protocol for Can1 described earlier can be modified to study the downregulation of membrane proteins at any endocytic organelle. In addition, we envision that RapiDeg can be useful to identify downstream sorting machinery, circumvent trafficking defects of the ubiquitination machinery in mutant backgrounds, and provide methods for temporally precise degradation and inactivation of proteins in the endomembrane system.

3.4.1 Screening for sorting machinery using RapiDeg

RapiDeg can be easily incorporated into high-throughput screening schemes to identify the machinery required to sort ubiquitinated cargo proteins. Additionally, replacing GFP with the pH-sensitive variant pHluorin would allow screening for mutants defective for vacuolar sorting using flow cytometry. Our lab and others have successfully employed flow cytometry to screen the yeast deletion collection using pHluorin fusion-based reporters upon substrate stimulation (Henne et al., 2015; Wrasman et al., 2018); RapiDeg provides the advantage of being able to design such a screen using any plasma membrane transporter, even if its substrate is unknown.

3.4.2 Bypassing adaptor trafficking defects using RapiDeg

ARTs are soluble proteins that can be recruited to the PM to ubiquitinate their cargo (Lin et al., 2008). However, some Rsp5 adaptor proteins, such as Ssh4, are membrane embedded and thus require trafficking through the endomembrane system to reach their functional localizations (Li, Rong, Chuang, Peng, & Emr, 2015). As such, mutants that appear to block vacuolar sorting of the cargo may, instead, block trafficking of the ubiquitination machinery. The RapiDeg assay provides a handy tool to distinguish such direct vs indirect effects on cargo sorting. Using RapiDeg-based degradation of the vacuole membrane lysine transporter, Ypq1, we demonstrated that although vacuole fusion mutants block Ypq1-GFP sorting to the lumen, they had no effect on Ypq1-GFP-2xFKBP sorting (Zhu et al., 2017). These mutants are instead defective for the correct localization of the vacuole membrane Rsp5 adaptor, Ssh4, which is required for Ypq1 ubiquitination.

3.4.3 *Protein inactivation using RapiDeg*

While null mutants have been extremely useful in characterizing gene function, these are often complicated by the accumulation of secondary mutations (Teng et al., 2013), or the inviability of the mutant. Thus, rapid inactivation of target proteins is preferable to avoid secondary effects. Temperature-sensitive alleles, when available, have been invaluable in this regard. For proteins where temperature sensitive alleles have not been generated, a temperature-sensitive degron can be fused to induce down-regulation (Dohmen, Wu, & Varshavsky, 1994). Furthermore, the Aux/IAA-induced degron (AID) system has also been used for efficient proteasomal degradation of cytoplasmic proteins, as well as ER and mitochondrial membrane proteins (Nishimura, Fukagawa, Takisawa, Kakimoto, & Kanemaki, 2009; Ramos, Zenser, Leyser, & Callis, 2001). RapiDeg offers a complementary approach for rapid inactivation of membrane proteins in the endomembrane system. Our lab has successfully employed this approach for targeting PM, Golgi, endosome as well as vacuole membrane proteins for vacuolar degradation (Zhu et al., 2017).

4. Concluding remarks

Our understanding of the varied physiological mechanisms facilitated by protein ubiquitination has expanded significantly in the last few decades. In the endocytic pathway, ubiquitination can guide endocytosis and trafficking, as well as protein sorting and degradation. Additionally, ubiquitin regulates both the fate of the cargo and the machinery required for cargo sorting. The protocols presented here offer the ability to employ clever genetic selections to carefully study these endocytic processes. While we understand the sorting determinants of a few model cargos, further studies that expand this knowledge for multiple cargos will be required to finely dissect the underlying mechanisms. Although these methods are described for studies in yeast, the general concepts are applicable to other model systems as well. These assays can also easily be adapted to study mechanisms of transmembrane protein sorting on any endocytic organelle in yeast.

Acknowledgment

Work in the authors' laboratory is supported by a Cornell University Research Grant (CU563704) to S.D.E.

References

Alvaro, C. G., Aindow, A., & Thorner, J. (2016). Differential phosphorylation provides a switch to control how alpha-arrestin Rod1 down-regulates mating pheromone response in Saccharomyces cerevisiae. *Genetics*, *203*, 299–317.

Banaszynski, L. A., Liu, C. W., & Wandless, T. J. (2005). Characterization of the FKBP.rapamycin.FRB ternary complex. *Journal of the American Chemical Society*, *127*, 4715–4721.

Bankaitis, V. A., Johnson, L. M., & Emr, S. D. (1986). Isolation of yeast mutants defective in protein targeting to the vacuole. *Proceedings of the National Academy of Sciences of the United States of America*, *83*, 9075–9079.

Belshaw, P. J., Ho, S. N., Crabtree, G. R., & Schreiber, S. L. (1996). Controlling protein association and subcellular localization with a synthetic ligand that induces heterodimerization of proteins. *Proceedings of the National Academy of Sciences of the United States of America*, *93*, 4604–4607.

Bessa, D., Pereira, F., Moreira, R., Johansson, B., & Queiros, O. (2012). Improved gap repair cloning in yeast: Treatment of the gapped vector with Taq DNA polymerase avoids vector self-ligation. *Yeast*, *29*, 419–423.

Dohmen, R. J., Wu, P., & Varshavsky, A. (1994). Heat-inducible degron: A method for constructing temperature-sensitive mutants. *Science*, *263*, 1273–1276.

Ellgaard, L., Molinari, M., & Helenius, A. (1999). Setting the standards: Quality control in the secretory pathway. *Science*, *286*, 1882–1888.

Fernandez-Murray, J. P., Ngo, M. H., & McMaster, C. R. (2013). Choline transport activity regulates phosphatidylcholine synthesis through choline transporter Hnm1 stability. *The Journal of Biological Chemistry*, *288*, 36106–36115.

Gietz, R. D. (2014). Yeast transformation by the LiAc/SS carrier DNA/PEG method. *Methods in Molecular Biology*, *1205*, 1–12.

Gietz, R. D., & Woods, R. A. (2001). Genetic transformation of yeast. *Biotechniques*, *30*, 816–820. 822–826, 828 passim.

Guiney, E. L., Klecker, T., & Emr, S. D. (2016). Identification of the endocytic sorting signal recognized by the Art1-Rsp5 ubiquitin ligase complex. *Molecular Biology of the Cell*, *27*, 4043–4054.

Haruki, H., Nishikawa, J., & Laemmli, U. K. (2008). The anchor-away technique: Rapid, conditional establishment of yeast mutant phenotypes. *Molecular Cell*, *31*, 925–932.

Hein, C., Springael, J. Y., Volland, C., Haguenauer-Tsapis, R., & Andre, B. (1995). NPl1, an essential yeast gene involved in induced degradation of Gap1 and Fur4 permeases, encodes the Rsp5 ubiquitin-protein ligase. *Molecular Microbiology*, *18*, 77–87.

Heitman, J., Movva, N. R., Hiestand, P. C., & Hall, M. N. (1991). FK 506-binding protein proline rotamase is a target for the immunosuppressive agent FK 506 in Saccharomyces cerevisiae. *Proceedings of the National Academy of Sciences of the United States of America*, *88*, 1948–1952.

Henne, W. M., Stenmark, H., & Emr, S. D. (2013). Molecular mechanisms of the membrane sculpting ESCRT pathway. *Cold Spring Harbor Perspectives in Biology*, *5*.

Henne, W. M., Zhu, L., Balogi, Z., Stefan, C., Pleiss, J. A., & Emr, S. D. (2015). Mdm1/Snx13 is a novel ER-endolysosomal interorganelle tethering protein. *The Journal of Cell Biology*, *210*, 541–551.

Herrador, A., Livas, D., Soletto, L., Becuwe, M., Leon, S., & Vincent, O. (2015). Casein kinase 1 controls the activation threshold of an alpha-arrestin by multisite phosphorylation of the interdomain hinge. *Molecular Biology of the Cell*, *26*, 2128–2138.

Hicke, L., & Dunn, R. (2003). Regulation of membrane protein transport by ubiquitin and ubiquitin-binding proteins. *Annual Review of Cell and Developmental Biology*, *19*, 141–172.

Hovsepian, J., Defenouillere, Q., Albanese, V., Vachova, L., Garcia, C., Palkova, Z., et al. (2017). Multilevel regulation of an alpha-arrestin by glucose depletion controls hexose transporter endocytosis. *The Journal of Cell Biology*, *216*, 1811–1831.

Hurley, J. H., & Emr, S. D. (2006). The ESCRT complexes: Structure and mechanism of a membrane-trafficking network. *Annual Review of Biophysics and Biomolecular Structure*, *35*, 277–298.

Ito, H., Fukuda, Y., Murata, K., & Kimura, A. (1983). Transformation of intact yeast cells treated with alkali cations. *Journal of Bacteriology*, *153*, 163–168.

Jones, E. W. (1977). Proteinase mutants of Saccharomyces cerevisiae. *Genetics*, *85*, 23–33.

Komander, D. (2009). The emerging complexity of protein ubiquitination. *Biochemical Society Transactions*, *37*, 937–953.

Lang, M. J., Martinez-Marquez, J. Y., Prosser, D. C., Ganser, L. R., Buelto, D., Wendland, B., et al. (2014). Glucose starvation inhibits autophagy via vacuolar hydrolysis and induces plasma membrane internalization by down-regulating recycling. *The Journal of Biological Chemistry*, *289*, 16736–16747.

Li, M., Rong, Y., Chuang, Y. S., Peng, D., & Emr, S. D. (2015). Ubiquitin-dependent lysosomal membrane protein sorting and degradation. *Molecular Cell*, *57*, 467–478.

Lin, C. H., MacGurn, J. A., Chu, T., Stefan, C. J., & Emr, S. D. (2008). Arrestin-related ubiquitin-ligase adaptors regulate endocytosis and protein turnover at the cell surface. *Cell*, *135*, 714–725.

Llopis-Torregrosa, V., Ferri-Blazquez, A., Adam-Artigues, A., Deffontaines, E., van Heusden, G. P., & Yenush, L. (2016). Regulation of the yeast Hxt6 hexose transporter by the Rod1 alpha-Arrestin, the Snf1 protein kinase, and the Bmh2 14-3-3 protein. *The Journal of Biological Chemistry*, *291*, 14973–14985.

Longtine, M. S., McKenzie, A., 3rd, Demarini, D. J., Shah, N. G., Wach, A., Brachat, A., et al. (1998). Additional modules for versatile and economical PCR-based gene deletion and modification in Saccharomyces cerevisiae. *Yeast*, *14*, 953–961.

MacDonald, C., Buchkovich, N. J., Stringer, D. K., Emr, S. D., & Piper, R. C. (2012). Cargo ubiquitination is essential for multivesicular body intralumenal vesicle formation. *EMBO Reports*, *13*, 331–338.

MacGurn, J. A., Hsu, P. C., & Emr, S. D. (2012). Ubiquitin and membrane protein turnover: From cradle to grave. *Annual Review of Biochemistry*, *81*, 231–259.

MacGurn, J. A., Hsu, P. C., Smolka, M. B., & Emr, S. D. (2011). TORC1 regulates endocytosis via Npr1-mediated phosphoinhibition of a ubiquitin ligase adaptor. *Cell*, *147*, 1104–1117.

Mumberg, D., Muller, R., & Funk, M. (1995). Yeast vectors for the controlled expression of heterologous proteins in different genetic backgrounds. *Gene*, *156*, 119–122.

Munro, S. (2004). Organelle identity and the organization of membrane traffic. *Nature Cell Biology*, *6*, 469–472.

Nathan, J. A., Kim, H. T., Ting, L., Gygi, S. P., & Goldberg, A. L. (2013). Why do cellular proteins linked to K63-polyubiquitin chains not associate with proteasomes? *The EMBO Journal*, *32*, 552–565.

Nikko, E., & Pelham, H. R. (2009). Arrestin-mediated endocytosis of yeast plasma membrane transporters. *Traffic*, *10*, 1856–1867.

Nikko, E., Sullivan, J. A., & Pelham, H. R. (2008). Arrestin-like proteins mediate ubiquitination and endocytosis of the yeast metal transporter Smf1. *EMBO Reports*, *9*, 1216–1221.

Nishimura, K., Fukagawa, T., Takisawa, H., Kakimoto, T., & Kanemaki, M. (2009). An auxin-based degron system for the rapid depletion of proteins in nonplant cells. *Nature Methods*, *6*, 917–922.

O'Donnell, A. F., Huang, L., Thorner, J., & Cyert, M. S. (2013). A calcineurin-dependent switch controls the trafficking function of alpha-arrestin Aly1/Art6. *The Journal of Biological Chemistry*, *288*, 24063–24080.

Odorizzi, G., Babst, M., & Emr, S. D. (1998). Fab1p PtdIns(3)P 5-kinase function essential for protein sorting in the multivesicular body. *Cell*, *95*, 847–858.
Odorizzi, G., Katzmann, D. J., Babst, M., Audhya, A., & Emr, S. D. (2003). Bro1 is an endosome-associated protein that functions in the MVB pathway in Saccharomyces cerevisiae. *Journal of Cell Science*, *116*, 1893–1903.
Paiva, S., Vieira, N., Nondier, I., Haguenauer-Tsapis, R., Casal, M., & Urban-Grimal, D. (2009). Glucose-induced ubiquitylation and endocytosis of the yeast Jen1 transporter: Role of lysine 63-linked ubiquitin chains. *The Journal of Biological Chemistry*, *284*, 19228–19236.
Piper, R. C., & Luzio, J. P. (2007). Ubiquitin-dependent sorting of integral membrane proteins for degradation in lysosomes. *Current Opinion in Cell Biology*, *19*, 459–465.
Ramos, J. A., Zenser, N., Leyser, O., & Callis, J. (2001). Rapid degradation of auxin/indoleacetic acid proteins requires conserved amino acids of domain II and is proteasome dependent. *Plant Cell*, *13*, 2349–2360.
Rivera, A. L., Magana-Ortiz, D., Gomez-Lim, M., Fernandez, F., & Loske, A. M. (2014). Physical methods for genetic transformation of fungi and yeast. *Physics of Life Reviews*, *11*, 184–203.
Robinson, J. S., Klionsky, D. J., Banta, L. M., & Emr, S. D. (1988). Protein sorting in Saccharomyces cerevisiae: Isolation of mutants defective in the delivery and processing of multiple vacuolar hydrolases. *Molecular and Cellular Biology*, *8*, 4936–4948.
Rothman, J. H., & Stevens, T. H. (1986). Protein sorting in yeast: Mutants defective in vacuole biogenesis mislocalize vacuolar proteins into the late secretory pathway. *Cell*, *47*, 1041–1051.
Sikorski, R. S., & Hieter, P. (1989). A system of shuttle vectors and yeast host strains designed for efficient manipulation of DNA in Saccharomyces cerevisiae. *Genetics*, *122*, 19–27.
Smirnov, M. N., Smirnov, V. N., Budowsky, E. I., Inge-Vechtomov, S. G., & Serebrjakov, N. G. (1967). Red pigment of adenine-deficient yeast Saccharomyces cerevisiae. *Biochemical and Biophysical Research Communications*, *27*, 299–304.
Stringer, D. K., & Piper, R. C. (2011). A single ubiquitin is sufficient for cargo protein entry into MVBs in the absence of ESCRT ubiquitination. *The Journal of Cell Biology*, *192*, 229–242.
Teng, X., Dayhoff-Brannigan, M., Cheng, W. C., Gilbert, C. E., Sing, C. N., Diny, N. L., et al. (2013). Genome-wide consequences of deleting any single gene. *Molecular Cell*, *52*, 485–494.
Weir, M., & Keeney, J. B. (2014). PCR mutagenesis and gap repair in yeast. *Methods in Molecular Biology*, *1205*, 29–35.
Wilson, D. S., & Keefe, A. D. (2001). Random mutagenesis by PCR. *Current Protocols in Molecular Biology*, *Chapter 8*, Unit8.3.
Wrasman, K., Alioto, S. L., Zhang, Y., Hoban, K., Khairy, M., Goode, B. L., et al. (2018). A flow cytometry-based phenotypic screen to identify novel endocytic factors in Saccharomyces cerevisiae. *G3 (Bethesda)*, *8*, 1497–1512.
Zhu, L., Jorgensen, J. R., Li, M., Chuang, Y. S., & Emr, S. D. (2017). ESCRTs function directly on the lysosome membrane to downregulate ubiquitinated lysosomal membrane proteins. *eLife*, *6*, e26403.

CHAPTER THIRTEEN

Detection of ubiquitinated targets in mammalian and *Drosophila* models

Elena Maspero[a,*], Valentina Fajner[a], Janine Weber[a], Simona Polo[a,b,*]
[a]IFOM, Fondazione Istituto FIRC di Oncologia Molecolare, Milan, Italy
[b]Dipartimento di oncologia ed emato-oncologia, Universita' degli Studi di Milano, Milan, Italy
*Corresponding authors: e-mail address: elena.maspero@ifom.eu; simona.polo@ifom.eu

Contents

Abstract

In this chapter, we describe techniques to detect ubiquitination events occurring in vivo. We focus on methodologies capable of preserving and detecting target protein ubiquitination in physiological conditions, without overexpressing a tagged version of ubiquitin. We provide detailed protocols for mammalian and *Drosophila melanogaster* systems using linkage-specific antibodies against ubiquitin. We analyze immunoblotting and immunofluorescence approaches as well as an ELISA-based quantitative approach, which allows comparing more samples and conditions. Finally, we describe the use of in vitro ubiquitination assays with purified proteins that allows a direct analysis of the molecular mechanisms behind specific E3 activity.

Methods in Enzymology, Volume 619
ISSN 0076-6879
https://doi.org/10.1016/bs.mie.2018.12.030
© 2019 Elsevier Inc.
All rights reserved.

Abbreviations

ATP adenosine 5′-triphosphate
BCA bicinchoninic acid
BSA bovine serum albumin
Cy5 cyanine5
DAPI 4,6-diamidino-2-phenylindole
DELFIA dissociation-enhanced lanthanide fluorescence immunoassay
DTT dithiothreitol
ECL enhanced chemiluminescence
EGF epidermal growth factor
EGFR epidermal growth factor receptor
ELISA enzyme-linked immunosorbent assay
FBS fetal bovine serum
GST glutathione *S*-transferase
IB immunoblot
IF immunofluorescence
IP immunoprecipitation
NEM *N*-ethylmaleimide
PBS phosphate-buffered saline
PLA proximity ligation assay
PVDF polyvinylidene fluoride
SDM Schneider's *Drosophila* medium
SDS sodium dodecyl sulfate
TBS Tris-buffered saline
TR-FRET time-resolved fluorescence resonance energy transfer
WT wild type

1. Introduction

Ubiquitin (Ub) is a versatile posttranslational modification that can change the fate, activity or localization of target proteins. Ubiquitination occurs in various ways, ranging from a simple monoubiquitination to polymeric ubiquitin chains with complex topologies. All these linkages have been detected in cells (Peng et al., 2003; Xu & Peng, 2008; Yau & Rape, 2016) and serve as distinct signals associated with various cellular functions (Komander & Rape, 2012; Oh, Akopian, & Rape, 2018; Park & Ryu, 2014).

The power of the Ub modification stems from its capacity to act as a protein–protein interaction module that targets substrates to a plethora of downstream effectors, generating Ub-regulated networks (Grabbe, Husnjak, & Dikic, 2011; Komander & Rape, 2012). It is thus not surprising that Ub is a key regulator of virtually all the processes that govern the fate of cells and organisms, starting from embryonic development up to the molecular networks underlying cell growth and damage responses. Aberrations of these systems underlie many diseases, including malignancies; inflammatory, immune and metabolic disorders; and neurodegeneration.

One Ub-regulated network pivots around the epidermal growth factor receptor (EGFR). EGF stimulation promotes ubiquitination of EGFR and of its endocytic adapters, providing a striking example of concerted regulation of signaling, ultimately regulating the route of EGFR internalization (Acconcia, Sigismund, & Polo, 2009; MacGurn, Hsu, & Emr, 2012; Polo, Di Fiore, & Sigismund, 2014; Sigismund et al., 2008). Upon EGF stimulation, EGFR becomes phosphorylated at different tyrosine residues; these in turn become docking sites for different effectors, among them the E3 ligase Cbl that catalyzes addition of multiple Ub molecules to the receptor (Levkowitz et al., 1999). By targeting the receptor, Ub mediates its internalization and degradation, leading to signal termination (Haglund et al., 2003; MacGurn et al., 2012; Sigismund et al., 2013). On the other hand, by modifying the endocytic machinery, Ub also contributes to the "endocytic regulation" of signaling in time and space. Indeed, ubiquitination recruits the EGF receptor to an internalization pathway containing specific Ub-binding "route-controllers," such as eps15, epsin, and proteins of the ESCRT complex. These factors ferry the receptor toward a degradative fate in lysosomes and away from a pathway that mediates recycling back to the plasma membrane (Sigismund et al., 2008). This biological model illustrates an additional degree of complexity of Ub-based signaling, namely its dynamic nature, which depends upon its integration with other signaling networks present within the cell.

It is important to note that in such dynamic networks, the stoichiometry of ubiquitination is usually very low. In other words, only a small fraction of each target protein is in a ubiquitinated state. Thus, methods to preserve and optimize the ubiquitination signal are particularly important. Here, we describe a number of techniques we established in the lab using ubiquitin-mediated regulation of EGFR as a model, but protocols can be easily modified for the study of other proteins of interest.

2. In vivo detection of ubiquitinated targets

In this chapter, we describe reliable techniques to detect ubiquitination events occurring in vivo. We will focus on methodologies to preserve and detect endogenous target protein ubiquitination using antibodies against Ub, rather than overexpressing a tagged version of Ub. Given the fact that the Ub protein is highly conserved throughout eukaryotic evolution, these antibodies could work in virtually any eukaryotic system.

The simplest and most direct approach is to immunoprecipitate the target protein from cell lysate (often after giving cells a physiological stimulus), followed by Western blot analysis utilizing specific antibodies against Ub. This approach requires an antibody capable of immunoprecipitating the protein of interest. A second approach relies on immunoprecipitation (IP) of bulk ubiquitinated proteins from total cell lysate followed by detection of the protein of interest using a target-specific antibody. This second method relies on the availability of antibodies against Ub that are highly specific and can precipitate it efficiently. A variant of these procedures is represented by an ELISA-based assay that is particularly well suited when many different samples need to be compared quantitatively. The final antibody-based protocol discussed here relies on immunofluorescence (IF) analysis to visualize the Ub-modified proteins in situ. This protocol requires anti-Ub antibodies that are able to distinguish specific polyUb signals [e.g., polyUb chains with Ub-Ub amide linkages utilizing either the K63 or K48 side chains of Ub (Newton et al., 2012, 2008)]. These can be used for colocalization studies along with antibodies that recognize the target protein. An even more sensitive implementation of this kind of IF analysis, not discussed further here, is the proximity ligation assay (PLA), which is based on signal amplification from oligonucleotides coupled to two different antibodies (Alam, 2018; Fredriksson et al., 2002; Sahin, Jollivet, Berthier, de The, & Lallemand-Breitenbach, 2016). We provide detailed protocols for both mammalian and *Drosophila* systems for all these approaches to in vivo ubiquitin-protein conjugate detection.

2.1 Immunoprecipitation/immunoblot analysis of cultured mammalian cells

Ligand stimulation induces the rapid ubiquitination of the EGFR (peak at 2 min) followed by its endocytosis (10–30 min) and lysosomal degradation (60–120 min). In this experiment, HeLa cells (a single 60%–70% confluent 10 cm plate for each condition) are serum starved for 16 h and then stimulated with 100 ng/mL of EGF for 0, 2, 10, 30, and 120 min. To control for the specificity of the anti-Ub signal, a parallel IP with lysates of cells knocked down for the protein of interest (in this case, HeLa cells knocked down for the EGFR) should be performed. An IP with an unrelated antibody of the same isotype as the antibody against the protein of interest is also recommended.

Since EGFR traffics through the *endo*-lysosomal compartment, a way to accumulate the modified receptor is to block this degradative pathway using, lysosomal inhibitors such as chloroquine or bafilomycin A1 (added 1 h before EGF stimulation); these lysosomotropic agents prevent endosomal acidification, which is needed for the acid proteases in the lysosome to be active. It has to be noted, however, that chloroquine was recently shown to impair endosome/autophagosome fusion with the lysosome in addition to affecting the acidity and/or degradative activity of this latter organelle (Mauthe et al., 2018).

2.1.1 Solutions and reagents

- RIPA lysis buffer: 50 m*M* Tris–HCl pH 8, 150 m*M* NaCl, 1 m*M* EDTA, 1% Triton X-100, 1% sodium deoxycholate, 0.1% SDS, supplemented with a cocktail of protease, phosphatase, and DUB inhibitors freshly added to the buffer prior to lysis: 20 m*M* sodium pyrophosphate pH 7.5, protease cocktail CALBIOCHEM (500 × reconstituted stock solution), 50 m*M* NaF, 2 m*M* PMSF, 10 m*M* sodium vanadate in HEPES pH 7.5, 5 m*M* NEM, or 25 μ*M* of the cell permeable broad spectrum deubiquitylating enzymes inhibitor PR619 (Sigma).
- Laemmli buffer 2 ×: 4% SDS, 125 m*M* Tris–HCl pH 6.8, 20% glycerol, 0.002% saturated bromophenol blue, 10% (v/v) β-mercaptoethanol (14 *M* stock).
- Denaturing solution: 6 *M* guanidium chloride, 20 m*M* Tris–HCl pH 7.5 with freshly added 1 m*M* PMSF and 5m*M* β-mercaptoethanol.
- EGF from INALCO (cat. AF-100-15). Mouse monoclonal anti-Ub antibodies: P4D1 (Santa Cruz, cat. sc-8017), FK2 (Enzo Life Sciences, cat. BML-PW8805); anti-EGFR antibodies: rabbit polyclonal anti-EGFR, which recognizes amino acid residues 1172–1186 (intracellular) of human EGFR (Eurogentech), and mouse monoclonal anti-EGFR [m108 hybridoma, directed against the extracellular domain of human EGFR, ATCC HB-9764]; recombinant Protein G-Sepharose 4B (Thermo Fischer Scientific, cat. 10-1243); goat anti-mouse HRP-conjugated antibody and goat anti-rabbit HRP-conjugated antibody (Bio-Rad, cat. 170-6516 and 170-6515).

2.1.2 Procedure A

1. Wash the 10 cm plates twice in cold PBS. Fully drain the PBS from plates by leaving them tilted for 2–3 min.
2. Lyse the cells directly on the plate by scraping with 150–200 μL/plate of RIPA buffer plus inhibitors. When using RIPA with 1% SDS

(see practical considerations below), use the minimal amount of buffer required to cover the plate (≤80–100 μL/plate).

3. Incubate the lysates on ice for 10 min then centrifuge at 15,000 × *g* for 15 min at 4°C (standard RIPA) or ultracentrifuge at 120,000 × *g* for 45 min at 4°C (RIPA 1% SDS).
4. Collect the supernatants and determine protein concentrations using Bradford protein assay (not ideal for lysis buffer based on SDS) or bicinchoninic acid (BCA) assay. Adjust the concentration to 1 μg/mL using lysis buffer.
5. Incubate the samples in the presence of anti-EGFR intracellular polyclonal antibody with gentle agitation at 4°C for 2 h (250 μg of HeLa lysate is sufficient to visualize EGFR ubiquitination).
6. Add 30 μL of 50% slurry Protein G-conjugated Sepharose beads and incubate at 4°C for an additional 2 h.
7. Wash the beads four times with 0.5 mL of RIPA buffer (centrifuge for 1 min at 400 × *g*).
8. Add 20 μL 2 × Laemmli buffer to the beads (previously drained of wash buffer), heat for 5 min at 95°C and spin for 1 min at 15,000 × *g* before loading on an SDS-PAGE gel. Keep 1/10 of immunopurified samples for anti-EGFR immunoblot analysis using a separate gel; although this immunoblotting will be done in parallel with anti-Ub blotting, nitrocellulose is recommended for anti-EGFR, whereas the anti-Ub blots are best done with PVDF.
9. After resolving the remainders of the IP samples by SDS-PAGE, transfer proteins to a PVDF membrane that had been previously wetted by incubation in 100% methanol for 5 min at RT and then extensively rinsed in TBS-T buffer (TBS, 0.1% Tween20).
10. After transfer, treat the filters in denaturing solution for 30 min at 4°C.
11. Wash the filters five times in TBS-T (5 min each) and incubate the filters overnight at 4°C in 5% BSA/TBS-T.
12. Incubate the filters with anti-Ub P4D1 antibody diluted 1:1000 in 5% BSA/TBS-T for 1 h at room temperature.
13. Wash the filters three times in TBS-T (10 min each) and incubate the filters with the anti-mouse horseradish peroxidase-conjugated secondary antibody diluted in 5% BSA/TBS-T for 30 min at room temperature.
14. Wash the filters three times in TBS-T (5 min each). Signal is revealed using ECL reagents from Amersham. Parallel electro-transfer and immunoblotting should be done with anti-EGFR except transfer proteins to nitrocellulose.

2.1.3 Procedure B

Direct IP of the target protein followed by anti-Ub detection relies on the availability of a good antibody capable of immunoprecipitating the desired protein. Since such an antibody is not always available, an alternative approach is to IP the entire pool of free Ub and ubiquitinated proteins from cellular lysates using a pan-specific immunoprecipitating anti-Ub antibody—such as the FK2 mAb-Agarose from MBL—that recognizes both monoUb and polyUb chains of different types, followed by Western blot analysis of the protein of interest. This protocol is simply the reciprocal procedure to the IP-IB using in Procedure A, and as such can also be used as a control for that procedure even if an IP-capable antibody against the protein of interest were available.

1. Incubate 500 μg of lysate from cells subjected to each treatment (such as EGF, MG132, or other), as well as from untreated control cells, with 10 μL of anti-FK2 monoclonal antibody conjugated to agarose beads for 2 h (up to overnight) at 4°C, rocking.
2. Wash the beads four times with 0.5 mL of RIPA buffer, with centrifugations for 1 min at 400 × *g* to pellet the beads.
3. Add 20 μL of 2 × Laemmli buffer to the drained beads, heat for 5 min at 95°C, spin for 1 min at 15,000 × *g* and loaded the samples on an SDS-PAGE gel.
4. Proteins from the gel are transferred to nitrocellulose, and immunoblotting is performed against the protein of interest (EGFR in this case). Good positive controls for substrate ubiquitination would be eps15 (monoubiquitinated upon EGF), EGFR (polyubiquitinated upon EGF), or β-catenin (stabilized and ubiquitinated upon MG132).

2.1.4 Practical considerations

- As mentioned before, EGF stimulation induces ubiquitination not only of the receptor itself but also of other accessory adaptor proteins important for EGFR endocytic trafficking (Polo et al., 2002). In order to exclude interacting proteins from the IP, lysates can be prepared in modified RIPA buffer containing 1% of SDS. Lysates are subjected to ultracentrifugation at 120,000 × *g* for 45 min at 4°C, and supernatants are collected. Dilution to reach 0.2% SDS (using RIPA buffer w/o SDS) is performed prior to protein quantification and the IP reactions. The ability of the antibody to recognize the target protein under these lysis and dilution conditions should be tested before proceeding with more extensive analysis. The Bradford protein assay quantitation method is not ideal when using SDS. We recommend to use the BCA method and to dilute the lysate to a final concentration of 0.2% SDS prior to quantification.

- Cold PBS and incubation on ice during cell lysis is important to block EGFR internalization and preserve the cellular lysate. NEM or other DUBs inhibitors are required in the lysis buffer as DUBs are active even at low temperature.
- Staining the PVDF membrane with Ponceau solution is not recommended, as this might interfere with antibody recognition Prestained molecular weight markers are used to check transfer efficiency.
- It is particularly important to avoid drying the PVDF membrane during the denaturing treatments required for the anti-Ub blotting. In case of drying, it is possible to rehydrate the PVDF membrane again by treating with methanol for 2 min followed by three washes (5 min each) in TBS-T buffer.
- As a control for nonspecific binding of the protein of interest to FK2-Protein G-Sepharose beads, a parallel IP with an unrelated antibody raised in the same species and of the immunoglobulin isotype as the FK2 antibody (mouse IgG1) is recommended.
- A variant of this procedure B relies on the use of linkage-specific antibodies [such as ones available against Ub with K63, K48, or K11 linkages that were generated by Genentech (Newton et al., 2012, 2008)]. Using these antibodies, we discovered that EGF mostly induces K63-linked ubiquitination of a variety of proteins (Argenzio et al., 2011).
- Overexpression of the protein of interest and/or tagged version of Ub should be avoided since it may lead to artifacts. Deregulated expression of the protein of interest may lead to misfolded protein that is targeted for degradation and therefore ubiquitinated. Artificially overexpression of tagged Ub may affect the physiology and bioactivity of the Ub molecule (e.g., preventing the action of deubiquitinated enzymes).

2.2 Immunoprecipitation in *Drosophila melanogaster*

Here below we report related protocols we have optimized for immunological detection of ubiquitin-modified proteins in fruit fly ovaries. The protocol can be easily scaled according to different starting volumes. Ovaries are favorable for experimental analysis for several reasons. The large size and relatively simple organization of the organs make them easy to dissect and suitable for microscopy using either living and fixed whole mount tissue.

In addition, tissues can be treated ex vivo with drugs (e.g., MG132 at 50 μ*M*, Chloroquine at 50 μ*M*, bafilomycin-A at 200 n*M*), avoiding possible toxicity due to animal feeding. To preserve the organ during these treatments, it is recommended to use *Drosophila* cell culture medium supplemented with FBS and to not exceed the 4 h incubation, although optimal treatment duration has to be experimentally verified.

Ovaries of adult females consist in 16–20 parallel ovarioles containing developing egg chambers arranged in a linear array of progressive developmental stages (from 1 to 14). Three-day-old WT female flies, previously fed on fresh yeast for 48 h, were dissected in Schneider's *Drosophila* medium (SDM) [for dissection technique, see the protocol from Wong & Schedl (2006)]. To enrich samples for ubiquitinated species that are subject to proteasome-mediated degradation, we treated the ovaries with the proteasome inhibitor MG132; this will enhance, for example, the accumulation of Armadillo, the fly ortholog of human β-catenin (Fig. 1A).

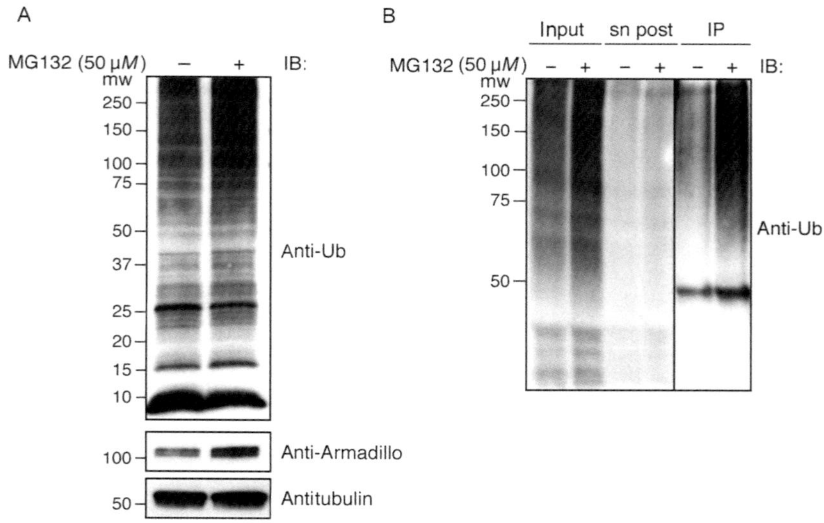

Fig. 1 Biochemical analysis of ubiquitinated proteins from *Drosophila* ovaries treated with MG132. (A) Lysates from fly ovaries were IB with the indicated antibodies. Upon proteasome inhibition with MG132 there is an increase of total Ub-conjugated species (*upper panel*) and of Armadillo (*middle panel*). Antitubulin blot is used as a normalizer. (B) Immunoprecipitated Ub-conjugated species are analyzed by IB using anti-Ub antibody. Anti-FK2 is quite efficient in depleting ubiquitinated proteins as visible comparing 50 μg of lysate before and after IP (sn post, supernatant post-IP). Immunoprecipitated species are clearly enriched upon MG132 treatment.

2.2.1 Solutions and reagents

- RIPA lysis buffer (Section 2.1.1).
- Laemmli buffer (Section 2.1.1).
- Denaturing solution (Section 2.1.1).
- SDM 1× with L-Glutamine (Gibco from Life Technologies, cat. 21720024). MG132 (Sigma, cat. 4774787), Fetal Bovine Serum (FBS, Euroclone). Mouse monoclonal anti-Ub antibodies FK2 (Enzo Life Sciences, cat. BML-PW8805), pan-anti-ubiquitin FK2 mAb-Agarose (MBL, cat. D058-8); mouse monoclonal anti-Armadillo/β-catenin antibody (Developmental Studies Hybridoma Bank clone N2 7A1); mouse monoclonal antitubulin (Sigma, cat T5168); Goat anti-mouse HRP-conjugated antibody (Bio-Rad, cat. 170-6516); recombinant Protein G-Sepharose 4B (Thermo Fischer Scientific, cat. 10-1243); Amersham ECL Western Blotting Detection Reagent (cat. GE Healthcare RPN2106).

2.2.2 Procedure

1. Dissect fly ovaries in SDM following the technique described in (Wong & Schedl, 2006). Once the ovaries are removed from the carcass, tease apart the anterior portion of the ovarioles, leaving them attached at the posterior, using either sharp forceps or stiff wires (e.g., tungsten needles).
2. Divide the dissected ovaries of each sample (control and MG132-treated in this example) between two microcentrifuge tubes (80 ovaries/1.5 mL tube) and treat them ex vivo with 50 μ*M* MG132 (or DMSO for the control) in 500 μL SDM medium supplemented with 10% FBS, for 2 h at room temperature while rocking.
3. After treatment, wash the samples twice in 500 μL of cold PBS. At the end of the second wash, spin the samples for 1 min at 8000 × *g* and carefully remove the remaining PBS using a microcapillary loading gel tip.
4. Lyse each pellet (80 ovaries) in 450 μL of RIPA buffer. First add 50–60 μL of buffer and mechanically lyse it with a pestle for 1–2 min, in ice. Then add buffer up to 450 μL and pipette thoroughly.
5. Leave the lysate on ice for 20 min and then spin at 15,000 × *g* for 10 min at 4°C.
6. Collect the supernatants, carefully avoiding the surface layer of lipids.
7. Repeat steps 5 and 6 until the collected supernatants are clear.
8. Protein content is determined with the BCA assay.
9. Proceed with anti-Ub IP as indicated in Section 2.1.3.

10. The immunoprecipitated species in Laemmli loading buffer are ready to be loaded for SDS-PAGE.
11. Anti-Ub Western blotting is performed as described for mammalian proteins (Section 2.1.2).

2.2.3 Practical considerations

- Ovary dissection is performed in SDM medium, and the collected samples are kept on ice until they are ready for treatment, in order to preserve organ integrity.
- Female reproduction is optimized for high fecundity under ideal environmental conditions, and in the absence of sufficient nutrients, egg production is halted. Thus, female flies should be maintained on fresh yeast paste for at least 2 days prior to dissections to ensure proper feeding of the flies and large and healthy ovaries for the experimental procedure.
- 160 ovaries correspond to about 3 mg of protein lysate. The number of ovaries dissected can vary according to amount of final lysate required for the experiment.
- For anti-Armadillo Western blotting, the monoclonal antibody from hybridoma clone N2 7A1 was tested on PVDF and nitrocellulose membrane and it performs well with both filters, but this should be tested for each protein of interest.

2.3 ELISA-based immunoassay

To assess substrate ubiquitination from low amounts of sample, a robust and reliable method is the ELISA assay. This technology is based on a method of "sandwich-recognition" of a target protein using a capture antibody and a detection antibody. After immobilization of the capture antibody (e.g., anti-Ub) on a solid surface (the bottom of a microwell), the analyte (appropriate cellular lysate) is incubated on the plate. After removal of the unbound analyte, incubation with the detection antibody against the substrate (e.g., the EGFR) will produce a specific signal in those samples where the substrate is ubiquitinated. The procedure could also be reversed with the antisubstrate coated on the plate and the anti-Ub as revealing antibody (Savio et al., 2016).

An optimized version of ELISA is represented by the DELFIA (Dissociation-Enhanced Lanthanide Fluoroimmunoassay) technology from PerkinElmer. The signal relies on a Europium (Eu)-conjugated antibody that is able to produce a fluorescent signal when incubated with acidic enhancement buffer. The lanthanide ions released into solution rapidly form a fluorescent species that is stable for up to 8 h.

2.3.1 Solutions and reagents

- The commercial DELFIA kit (PerkinElmer) includes: microwell plates, wash buffer, assay buffer, Europium-labeled secondary antibodies (anti-mouse cat. AD0124; anti-rabbit cat. AD0105), and enhancement solution.
- Plate coating buffer: 3.03 g $NaCO_3$, 6.0 g $NaHCO_3$ dissolved in 1 L H_2O (pH 9.6).
- PerkinElmer EnVision Multilabel Plate Reader.

2.3.2 Procedure

1. Coat the DELFIA microwell plate with the capturing antibody: dilute the anti-Ub (FK2) antibody to 5 μg/mL in coating buffer; add 100 μL to each well and incubate plate overnight at 4°C with gentle shaking.
2. Wash each well once with 100 μL of coating buffer and block the microwell plate by incubating with 200 μL of 2% BSA/PBS for either 1–2 h at RT or O/N at 4°C.
3. Wash the plate once in wash buffer. Use 150–200 μL/well of buffer to wash the wells.
4. Incubate 25–50 μg of lysate/well (in triplicate) overnight at 4°C. Dilute the samples in assay buffer to reach a volume of 100 μL/well
5. Wash the plate three times in wash buffer.
6. Incubate the plate with detection antibody for 1 h at room temperature: dilute the antibody against the protein of interest to 1 μg/mL in 100 μL of assay buffer.
7. Wash the plate three times in wash buffer.
8. Add the selected Eu-labeled secondary antibody (1 μg/mL in assay buffer, 100 μL/well) and incubate for an additional 1 h at room temperature.
9. Wash the plate three times in wash buffer.
10. Add 100 μL/well of Enhancement Solution to release and activate Eu^{3+} ions and incubate for 15 min without shaking or 5 min with shaking at room temperature.
11. Signal is detected using the PerkinElmer EnVision Plate Reader (excitation at 340 nm and emission at 615 nm).

2.3.3 Practical considerations

- As a negative control, include a lysate from a cell line devoid of your target protein (or at least a "no-lysate" condition).
- Typically, the total signal and sensitivity is better using plates directly coated with antibody rather than with streptavidin. However, antibody-consumption is significantly higher in direct coating

(e.g., 0.5 μg of antibody/well) than in indirect coating (e.g., 200 ng of biotinylated antibody/well).

- In principle, coated plates could be stored at 4°C until use, but better results are obtained with freshly prepared plates.
- In the case of a time course or a titration experiment, it is advisable to evaluate total level of the protein of interest in the same samples. To this end, it is possible to set up a second DELFIA assay capturing the substrate with an antibody targeting an epitope different from the one used in the detection. DELFIA assays can also be multiplexed. Dual-label assays are performed with Eu and Samarium (Sm, 643 n*M*) or Terbium (Tb, 545 n*M*). The choice is dictated by the required sensitivity and assay protocol.

2.4 Immunofluorescence detection of ubiquitinated proteins in cultured mammalian cells

This method allows one to directly visualize ubiquitin chains in cells through the use of specific anti-Ub antibodies. Since ubiquitinated proteins are expressed throughout the cell, anti-Ub antibodies stain both nucleus and cytoplasm. However, upon specific signal induction (i.e., EGF stimulation or DNA damage) relocalization of the Ub signal may be evident. In the case of the EGFR trafficking, EGF causes rapid relocalization of the K63-Ub chain signal from the cytosol to EGFR-positive foci. By following EGFR with fluorescent EGF, EGFR is found to colocalize with the anti-K63-Ub chain antibody but not with anti-K48 chain antibody (Argenzio et al., 2011). The opposite behavior is seen when cells are treated with MG132 that rapidly accumulate K48-modified proteins in multiple small cytoplasmic foci (Fig. 2). In case the signal of interest does not induce massive relocalization of the Ub signal, PLA analysis is recommended for visualizing the presence of ubiquitin on the protein of interest (Alam, 2018; Fredriksson et al., 2002; Sahin et al., 2016).

2.4.1 Buffers and reagents

- Binding Buffer: Minimum Essential Media (MEM), containing 1 m*M* sodium pyruvate, 0.1 m*M* nonessential amino acids, 20 m*M* HEPES pH 7.5, and 0.1% BSA to saturate nonspecific antibody-binding sites.
- Paraformaldehyde solution 4% in PBS (HiMedia lab, cat. TCL119).
- Permeabilization solution (PBS, 0.1% BSA, 0.1% Triton X-100).
- 10% goat serum (Jackson Immunoresearch Labs, cat. 005-000-121).
- 2% BSA dissolved in PBS.

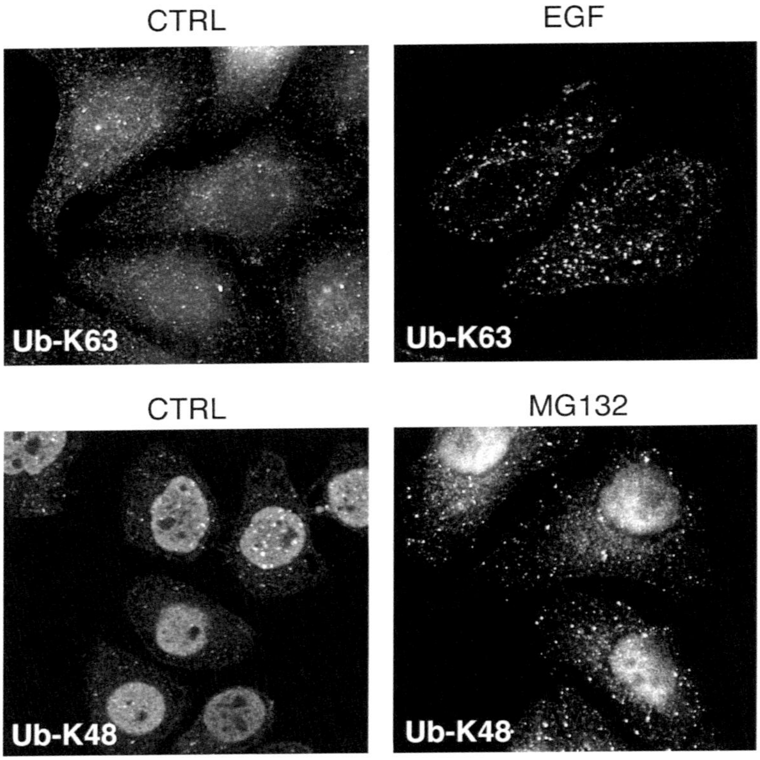

Fig. 2 Immunofluorescence analysis with anti-Ub specific antibodies. In the *upper panels*, IF analysis of HeLa cells serum-starved (*left panel*) or treated for 10 min with rhodamine-EGF. Ub-K63 specific signal accumulates upon EGF stimulation in discrete dots (*right panel*). In the *lower panels*, IF analysis of untreated Hela cells (*left panel*) or treated for 4 h with 5 μ*M* MG132. Ub-K48 antibody signal accumulates in multiple dots upon treatment (*right panel*).

- Rhodamine-EGF (Thermo Fisher Scientific, cat. E3481); Alexa-EGF 555 (Thermo Fisher Scientific, cat. E35350); Anti-Ub K63 (human Apu3 clone, Genentech) or anti-Ub K63 (rabbit Merck Millipore, cat. 05-1308); anti-Ub K48 (human Apu2 clone, Genentech) or anti-Ub K48 (rabbit Merck Millipore, cat. 05-1307); Alexa-488 goat antihuman (Thermo Fisher Scientific, cat. A-11013); Alexa-488 donkey anti-rabbit (Thermo Fisher Scientific, cat. A-21206); Alexa-647 donkey anti-mouse (Thermo Fisher Scientific, cat. A-31571); DAPI (Sigma-Aldrich, cat. D9542).
- Dako pen (Agilent Technologies, cat. S2002).
- Mounting medium (70% glycerol in PBS, 1.5% DABCO (1,4-Diazabicyclo[2.2.2]octane, Sigma, cat. D2522).

2.4.2 Procedure

1. Plate on glass coverslips the correct number of cells in order to have ~60% confluence the day of the experiment. One coverslip will be used for each specific time point in the time course below.
2. Distribute one coverslip/each well of a 24-well plate.
3. Wash and place the coverslips in 500 μL of binding buffer for 4 h.
4. Incubate the coverslips at 37°C in the presence of a fluorophore-conjugated EGF replacing the medium with 500 μL of Alexa 555-EGF 1:500 or Rhodamine-EGF 1:100 in binding buffer, starting from the longer incubation time (from 60 to 2 min).
5. At the end of the EGF stimulation, place the 24-well plate on ice and wash each well twice with cold PBS.
6. Fix the cells for 10 min in 4% paraformaldehyde at room temperature.
7. Wash three times with PBS at room temperature.
8. Treat with permeabilization solution for 10 min.
9. Wash three times with PBS.
10. Block in 10% goat serum diluted in 2% BSA/PBS for 30 min at room temperature.
11. Incubate coverslips with primary antibody in 1% BSA/PBS for 1 h at room temperature (either human K48 or K63 or FK2 at 1 μg/mL).
12. Wash three times with PBS.
13. Incubate coverslips with Alexa-488 goat antihuman or anti-mouse (1:400) in 1% BSA/PBS for 45 min.
14. Add DAPI solution (0.5 μg/mL) for nuclei staining and incubate for 10 min.

10\. Wash three times with PBS.

11\. Mount coverslips in glycerol for confocal analysis. For this, draw a circle (around 13 mm) on a microscope slide with a Dako pen to create a spreading barrier for the stained specimen, and put 3 μL of glycerol in the center of the circle.

12\. Dip the coverslip in distilled water, remove the excess of water with filter paper placed on the edge of the coverslip and carefully lower the coverslip onto the glycerol drop in order to prevent the formation of bubbles.

13\. Seal the coverslips with nail polish before microscope acquisition.

2.4.3 Practical considerations

- The incubation time with binding buffer in the absence of serum minimizes basal EGFR activity localizing most of the EGFR at the cell surface. Starvation is typically performed in binding buffer for 4–16 h.

- Cold PBS washes at step 5 are required after EGF stimulation in order to remove the excess of labeled EGF and, at the same time, stop the internalization process.
- Upon stimulation of cells with fluorescently labeled EGF, the anti-Ub-K63 signal is relocalized to active EGFR-containing foci (Fig. 2). It is noteworthy that K63 signal could also be ascribed to ubiquitination of EGFR-associated proteins.

2.5 Immunofluorescence detection of ubiquitinated proteins in *Drosophila melanogaster*

Fly ovaries are a powerful tool to monitor Ub-conjugated species formed in response to certain stimuli (i.e., starvation-induced autophagy). The ovary is a nutrient-sensitive organ that quickly responds to food deprivation by inducing autophagy in previtellogenic egg chambers (stage 1–7) (Barth, Szabad, Hafen, & Kohler, 2011). Ubiquitinated proteins or organelles are recognized by autophagy receptors (called p62/SQSTM1 or Ref(2)p in the fly) and then engulfed in autophagosomes for degradation. Autophagosomes are key components of the autophagic machinery and consist of double-membrane vesicles that eventually fuse with the lysosome for degradation of their contents (Glick, Barth, & Macleod, 2010). These structures can be easily detected by IF using ATG8 as specific marker (not shown here) and are clearly labeled by Ub (Fig. 3). For this experiment, 3-day-old flies are fed on fresh yeast (rich nutrient conditions) for 48 h and then moved to agar-molasses plates (food deprivation conditions) for the following 48 h.

2.5.1 *Buffers and reagents*

- Agar-molasses plate: 36 mL of Golden A molasses (Natural American Foods, cat. OT368088), 8.8 g of Agar (BD Bacto Dehydrated Agar, cat. 214010, Thermo Fisher Scientific), 10% in 95% ethanol Tegosept (Sigma, cat. H3647); bring up to 1 L with water.
- Paraformaldehyde solution 4% in phosphate-buffered saline (HiMedia lab, cat. TCL119).
- Washing solution (PBS, 0.1% Triton X-100).
- Permeabilization solution (PBS, 1% Triton X-100).
- 5% BSA dissolved in PBS 0.1% Triton X-100.
- DAPI (Sigma-Aldrich, cat. D9542).
- Mouse monoclonal anti-Ub antibodies FK2 (Enzo Life Sciences, cat. BML-PW8805); rabbit anti-Ref(2)p antibody (kindly provided by

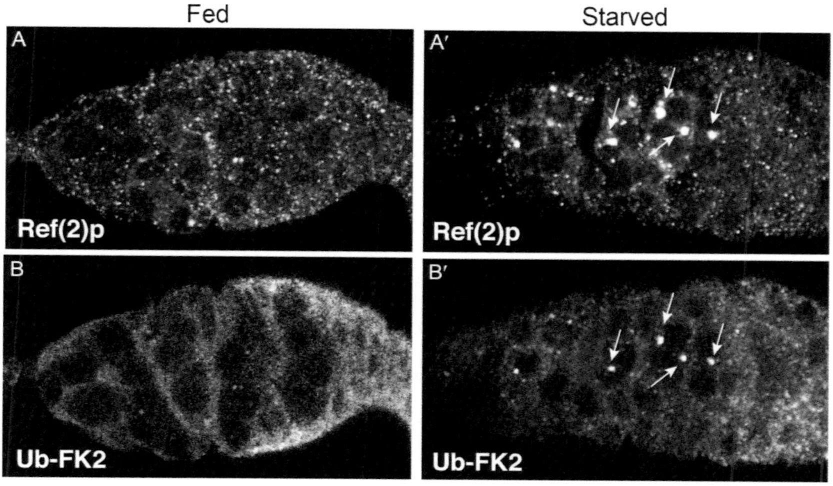

Fig. 3 Immunofluorescence analysis of starvation-induced autophagy in fly ovaries. During oogenesis, nutrient depletion induces autophagy at several checkpoints. The first checkpoint is in the germarium, where autophagy can be detected during normal development, but are increased under starvation conditions. Germaria are smaller and thinner than the rest of egg chambers thus antibody penetration is generally easier at this stage. IF images show germaria of fed (A, B) and starved (A′, B′) flies. Three-day-old flies were fed on fresh yeast for 48 h and then starved on agar-molasses plate for the following 48 h. Egg chambers from starved flies accumulate the autophagy adaptor Ref(2)p, in which the ubiquitin signal is relocalized (*white arrows*).

Prof. T. E. Rusten, University of Oslo); Alexa-647 donkey anti-mouse antibody (Thermo Fisher Scientific, cat. A-31571); Alexa-488 donkey anti-rabbit antibody (Thermo Fisher Scientific, cat. A-21206).

- Mounting medium (70% glycerol in PBS, 1.5% DABCO (1,4-Diazabicyclo[2.2.2]octane, Sigma, cat. D2522).

2.5.2 Procedure

1. Dissect fly ovaries in PBS (10 ovaries/condition) following the technique described in (Wong & Schedl, 2006). Once the ovaries are removed from the carcass, tease apart the anterior portion of the ovarioles, leaving them attached at the posterior, using either sharp forceps or stiff wires (e.g., tungsten needles).
2. Once all the ovaries have been isolated, transfer them to a 0.5-mL tube with 300 μL 4% paraformaldehyde solution and incubate 20 min at room temperature. From this point, perform all the following steps of incubation and washes while rocking.

3. Allow the ovaries to settle to the bottom of the tube and carefully remove the PFA using a P200 tip to avoid taking up the ovaries.
4. Wash three times with 500 μL PBS-Triton (PBS-T) 0.1% for 5 min each at room temperature.
5. Permeabilize the ovaries by incubating them with 500 μL PBS-T 1% for 20 min.
6. Block in 5% BSA prepared in PBS-T 0.1% for at least 30 min at room temperature (up to overnight at 4°C).
7. Incubate the ovaries overnight at 4°C with the primary antibodies diluted in 300 μL of 5% BSA–PBS-T 0.1%. Optimal antibody dilution must be determined empirically.
8. Remove the primary antibodies and perform three 5-min washes as in step 4.
9. Incubate with secondary antibodies diluted in 300 μL PBS-T 0.1% for 2 h at room temperature.
10. Remove the secondary antibodies and wash in PBS-T 0.1% for 5 min three times.
11. Add DAPI for nuclei staining and incubate for 10–15 min.
12. Wash three times in 500 μL PBS-T 0.1% for 5 min each.
13. Carefully remove all the PBS-T 0.1% and add PBS.
14. Using forceps, transfer the ovaries into a drop of PBS on a microscope slide between two pieces of transparent tape to form a bridge. This will prevent the organs from becoming too compressed under the coverslip.
15. Using forceps and tungsten needle, carefully separate all the ovarioles and remove the mature eggs that could create an excessive thickness of the sample allowing the movement of the smaller egg chambers during microscope image acquisition.
16. Remove the excess of PBS and add 16 μL of mounting medium.
17. Gently cover with a fresh cover slip (18 × 18 mm) avoiding bubbles formation and seal it with nail polish.

2.5.3 Practical considerations

- During dissection, before fixation is important to separate the single ovarioles in order to allow the proper penetration of fixative and antibodies in all the egg chambers. The ovarioles separation can be achieved also pipetting up and down the ovaries during the step of washing and blocking. This approach is not recommended for the analysis of the early egg chambers and germaria, since they can detach during the procedure.

- Do not prolong the dissection for more than 30 min per sample, in order to avoid organ deterioration.
- Permeabilization time can be increased up to 30 min. For thinner tissue, such as larval imaginal discs, 10 min permeabilization is sufficient for proper antibody penetration.
- To mount coverslips for confocal analysis, a Dako pen can be used in place of tape as described in Section 2.4.2.

3. In vitro detection of ubiquitinated targets

The development of an in vitro system with purified components is a useful method to validate the data observed in vivo and study at the molecular level the ubiquitin-ligation machinery responsible for the ubiquitination of a specific protein. It also allows one to determine the minimal components required for the ubiquitination process (including adaptors, allosteric activators, or posttranslational modifications) and evaluate possible enzyme inhibitors. Moreover, the nature of the Ub chains formed during the reaction can be investigated using mass spectrometry analysis with AQUA peptides (Phu et al., 2011). Our group has previously characterized in detail the ubiquitin-protein ligase (E3) Nedd4, including studying its substrates (Polo et al., 2002), its regulation (Woelk et al., 2006) and the molecular mechanism behind its catalytic activity (Maspero et al., 2011, 2013). Nedd4 belongs to the HECT class of E3 ligases (Rotin & Kumar, 2009), which form an intermediate thioester bond with Ub through an E3 cysteine residue, before catalyzing the covalent attachment of ubiquitin to its specific substrate or, for Ub-chain formation, to ubiquitin itself.

In this section, we describe a protocol designed to investigate the activity of the E3 ligase of choice on a specific substrate. The E1 (ubiquitin-activating enzyme), E2 (ubiquitin-conjugating enzyme), and E3 (ubiquitin ligase) enzymes used in these assays are produced and purified in-house according to protocols previously described (Maspero & Polo, 2016).

3.1 Substrate ubiquitination and Ub-chain formation

3.1.1 Buffers and reagents

- His-tagged human E1 enzyme (plasmid: Addgene clone 34965, expressed and purified in house (Berndsen & Wolberger, 2011); human E2 enzyme Ube2D3 (expressed and purified in-house); human E3 enzyme Nedd4 (expressed as a GST fusion protein and then cleaved from the GST tag and purified as previously described (Maspero & Polo, 2016)); GST- or MBP-tagged substrate of interest. In the specific protocol described below, we use purified GST-gammaENaC expressed and purified in our lab).

- 0.1 m*M* ATP stock solution: Dissolve ATP in H_2O, adjust the pH to 7.0, aliquot and store at −80°C.
- Ubiquitination Buffer 10 ×: 250 m*M* Tris–HCl pH 7.6, 50 m*M* $MgCl_2$, and 1 *M* NaCl. 2 m*M* ATP (Sigma) and 0.2 μ*M* DTT (Sigma) are freshly added to the buffer prior to reaction.
- RIPA buffer (Section 2.1.1).
- Denaturing Solution (Section 2.1.1).
- Laemmli Buffer (Section 2.1.1).
- Glutathione Sepharose 4B beads (GE Healthcare, cat. 17-0756-05).
- Imperial protein stain (Thermo Fisher Scientific, cat. 24615); Anti-Ub P4D1 antibody (Santa Cruz, cat. sc-8017).

3.1.2 Procedure

1. Set up reactions with final concentrations of 20 n*M* E1, 250 n*M* $HECT^{Nedd4}$, 300 n*M* GST-substrate (gamma-ENaC) bound to glutathione beads, 2 m*M* ATP, 0.2 μ*M* DTT, 1 μ*M* Ub (WT) and 1 × Ubiquitination buffer. Reactions of 50 μL are set up for each experimental condition.
2. Transfer 50 μL of the common mix in a separate tube, this represents the negative control of the reaction.
3. Start the reaction by addition of purified His6-tagged Ube2D3 (UbcH5c) to a final concentration of 250 n*M* in the above reaction mixes.
4. Incubate the samples at 37°C for 10, 30, and 60 min. At the end of the incubation time, put the vial on ice.
5. Centrifuge (400 × *g* for 2 min) at 4°C to separate the glutathione-bound ubiquitinated substrate from free Ub chains and enzymes present in the supernatant.
6. Add 15 μL of Laemmli buffer (2 ×) to 15 μL of the supernatants and load on 4%–20% gradient gel. These will allow good separation of the polyUb chains (and ubiquitinated enzymes) generated during the reaction.
7. Wash the pellet containing the ubiquitinated substrate linked to the beads four times in RIPA buffer, then add 2 × Laemmli buffer to the drained beads.
8. Heat for 5 min at 95°C and centrifuge for 1 min at 15,000 × *g*. Supernatants are resolved in an 8% SDS-PAGE gel.

9. Protein ubiquitination is evaluated by immunoblotting with anti-Ub (Section 2.1.2).
10. After the IB, the PVDF membrane can be stained with Imperial Protein Stain solution for 10 min at room temperature.
11. Wash the membrane extensively with water in order to remove excess stain.
12. Dry the membrane completely to visualize proteins.

3.1.3 Practical considerations

- It is recommended to avoid multiple freeze/thaw cycles for purified proteins. Make small aliquots of each purified enzyme and store at −80°C.
- It is recommended to follow substrate ubiquitination kinetics. The precise time points used need to be experimentally tested considering as end point reaction the condition in which free monomeric ubiquitin is totally consumed.
- In the case of HECT E3 enzymes free polyUb chains are formed during the reaction and can be considered as a proxy of the enzyme activity in the absence of substrates.
- If pellet is not visible after centrifugation (step 7), 15 μL of Glutathione beads washed in ubiquitination buffer can be added.
- Protein modification by a single Ub results in a ∼8 kDa shift in apparent molecular mass. Multiple monoUb addition or polyUb chain addition results in a smear running above the position of the unmodified substrate protein. Percentage of the gel may change according to the molecular weight of the substrate.
- To normalize the reactions, instead of staining the membrane with Imperial Stain solution, substrates can also be recognized with a specific antibody or antitag antibody (usually GST or MBP).
- Avoid, if possible, to use a His-tagged version of the protein substrate since E2 enzymes are usually produced with an uncleavable His-tag and this will result in E2 precipitation in step 5.

3.2 Time-resolved ubiquitin fluorescence resonance energy transfer (TR-Ub-FRET) assay

The TR-Ub-FRET assay is a ubiquitination assay that enables the detection of polyubiquitin chain formation in solution in real-time or in an end-point configuration without the need of running and staining an SDS-PAGE gel or performing a Western blot.

The assay measures ubiquitin-chain formation by the ubiquitination machinery using a mix of labeled and unlabeled monoUb. Specifically, the

100 × TRF-Ubiquitin-mix (South Bay Bio; see also https://southbaybio.com/products/100x-trf-ubiquitin-mix.html) consists of a mixture with no specified ratio of Europium Cryptate-labeled ubiquitin, the acceptor fluorophore cyanine5 (Cy5)-labeled ubiquitin, and unlabeled ubiquitin. When Ub-chain formation is initiated by adding ATP, unlabeled, Cryptate- and Cy5-labeled ubiquitin monomers become incorporated in the growing polyUb chain. As donor and acceptor Ubs become incorporated into chains they come into close proximity, resulting in energy transfer and signal detection. The longer the Ub chain is growing the more signal is detected. The presence of unlabeled Ub prevents FRET signal from Ub dimers only. With homemade enzymes, an initial titration of the E1, E2, and E3 is required to get the best signal/background ratio. This assay can be used to test the kinetics and processivity of E3 ubiquitin ligases and allows high throughput screenings for inhibitors.

Here we give instructions for activity measurements and data analysis using the HECT domain of the Nedd4 E3 ligase as an example (Fig. 4).

3.2.1 Buffers and reagents

- Buffer: 25 m*M* Tris–HCl pH 7.6, 100 m*M* NaCl, 5 m*M* $MgCl_2$, 0.01% Tween20.
- 100 m*M* ATP stock solution.
- E1, E2, and E3 enzymes (see Section 3.1.1). For the E3 enzyme HECT Nedd4, cleavage of the GST-tag and purification of the enzymatic domain is performed according to a previously described protocol (Maspero & Polo, 2016).
- 100 × TRF-Ubiquitin-Mix (South Bay Bio, cat. SBB-TR0051-2K).
- OptiPlate-384 well Microplate, White Opaque (PerkinElmer, cat. 6007290).
- PerkinElmer EnVision Multilabel Plate Reader.

3.2.2 Procedure

1. Set the reader as follows: Excitation Filter = UV2 (TRF) 320 nm; Emission Filter 1 = APC 665 nm; Emission Filter 2 = Europium 615 nm; Delay Time = 50 μs; Window Time = 400 μs.
2. Prepare a mix of E1 (15 n*M*), E2 (350 n*M*), E3 $HECT^{Nedd4}$ (100 n*M*) and the TRF-Ubiquitin-mix (1 ×) in TR-Ub-FRET-buffer and aliquot 49 μL in each well of the Opti-384 well plate.
3. Set up control reaction wells adding 1 μL of water (for background subtractions and normalization).

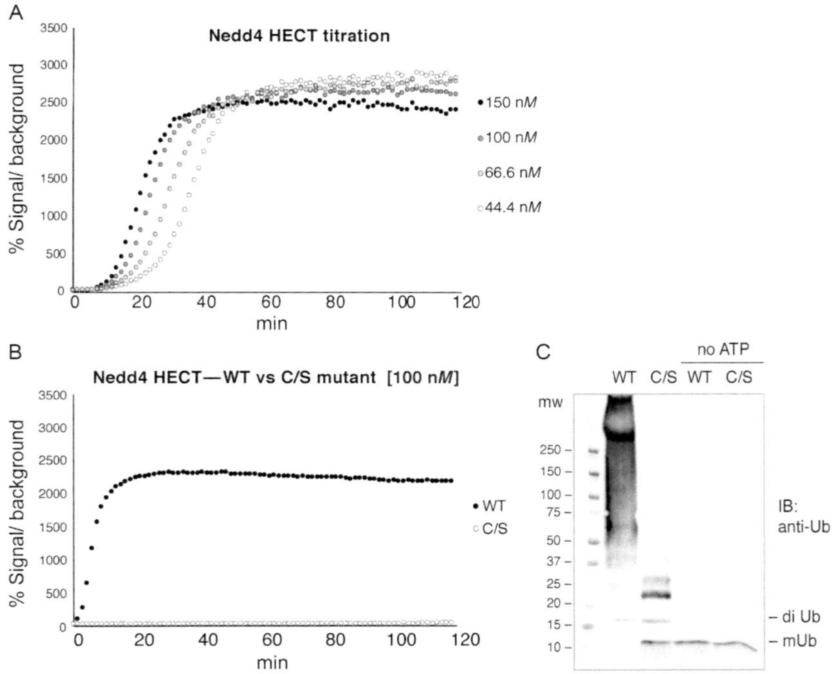

Fig. 4 TR-FRET ubiquitination assay. (A) Optimization of E3 ligase concentration. Bacterially expressed E1, E2 (His-tagged UBE2D3) and serial dilutions of the Nedd4 HECT domain were incubated with the Ub mix of Europium Cryptate ubiquitin, Cy5-ubiquitin, and WT ubiquitin in an optimized ratio. Emission was immediately recorded. (B) Same as in (A), but using Nedd4 WT and catalytically inactive (C/S) mutant at 100 n*M* concentration. (C) Western blot of samples from the TR-FRET assay shown in (B). At the end of the measurement (120 min) samples are loaded on a gel, transferred to a filter and detected with antiubiquitin antibody as described in Section 2.1.2.

4. Start the reaction in the other wells by adding 1 µL of 100 m*M* ATP to the enzymes/Ub mix.
5. Start measuring the Opti-384 well plate using the PerkinElmer EnVision Multilabel Plate Reader.
6. Quantify the raw signal by calculating the product of the ratio of acceptor to donor emission signal for all reactions (including the control reactions). Raw Signal [*r*(Signal) = (Emission (620 nm)/Emission (665 nm)] $\times 10^4$.
7. Subtract the background raw signal (no ATP) from the experimental signal (with ATP). Delta Signal (*r*(Delta)) = *r*(Signal reaction) − *r*(Signal background).

8. Divide the Delta Signal you obtained with the background raw signal and normalize it to 100%. Signal/Background % = r(Delta)/r(Signal background) * 100%.
9. Plot the time (minutes) against Signal/Background % with a scatter plot in Excel.

3.2.3 Practical considerations

- It is preferable to run the reactions in duplicates, even if results obtained with the same batch of enzymes and Ub mix are, in our hands, extremely reproducible.
- At the end of the measurements, the samples used in the Opti-384 well plate can be resolved by SDS-PAGE and analyzed by anti-Ub antibody immunoblotting as described above (Section 2.1.2). A representative Western blot is shown in Fig. 4C.

4. Concluding remarks

The regulatory influence exerted by Ub pervades virtually all areas of cellular biology; yet, its analysis is often difficult and requires specialized knowledge. Since only a small fraction of the protein of interest is in the ubiquitinated form, experiments to demonstrate covalent modification in vivo are quite challenging. In case of polyubiquitination events that drive protein degradation (e.g., K48/K11 chains), modified proteins are rapidly degraded and thus poorly detectable, but the use of proteasome inhibitors may solve the issue. In case of mono/polyubiquitination events that drive alternative localization or promote signaling (e.g., K63), the identification of the right stimulus and its kinetics is a priority. Strategies for identification of modification sites or modifying enzymes vary depending on the type of target, and in vitro analysis requires proper isolation and handling of the relevant enzymes. Whatever the case, cross-validation with more than one approach is highly recommended in order to prove that the protein of interest is modified by ubiquitin.

References

Acconcia, F., Sigismund, S., & Polo, S. (2009). Ubiquitin in trafficking: The network at work. *Experimental Cell Research*, *315*(9), 1610–1618. https://doi.org/10.1016/j.yexcr.2008.10.014. S0014-4827(08)00434-5 [pii].

Alam, M. S. (2018). Proximity ligation assay (PLA). *Current Protocols in Immunology*, *123*(1), e58. https://doi.org/10.1002/cpim.58.

Argenzio, E., Bange, T., Oldrini, B., Bianchi, F., Peesari, R., Mari, S., et al. (2011). Proteomic snapshot of the EGF-induced ubiquitin network. *Molecular Systems Biology*, 7, 462. https://doi.org/10.1038/msb.2010.118.

Barth, J. M., Szabad, J., Hafen, E., & Kohler, K. (2011). Autophagy in Drosophila ovaries is induced by starvation and is required for oogenesis. *Cell Death and Differentiation*, *18*(6), 915–924. https://doi.org/10.1038/cdd.2010.157.

Berndsen, C. E., & Wolberger, C. (2011). A spectrophotometric assay for conjugation of ubiquitin and ubiquitin-like proteins. *Analytical Biochemistry*, *418*(1), 102–110. https://doi.org/10.1016/j.ab.2011.06.034.

Fredriksson, S., Gullberg, M., Jarvius, J., Olsson, C., Pietras, K., Gustafsdottir, S. M., et al. (2002). Protein detection using proximity-dependent DNA ligation assays. *Nature Biotechnology*, *20*(5), 473–477. https://doi.org/10.1038/nbt0502-473.

Glick, D., Barth, S., & Macleod, K. F. (2010). Autophagy: Cellular and molecular mechanisms. *The Journal of Pathology*, *221*(1), 3–12. https://doi.org/10.1002/path.2697.

Grabbe, C., Husnjak, K., & Dikic, I. (2011). The spatial and temporal organization of ubiquitin networks. *Nature Reviews. Molecular Cell Biology*, *12*(5), 295–307. https://doi.org/10.1038/nrm3099.

Haglund, K., Sigismund, S., Polo, S., Szymkiewicz, I., Di Fiore, P. P., & Dikic, I. (2003). Multiple monoubiquitination of RTKs is sufficient for their endocytosis and degradation. *Nature Cell Biology*, *5*(5), 461–466. https://doi.org/10.1038/ncb983.

Komander, D., & Rape, M. (2012). The ubiquitin code. *Annual Review of Biochemistry*, *81*, 203–229. https://doi.org/10.1146/annurev-biochem-060310-170328.

Levkowitz, G., Waterman, H., Ettenberg, S. A., Katz, M., Tsygankov, A. Y., Alroy, I., et al. (1999). Ubiquitin ligase activity and tyrosine phosphorylation underlie suppression of growth factor signaling by c-Cbl/Sli-1. *Molecular Cell*, *4*(6), 1029–1040.

MacGurn, J. A., Hsu, P. C., & Emr, S. D. (2012). Ubiquitin and membrane protein turnover: From cradle to grave. *Annual Review of Biochemistry*, *81*, 231–259. https://doi.org/10.1146/annurev-biochem-060210-093619.

Maspero, E., Mari, S., Valentini, E., Musacchio, A., Fish, A., Pasqualato, S., et al. (2011). Structure of the HECT: Ubiquitin complex and its role in ubiquitin chain elongation. *EMBO Reports*, *12*(4), 342–349. https://doi.org/10.1038/embor.2011.21.

Maspero, E., & Polo, S. (2016). In vitro ubiquitination: Self-ubiquitination, chain formation, and substrate ubiquitination assays. *Methods in Molecular Biology*, *1449*, 153–160. https://doi.org/10.1007/978-1-4939-3756-1_7.

Maspero, E., Valentini, E., Mari, S., Cecatiello, V., Soffientini, P., Pasqualato, S., et al. (2013). Structure of a ubiquitin-loaded HECT ligase reveals the molecular basis for catalytic priming. *Nature Structural & Molecular Biology*, *20*(6), 696–701. https://doi.org/10.1038/nsmb.2566.

Mauthe, M., Orhon, I., Rocchi, C., Zhou, X., Luhr, M., Hijlkema, K. J., et al. (2018). Chloroquine inhibits autophagic flux by decreasing autophagosome-lysosome fusion. *Autophagy*, *14*(8), 1435–1455. https://doi.org/10.1080/15548627.2018.1474314.

Newton, K., Matsumoto, M. L., Ferrando, R. E., Wickliffe, K. E., Rape, M., Kelley, R. F., et al. (2012). Using linkage-specific monoclonal antibodies to analyze cellular ubiquitylation. *Methods in Molecular Biology*, *832*, 185–196. https://doi.org/10.1007/978-1-61779-474-2_13.

Newton, K., Matsumoto, M. L., Wertz, I. E., Kirkpatrick, D. S., Lill, J. R., Tan, J., et al. (2008). Ubiquitin chain editing revealed by polyubiquitin linkage-specific antibodies. *Cell*, *134*(4), 668–678. https://doi.org/10.1016/j.cell.2008.07.039.

Oh, E., Akopian, D., & Rape, M. (2018). Principles of ubiquitin-dependent signaling. *Annual Review of Cell and Developmental Biology*, *34*, 137–162. https://doi.org/10.1146/annurev-cellbio-100617-062802.

Park, C. W., & Ryu, K. Y. (2014). Cellular ubiquitin pool dynamics and homeostasis. *BMB Reports*, *47*(9), 475–482.

Peng, J., Schwartz, D., Elias, J. E., Thoreen, C. C., Cheng, D., Marsischky, G., et al. (2003). A proteomics approach to understanding protein ubiquitination. *Nature Biotechnology*, *21*(8), 921–926. https://doi.org/10.1038/nbt849.

Phu, L., Izrael-Tomasevic, A., Matsumoto, M. L., Bustos, D., Dynek, J. N., Fedorova, A. V., et al. (2011). Improved quantitative mass spectrometry methods for characterizing complex ubiquitin signals. *Molecular & Cellular Proteomics*, *10*(5). https://doi.org/10.1074/mcp.M110.003756. M110 003756.

Polo, S., Di Fiore, P. P., & Sigismund, S. (2014). Keeping EGFR signaling in check: Ubiquitin is the guardian. *Cell Cycle*, *13*(5), 681–682. https://doi.org/10.4161/cc.27855.

Polo, S., Sigismund, S., Faretta, M., Guidi, M., Capua, M. R., Bossi, G., et al. (2002). A single motif responsible for ubiquitin recognition and monoubiquitination in endocytic proteins. *Nature*, *416*(6879), 451–455. https://doi.org/10.1038/416451a.

Rotin, D., & Kumar, S. (2009). Physiological functions of the HECT family of ubiquitin ligases. *Nature Reviews. Molecular Cell Biology*, *10*(6), 398–409. https://doi.org/10.1038/nrm2690.

Sahin, U., Jollivet, F., Berthier, C., de The, H., & Lallemand-Breitenbach, V. (2016). Detection of protein SUMOylation in situ by proximity ligation assays. *Methods in Molecular Biology*, *1475*, 139–150. https://doi.org/10.1007/978-1-4939-6358-4_10.

Savio, M. G., Wollscheid, N., Cavallaro, E., Algisi, V., Di Fiore, P. P., Sigismund, S., et al. (2016). USP9X controls EGFR fate by deubiquitinating the endocytic adaptor Eps15. *Current Biology*, *26*(2), 173–183. https://doi.org/10.1016/j.cub.2015.11.050.

Sigismund, S., Algisi, V., Nappo, G., Conte, A., Pascolutti, R., Cuomo, A., et al. (2013). Threshold-controlled ubiquitination of the EGFR directs receptor fate. *The EMBO Journal*, *32*(15), 2140–2157. https://doi.org/10.1038/emboj.2013.149.

Sigismund, S., Argenzio, E., Tosoni, D., Cavallaro, E., Polo, S., & Di Fiore, P. P. (2008). Clathrin-mediated internalization is essential for sustained EGFR signaling but dispensable for degradation. *Developmental Cell*, *15*(2), 209–219. https://doi.org/10.1016/j.devcel.2008.06.012.

Woelk, T., Oldrini, B., Maspero, E., Confalonieri, S., Cavallaro, E., Di Fiore, P. P., et al. (2006). Molecular mechanisms of coupled monoubiquitination. *Nature Cell Biology*, *8*(11), 1246–1254.

Wong, L. C., & Schedl, P. (2006). Dissection of Drosophila ovaries. *Journal of Visualized Experiments*, (1), e52. https://doi.org/10.3791/52.

Xu, P., & Peng, J. (2008). Characterization of polyubiquitin chain structure by middle-down mass spectrometry. *Analytical Chemistry*, *80*(9), 3438–3444. https://doi.org/10.1021/ac800016w.

Yau, R., & Rape, M. (2016). The increasing complexity of the ubiquitin code. *Nature Cell Biology*, *18*(6), 579–586. https://doi.org/10.1038/ncb3358.

CHAPTER FOURTEEN

Western blot analysis of the autophagosomal membrane protein LGG-1/LC3 in *Caenorhabditis elegans*

Alexander Springhorn, Thorsten Hoppe*
Institute for Genetics and CECAD Research Center, University of Cologne, Cologne, Germany
*Corresponding author: e-mail address: thorsten.hoppe@uni-koeln.de

Contents

Abstract

Autophagy is being studied intensively in *Caenorhabditis elegans* in the context of protein homeostasis and aging. However, in contrast to the yeast and mammalian autophagosomal membrane proteins Atg8 and LC3, lipidation of the *C. elegans* ortholog LGG-1 with phosphatidylethanolamine has rarely been investigated by western blotting. We attribute this shortcoming to technical problems with separating the non-lipidated from the lipidated LGG-1 protein by gel electrophoresis. Our new protocol for Western blot analysis is applicable for both the detection of transgenic and endogenous LGG-1 proteins and provides a quantifiable method to assess autophagic flux. As a proof of principle, we use this assay to analyze the role of the transcriptional master regulator HLH-30/TFEB in starvation-induced autophagy.

Methods in Enzymology, Volume 619
ISSN 0076-6879
https://doi.org/10.1016/bs.mie.2018.12.034
© 2019 Elsevier Inc.
All rights reserved.

1. Introduction

From yeast to mammals, food deprivation triggers an evolutionarily conserved autophagic response to supply the cell with energy through breakdown and recycling of its own cellular components (Singh & Cuervo, 2011). In mammalian cells, low nutrient levels inhibit the master cell growth regulator kinase mammalian target of rapamycin (mTOR) and activate AMP-activated protein kinase (AMPK) (Dikic & Elazar, 2018). The changed activity of both energy sensors triggers induction of the protein complexes Unc-51-like kinase 1 (ULK1) and Beclin1, thereby establishing the basis for autophagosome formation through local phosphatidylinositol 3-phosphate (PI3P) production at the ER and recruitment of autophagy-related 9 (ATG9)-positive membrane vesicles. WD-repeat protein interacting with phosphoinositides (WIPI) proteins act as PI3P effectors and promote the elongation of the phagosomal membrane (Fig. 1A) (Bakula et al., 2017; Dikic & Elazar, 2018; Mercer, Gubas, & Tooze, 2018; Nakatogawa, Suzuki, Kamada, & Ohsumi, 2009). This is partially achieved by retrieving the transmembrane protein ATG9 from the growing autophagosome to maintain ATG9 trafficking toward and away from the site of autophagosome formation (Fig. 1A) (Lu et al., 2011; Mercer et al., 2018; Orsi et al., 2012; Reggiori, Tucker, Stromhaug, & Klionsky, 2004; Saitsu et al., 2013). ATG9 trafficking is considered an important mechanism through which membrane material needed for the elongation process is transported from distant sites (Noda, 2017; Xie & Klionsky, 2007).

Additionally, WIPIs have been implicated in the recruitment of microtubule-associated protein light chain 3 (LC3) family proteins (LC3s) (Fig. 1A), which expand the phagosomal membrane through their membrane fusion activity (Dikic & Elazar, 2018; Dooley et al., 2014; Nakatogawa, Ichimura, & Ohsumi, 2007; Polson et al., 2010; Weidberg et al., 2011). LC3 precursors are cleaved constitutively near the C-terminus by ATG4 proteases, exposing a glycine residue, which serves as a site for subsequent covalent linkage to phosphatidylethanolamine (PE) (Fig. 1A). This lipidation step occurs at basal rates under nutrient-rich conditions, but is significantly enhanced upon starvation—potentially by (i) bringing components of the lipidation machinery in close proximity at the growing autophagosome, (ii) changing lipid composition, or (iii) by supplying more nonlipidated substrate (Dooley et al., 2014; Fujita et al., 2008; Ichimura et al., 2004; Mizushima, Yoshimori, & Levine, 2010). PE-conjugation

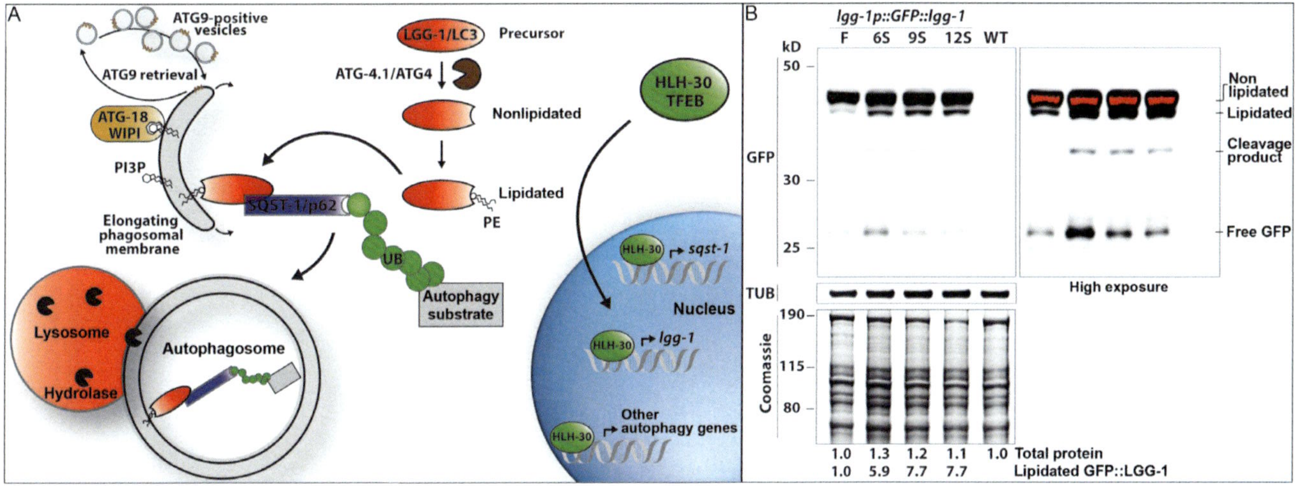

Fig. 1 (A) Overview of a subset of autophagy-related events occurring during starvation. Mammalian autophagy factors and the corresponding *C. elegans* orthologs are indicated. Food depletion triggers spatially confined synthesis of PI3P at the endoplasmic reticulum, which marks the site of autophagosome formation. WIPI proteins bind PI3P at this site and help elongate the phagosomal membrane, partly by regulating the delivery of ATG9-positive lipid vesicles, or by promoting the membrane fusion activity of LC3s. Membrane association of LC3s is achieved by conjugating PE to the C-terminal glycine residue, which is exposed upon cleavage of LC3 precursors by ATG4 proteases. p62/SQST-1 recruits poly-ubiquitylated (UB) substrates into the growing autophagosome by interacting with phagosomal membrane-bound LC3s. After completion of phagosomal membrane elongation to form a closed autophagosome, fusion of mature autophagosomes with acidic lysosomes leads to the degradation of autophagosomal substrates by lysosomal hydrolases. Food-depletion also triggers nuclear translocation of the transcriptional regulator TFEB/HLH-30, which boosts expression of several autophagy-related genes. (B) Western blot-based detection of GFP::LGG-1. Day 1 adult worms from strain DA2123 expressing GFP-tagged LGG-1 under the control of the *lgg-1* promoter were put on plates with (F) or without food for 6, 9, or 12 h (6S, 9S, 12S). Fed untagged wild-type (WT) worms served as GFP-negative control. After protein quantification of worm lysates, 15 μg of total protein per sample were either used for Coomassie Blue staining or Western blotting against GFP and tubulin. After digital imaging, quantification of lipidated GFP::LGG-1 band intensities and of Coomassie staining was done using ImageJ. For Coomassie staining, band intensities were measured over the whole lane (190–15 kD). Numbers below the blot indicate relative amounts of total protein loaded and of lipidated GFP::LGG-1. Overexposed areas are marked in *red*.

anchors LC3s to the phagosomal membrane, where they not only drive membrane expansion but, importantly, also confer specificity to the autophagosomal mode of protein degradation by acting as docking sites for autophagosomal cargo receptors like sequestosome 1 (p62) (Fig. 1A) (Dikic & Elazar, 2018). Through this interaction, ubiquitylated autophagy substrates, which are recognized by p62 via their ubiquitin conjugates, are incorporated into the autophagosome. Subsequent fusion of the mature autophagosome with acidic lysosomes results in the breakdown of sequestered autophagosomal cargo by lysosomal hydrolases (Fig. 1A).

While the core autophagy pathway is based on a multitude of posttranslational modifications and protein interactions, a range of transcription factors have also been implicated in boosting the autophagic response. The transcription factor EB (TFEB), for example, translocates to the nucleus upon starvation and upregulates expression of several genes encoding components of the autophagy machinery (Fig. 1A) (Settembre et al., 2013, 2011). In the absence of TFEB, total LC3 levels decrease and starvation-induced increases in lipidated LC3 are severely dampened. Conversely, TFEB gain-of-function induces LC3 lipidation and autophagosome formation. Thus far, it is unclear whether TFEB stimulates autophagic flux by providing higher levels of multiple autophagy factors, or whether it operates through specific pathways, e.g., to inhibit TOR activity or selectively boost the lipidation rate of LC3s.

In mammalian systems, Western blot-based analysis of LC3 levels serves as a commonly used readout to assess autophagic activity (Kabeya, 2000; Klionsky et al., 2016; Mizushima et al., 2010). Nonlipidated and lipidated LC3 can be resolved by sodium dodecyl sulfate polyacrylamide gel electrophoresis (SDS-PAGE), with the lipidated form running slightly faster. As mentioned above, lipidation of LC3 increases upon autophagy induction. However, since lipidated LC3 is an autophagy substrate by itself, its level can also be increased when autophagic flux is blocked by defective fusion of autophagosomes with lysosomes or defective lysosomal hydrolase function. Therefore, additional hints on whether a certain treatment induces or blocks autophagy are inferred from measuring levels of lipidated LC3 in the presence and absence of the chemical drugs bafilomycin or chloroquine, which neutralize the lysosomal pH and thereby render its hydrolases inactive (Klionsky et al., 2016; Mizushima et al., 2010). Under the assumption that these inhibitors cause a complete block in the last step of autophagic degradation, any induction of the pathway upstream of the inhibitory block should result in an additive increase of lipidated LC3 compared with the single conditions. Conversely, treatments conferring an additional block would not lead to further increases of this autophagy marker.

Due to a lack of commercially available antibodies against LC3, GABARAP, and GATE-16 family 1 (LGG-1) (Fig. 1A), the major LC3-like protein in *Caenorhabditis elegans*, autophagy research in worms relies mainly on transgenic expression of green fluorescent protein (GFP)-tagged LGG-1 (Kang, You, & Avery, 2007; Meléndez et al., 2003). The transgene is often used to count GFP-labeled autophagosomes, which are mainly monitored in seam cells of adult worms (Meléndez et al., 2003). However, according to Klionsky et al., this approach is error prone due to a small number of GFP puncta per cell even under autophagy-inducing conditions, and due to high background levels in the GFP channel (Klionsky et al., 2016). Western blot-based detection of the fusion protein is often hampered by difficulties in resolving lipidated vs nonlipidated GFP::LGG-1 on the gel (Alberti, Michelet, Djeddi, & Legouis, 2010; Chapin, Okada, Merz, & Miller, 2015; Chen, Scarcelli, & Legouis, 2017). By contrast, free GFP, which is generated through lysosomal proteolysis of the fusion protein under autophagy inducing conditions, but not when autophagy is blocked, is readily detectable and serves as a good tool to monitor autophagic flux (Djeddi et al., 2012; Kozlowski, Garvis, Bedet, & Palladino, 2014; Mizushima et al., 2010; Possik et al., 2014; Shintani & Klionsky, 2004). *C. elegans* researchers also attempted to assess autophagic flux by adding bafilomycin or chloroquine to the nematode growth medium (NGM). However, the chemicals are only effective when injected directly into adult animals, as judged by increased GFP::LGG-1 puncta formation in seam cells (Chang, Kumsta, Hellman, Adams, & Hansen, 2017; Wilkinson et al., 2015).

We sought to study autophagy in *C. elegans* by Western blot analysis of endogenous LGG-1 levels. To this end, we generated an antibody against LGG-1 (available upon request) and established a protocol that allows one to reproducibly separate the lipidated from the nonlipidated form of the protein. The same conditions also make possible the detection of the lipidated and nonlipidated form of the commonly used fusion protein GFP::LGG-1. Following starvation-induced autophagy as a way to stimulate LGG-1 lipidation, we analyzed a *C. elegans* mutant lacking the WIPI protein family member ATG-18 to assess autophagic flux (Fig. 1A). In line with a role for WIPI family proteins in autophagosome formation, *atg-18* mutants are short-lived during starvation and display higher levels of damaged mitochondria and polyQ aggregates, both of which are thought to be degraded by autophagy (Hibshman et al., 2018; Jia, Hart, & Levine, 2007).

Direct evidence for defective autophagic breakdown was provided by Lu et al., who showed that lipidated LGG-1 and the *C. elegans* p62 homolog sequestosome related 1 (SQST-1) (Fig. 1A) accumulate in *atg-18* mutant embryos, which is consistent with the observation that both proteins are degraded via autophagy (Guo et al., 2014; Lu et al., 2011; Tian et al., 2010). However, despite these results, it remains unclear whether ATG-18 promotes autophagosome formation by regulating ATG-9 trafficking and/or LGG-1-mediated membrane fusion, or whether the protein functions through other mechanisms (Lu et al., 2011; Wu et al., 2014).

To further test our Western blot-based autophagy readout, we investigated the function of the *C. elegans* TFEB homolog helix loop helix 30 (HLH-30) (Fig. 1A). Similar to TFEB, HLH-30 translocates to the nucleus upon starvation, or inhibition of TOR, to boost expression of autophagy genes like *lgg-1* (Lapierre et al., 2013; O'Rourke & Ruvkun, 2013; Silvestrini et al., 2018). Moreover, GFP::LGG-1 puncta quantification suggested that activation of HLH-30 is beneficial for autophagosome formation (Lapierre et al., 2013; Silvestrini et al., 2018). Physiological evidence for the requirement of HLH-30 for efficient autophagic breakdown was provided by O'Rourke et al., who reported that *hlh-30* mutants are sensitive to long-term fasting (O'Rourke & Ruvkun, 2013). We intended to further corroborate these findings by analyzing the effects of HLH-30 loss-of-function on LGG-1 levels during starvation-induced autophagy.

2. Experimental procedure

After completion of embryogenesis, *C. elegans* development proceeds through four larval stages (L1–L4) and ends at the day 1 adult stage (d1) (Girard et al., 2007). Worms are cultured on NGM agar plates seeded with the *Escherichia coli* bacterial strain OP50 as a food source. Early L1 larvae arrest development in the absence of food, which allows for synchronization of worm cultures at the L1 stage. During this procedure, all cultured animals except for egg shell-protected embryos are killed with bleaching solution. Embryos are then grown in liquid medium without food (M9 buffer), where they complete embryogenesis and enter developmental arrest as L1 hatchlings (Girard et al., 2007).

2.1 Induction of autophagy by starvation

In the protocol below, the quantities of nematodes are sufficient for the generation of one control and one starvation lysate.

- synchronize ~4000 L1 larvae by bleaching ~500 gravid worms and allowing the eggs to develop in M9 buffer for 12–15 h
- distribute L1 larvae to eight OP50-seeded NGM L-plates (94 × 16 mm Petri dishes filled with 30 mL NGM and seeded with 800 μL of a saturated OP50 liquid culture) at a density of 300–500 worms/plate and allow to develop for ~72 h at 20°C until mid d1 adult stage
- wash worms off the plates with a glass pipette and M9 buffer (or tap water) and collect them in a single 15 mL falcon tube
- remove OP50 by washing the worm pellet 3 × (i.e., spin at 800 rpm for 30 s and replace the supernatant with fresh M9 buffer)
- after the last wash, leave worm pellet in ~0.5 mL M9 buffer and distribute worms equally on 2 OP50-seeded NGM L-plates and 2 empty NGM L-plates (starvation condition)
- incubate plates at 20°C for 6–24 h

Note: We detect an increase in lipidated LGG-1 levels at the earliest after 5–6 h of starvation. Often, this increase disappears after 24 h of starvation, which could be due to increased autophagic flux or loss of autophagy induction. Starving d1 adult worms for 24 h leads to internal egg hatching, which can be circumvented by starving the worms from the mid L4 stage on. However, as LGG-1 levels change during development, we recommend conducting time course experiments either entirely at the L4 stage or during d1 adulthood.

2.2 Worm collection

- wash worms off the plates with a glass pipette and M9 buffer (or tap water) and collect fed and starved worms in two separate 15 mL falcon tubes
- remove OP50 and hatched L1 larvae by washing the worm pellets 3 × (i.e., spin at 800 rpm for 30 s and replace supernatant with fresh M9 buffer)
- after final wash, remove as much supernatant as possible and add 1 mL of chilled lysis buffer to the worm pellets to lower the amount of M9 buffer in the final lysate and to facilitate transfer
- transfer worm pellets in chilled lysis buffer with a glass pipette into one 1.5 mL Eppendorf tube per condition and spin in a tabletop microcentrifuge at 6000 rpm for 20 s
- take off supernatant and add 30–200 μL of chilled lysis buffer to the worm pellets, then snap freeze the Eppendorf tubes in liquid nitrogen and store at −80°C

Note: In order to load 15 μg total protein in a total volume of 10 μL containing 2.5 μL of 4 × SDS protein sample buffer and 1 μL of 10 × DTT stock solution, final protein concentrations of worm lysates should not be lower than ~2.4 μg/μL. As a rule of thumb, add the same volume of lysis buffer to the worm pellet as the volume that remains in the Eppendorf tube after removal of the supernatant. In our hands, this remaining volume ranges from 30 to 200 μL.

2.3 Lysate generation

- incubate frozen samples for 5 min at 95°C, then snap freeze in liquid nitrogen
- sonicate samples continuously with an ultrasonic homogenizer (SONOPULS mini20 equipped with microtip/sonotrode MS 1.5, BANDELIN) for 2–3 min at an amplitude (Am) of 70% until worm pellets are completely dissolved, then snap freeze in liquid nitrogen
- incubate samples for 5 min at 95°C, then snap freeze again in liquid nitrogen and let thaw
- clear lysates by spinning for 10 min at 18500 RCF/4°C
- transfer cleared lysates into new 1.5 mL Eppendorf tubes; store at −80°C

2.4 Protein quantification

Worm lysates are quantified using reagents from the Pierce BCA Protein Assay Kit (ThermoFisher Scientific).

- pipette 6 μL of dilutions of a BSA standard (2.0, 1.0, 0.5, 0.25, 0.125, 0.0625, 0.03125 μg/μL), 6 μL lysis buffer as a blank and both 3 μL and 1.5 μL of each worm lysate into corresponding wells of a 96-well plate and bring the final volume to 100 μL with a 50:1 mixture of reagent A and B
- incubate 96-well plate on shaking incubator for 25 min at 60°C and measure absorbance in a plate reader

Note: To ensure that protein concentrations of worm lysates do not exceed the 2 μg/μL BSA absorbance limit, we include a 1:4 dilution (1.5 μL) besides the 1:2 dilution (3 μL) in our assay. Selected lysate concentrations are all derived from the same dilution.

2.5 SDS-PAGE and western blotting

- when loading 10 μL per well, mix 2.5 μL of 4× SDS protein sample buffer and 1 μL of a 10× DTT stock solution with 6.5 μL lysis buffer containing 15 μg of worm lysate

  - alternatively, use 4 × NuPAGE LDS Sample Buffer (ThermoFisher Scientific) and 10 × NuPAGE Sample Reducing Agent (ThermoFisher Scientific)
- mix samples well by pipetting up and down and incubate for 10 min at 70°C
- We have tested several home-made and commercially available precast gels with different running buffers and came to the conclusion that best resolution of lipidated vs nonlipidated LGG-1 is achieved when samples are run at 150 volts (V) in MOPS running buffer on 12% NuPAGE Bis-Tris Gels (ThermoFisher Scientific). Good resolution of lipidated vs nonlipidated GFP::LGG-1 is accomplished with the same conditions. To sharpen protein bands, we add 500 μL of NuPAGE Antioxidant (ThermoFisher Scientific) to the running buffer in the upper buffer chamber
- transfer proteins to a nitrocellulose membrane (pore size 0.45 μm) by wet electro-transfer at 4°C
  - using a TE22 Mighty Small Transfer Tank (Hoefer), we obtain good results blotting for 1.5 h at 70 V in 1 × NuPAGE Transfer Buffer (ThermoFisher Scientific) containing 10% methanol
- after transfer, block membrane in 1 × Roti-Block (ROTH)/PBS for 1–2 h, then incubate overnight with primary antibody in 1 × Roti-Block/PBS at 4°C:
  - mouse monoclonal antiliving colors Clone JL-8 (TaKaRa) (1:5000)
  - mouse monoclonal anti-α-tubulin Clone B-5-1-2 (Sigma-Aldrich) (1:10000)
  - rabbit anti-LGG-1 (1:3000)
  - rabbit anti-SQST-1 (1:3000)
  - due to the size differences of the detected proteins and the high specificity of the antibodies, the membrane can be incubated with all four antibodies at the same time
- rinse membrane 3 × and wash 3 × 10 min with PBS-Tween20 (0.1%), then incubate with secondary antibody in 1 × Roti-Block/PBS for 2 h at RT
  - Peroxidase-conjugated AffiniPure Goat antiMouse IgG (H + L) (Jackson ImmunoResearch) (1:10000)
  - Peroxidase-conjugated AffiniPure Mouse antiRabbit IgG (H + L) (Jackson ImmunoResearch) (1:10000)
- rinse and wash membrane with PBSTween (0.1%) as before, then incubate membrane with GE Healthcare Amersham ECL Prime Western Blotting Detection Reagent (ThermoFisher Scientific) and image

Note: Good results were also obtained with fluorescently labeled secondary antibodies. Western blots were imaged digitally with the ChemiDoc Imaging System from Bio-Rad.

3. Reagents

Lysis buffer
- 25 m*M* Tris–HCl, pH 7.6
- 150 m*M* NaCl
- 1 m*M* ethylenediaminetetraacetic acid (EDTA)
- 1% Triton X-100
- 1% sodium deoxycholate (w/v)
- 1% SDS (w/v)

4 × SDS protein sample buffer (pH 6.8)
- 40% glycerol
- 0.04% bromophenol blue (W/V)
- 240 m*M* Tris–HCl, pH 6.8
- 8% SDS (w/v)

10 × DTT stock solution
- 1 *M* DTT in H_2O

NuPAGE™ MOPS running buffer
- 50 m*M* 3-(*N*-morpholino)propanesulfonic acid (MOPS)
- 50 m*M* Tris-base
- 1 m*M* EDTA
- 0.1% SDS (W/V)

4. Experimental results

The *C. elegans* strain DA2123 carrying the integrated transgenic construct [*lgg-1p::GFP::lgg-1*] is often used to monitor autophagy by ccunting GFP-positive puncta in worms. Applying our protocol, we show that this line is also suitable for analyzing levels of lipidated GFP::LGG-1 by Western blotting (Fig. 1B). Compared to the fed control, levels of lipidated GFP::LGG-1 increase upon 6, 9, and 12 h of starvation. As judged by the concomitant increase in free GFP levels, this rise in lipidated GFP::LGG-1 is not due to a block in autophagic flux, but likely results from starvation-induced stimulation of LGG-1 lipidation. Note that despite quantification of protein lysates, we often see slight differences in the amounts of total protein loaded.

This is revealed more accurately by Coomassie Blue protein staining than by Western blotting against tubulin. To account for unequal loading, we typically normalize levels of lipidated LGG-1 to the total amount of protein loaded and then calculate changes in lipidated LGG-1 relative to the respective fed or untreated controls.

To test whether the results obtained with transgenic GFP::LGG-1 are reflecting the behavior of the endogenous protein, we generated an antibody against LGG-1. Western blotting reveals a discrete double band corresponding to the molecular weight of processed nonlipidated LGG-1 (14kD, 116 residues) and of its lipidated form, which migrates more rapidly despite its higher mass (Fig. 2A). The intensities of both bands increase upon starvation (Fig. 2A). Thus, not only levels of lipidated but also of nonlipidated LGG-1 rise under autophagy-inducing conditions. Increases in nonlipidated LGG-1 might either be due to transcriptional upregulation of *lgg-1* expression or to posttranslational regulatory mechanisms inhibiting protein degradation, which may serve to provide more substrate for the lipidation reaction (Ichimura et al., 2004). The specificity of the antibody was confirmed by probing lysates of both (1) worms in which *lgg-1* was knocked down by RNAi and (2) worms mutant for ATG-4.1, a *C. elegans* homolog of mammalian ATG4s, which are known to cleave precursors of LC3s constitutively (Wu, Li, Wang, Noda, & Zhang, 2012). Knockdown of *lgg-1* decreases the intensity of the double band, while lysates from *atg-4.1* mutants display a higher molecular weight band corresponding to the unprocessed LGG-1 precursor (14.8kD, 123 residues) (Fig. 2A).

Using this method for immunoblot analysis of endogenous LGG-1 protein, we analyzed LGG-1 levels in fed vs starved *atg-18* mutants. As noted above, lack of the PI3P-binding protein leads to phenotypes indicative of perturbed autophagosome formation, which should block autophagic flux similar to bafilomycin or chloroquine treatment in mammalian systems. In line with this assumption, fed *atg-18* mutants display elevated levels of lipidated LGG-1 (Fig. 2B), which is normally targeted for autophagic turnover. Notably, steady-state levels of lipidated LGG-1 are further increased upon starvation (Fig. 2B). This starvation-dependent upregulation under conditions of impaired autophagic degradation provides evidence for the well-established notion that food depletion induces rather than blocks autophagy. *atg-18* mutants might therefore serve as a reliable tool for analysis of autophagic flux. However, as opposed to a bafilomycin- or chloroquine-induced block of autophagic flux downstream of autophagosome formation, which predominantly stabilizes the lipidated but not the nonlipidated form

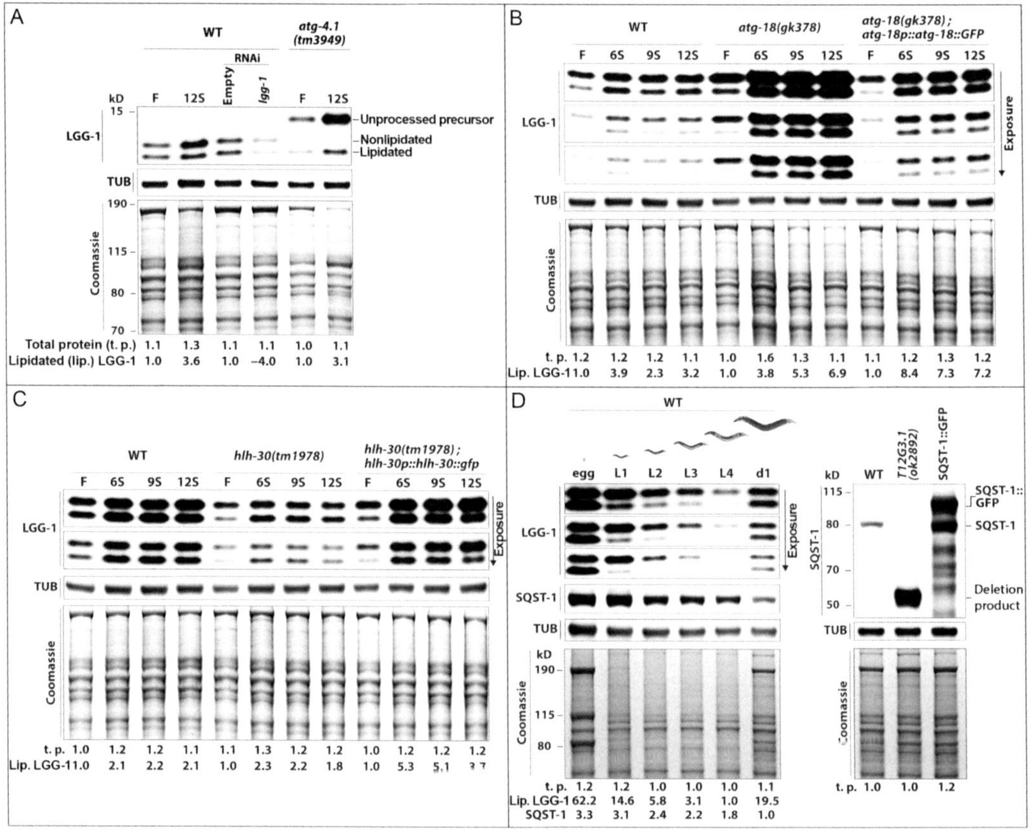

Fig. 2 See legend on opposite page.

of LC3 (Kabeya, 2000; Klionsky et al., 2016; Mizushima et al., 2010), fed *atg-18* mutants not only accumulate lipidated but also high levels of nonlipidated LGG-1 (Fig. 2B), which is not degraded by autophagy.

Bakula et al. reported similar observations in mammalian cells that had impaired function of WIPI3/4, while others have found selective increases of either lipidated or nonlipidated LC3 in the absence of WIPIs (Bakula et al., 2017; Polson et al., 2010; Saitsu et al., 2013). As high levels of nonlipidated LGG-1 could theoretically trigger an elevated rate of conversion of the nonlipidated into the lipidated form (Ichimura et al., 2004), it is unclear whether accumulation of lipidated LGG-1 truly results from impaired autophagic breakdown. Thus, to properly test *atg-18* mutants as a tool to monitor autophagic flux, early events of autophagosome formation need to be further dissected to understand why a block at this stage in the pathway leads to accumulation of nonlipidated LGG-1, and whether high levels of this form of the protein, rather than defective degradation, trigger a rise in lipidated LGG-1. Analyzing levels of free GFP generated from the GFP::LGG-1 fusion protein could provide additional evidence for defective breakdown, as free GFP levels should be lower in fed and starved *atg-18* mutants compared to the respective wild-type (WT) controls, if autophagic flux were reduced.

Fig. 2 (A) Western blot against endogenous LGG-1. Worms of the indicated genotypes were either allowed to develop to the day 1 (d1) adult stage and then put on plates with (F) and without food (12S) for 12 h, or cultivated on the indicated RNAi plates from the larval L1 stage on until d1 adulthood. Processing of the LGG-1 precursor is not completely abolished in *atg-4.1* mutants, which is why lipidation of residual amounts of processed nonlipidated LGG-1 still occurs. The nonlipidated LGG-1 band is detectable at higher exposures. Band intensities of lipidated LGG-1 were calculated relative to the respective fed or empty RNAi treated controls. (B, C) Western blot against endogenous LGG-1. Day 1 adult worms of the indicated genotypes were either put on plates with (F) or without food for 6, 9 or 12 h (6S, 9S, 12S). For each genotype, band intensities of lipidated LGG-1 were calculated relative to the fed controls. Accumulation of lipidated and nonlipidated LGG-1 in fed and starved *atg-18* mutants is largely suppressed by means of an *atg-18* rescue construct. Transgenic expression of HLH-30 rescues the striking loss of lipidated and nonlipidated LGG-1 in starved *hlh-30* mutants. (D) Western blot against endogenous LGG-1 and SQST-1. 15 μg of worm lysates of the indicated developmental stages were either probed with antibodies against LGG-1, SQST-1, or tubulin, or used for Coomassie Blue staining. Tubulin levels are also subject to developmental regulation, since loading differences between lanes do not account for strongly elevated amounts of tubulin in eggs relative to other developmental stages. Specificity controls for the antibody against SQST-1 are provided in the blot on the right. The antibody detects the 693-residue full-length protein in WT worms and a deletion product of 428 amino acids in worms carrying the *sqst-1* deletion *ok2892*. Additionally, the antibody recognizes GFP-tagged SQST-1 expressed from a transgene.

To study starvation-induced autophagy in the context of transcriptional regulation, we analyzed worms mutant for HLH-30 (Fig. 2C). Settembre et al. report that starvation-induced activation of TFEB is required to obtain sufficiently high levels of lipidated LC3 for optimal degradation of autophagic substrates (Settembre et al., 2011). In line with these findings, our data show that starved *hlh-30* mutants fail to accumulate either nonlipidated or lipidated LGG-1 to the extent that starved WT worms do (Fig. 2C). This defect can be rescued with a transgene expressing HLH-30 (Fig. 2C). The observed phenotype could simply result from downregulation of *lgg-1* expression in the absence of transcriptional inputs from HLH-30, which would explain why relative starvation-induced increases in lipidated LGG-1 are similar to those observed in wild-type worms (Fig. 2C). Alternatively, HLH-30 might bring about an increase in lipidated LGG-1 levels by providing a factor required for the stabilization (i.e., inhibition of degradation) of nonlipidated LGG-1, which could accompany an HLH-30-induced rise in lipidation activity. Regardless of how HLH-30 transcription factor activity alters lipidated LGG-1 levels, reduced levels of lipidated LGG-1 are likely to result in impaired autophagic breakdown of substrates, potentially explaining the high sensitivity of *hlh-30* mutants toward starvation (O'Rourke & Ruvkun, 2013).

As illustrated above, autophagy is often studied in the context of fully differentiated organisms to analyze its many roles in physiology. However, much can be learned from dissecting its functions during development (Tian et al., 2010). Fig. 2D shows that levels of lipidated LGG-1 are comparatively high in eggs, gradually decline as development proceeds through the four larval stages (L1–L4), and rise again in fully differentiated day 1 adult worms (d1). SQST-1 levels display a similar gradual decline from embryonic stages toward the larval L4 stage, but, unlike LGG-1, do not increase upon completion of development (Fig. 2D). While further studies are required to characterize the function and behavior of SQST-1 during autophagy in *C. elegans*, the quantitative differences in lipidated LGG-1 levels might be indicative of strong requirements for autophagy during embryogenesis and later on in adult life.

5. Conclusions

Studies of autophagy in *C. elegans* can be further substantiated by Western blot-based analyses of LGG-1 lipidation levels. Our protocol provides an improved method to achieve good resolution of the lipidated and nonlipidated form of both transgenic and endogenous LGG-1. The newly established protocol allowed us to obtain evidence for the requirement of HLH-30 for the generation of sufficiently high levels of lipidated LGG-1 during starvation-induced autophagy, and helped to characterize the *atg-18* mutant in the context of autophagic flux. Strikingly, both during starvation-induced autophagy and in *atg-18* mutants with impaired autophagosome formation, accumulation of lipidated LGG-1 is accompanied by an increase in nonmodified LGG-1. While the molecular basis for the accumulation of lipidated and nonlipidated LGG-1 in *atg-18* mutants remains unclear, starvation-induced increases in the nonlipidated form suggest that *C. elegans* boosts levels of lipidated LGG-1 by supplying more nonlipidated substrate, rather than by rendering PE conjugation more efficient. The latter might be the case in mammals, where autophagy induction often depletes rather than raises levels of nonlipidated LC3 (Kabeya, 2000; Klionsky et al., 2016; Mizushima & Yoshimori, 2007; Mizushima et al., 2010). Further dissecting the regulation of the lipidation process will help us to better understand the exact mode of action of HLH-30 and of its mammalian ortholog TFEB.

Taken together, our protocol facilitates the detailed analysis of the most commonly used autophagy marker in *C. elegans*, thereby offering a basis for gaining further insights into the mechanistic details and physiological impact of autophagy in this multicellular organism.

Acknowledgments

We thank the *Caenorhabditis* Genetics Center (funded by the NIH National Center for Research Resources), the Dana-Farber Cancer Institute, Addgene and Geneservice Ltd. for antibodies, plasmids, cDNAs, and strains. In particular, we thank Martina Liebl and Ricardo Marchante for raising LGG-1- and SQST-1-specific antibodies. Special thanks are extended to Shahida Moosa for critically reading the manuscript and to the Krüger lab for assistance with digital imaging and protein quantification. This work is supported by grants of the Deutsche Forschungsgemeinschaft (CECAD) and the European Research Council (consolidator grant 616499) to T.H.

References

Alberti, A., Michelet, X., Djeddi, A., & Legouis, R. (2010). The autophagosomal protein LGG-2 acts synergistically with LGG-1 in dauer formation and longevity in *C. elegans*. *Autophagy*, *6*, 622–633. https://doi.org/10.4161/auto.6.5.12252.

Bakula, D., Müller, A. J., Zuleger, T., Takacs, Z., Franz-Wachtel, M., Thost, A. K., et al. (2017). WIPI3 and WIPI4 β-propellers are scaffolds for LKB1-AMPK-TSC signalling circuits in the control of autophagy. *Nature Communications*, *8*, 15637. https://doi.org/10.1038/ncomms15637.

Chang, J. T., Kumsta, C., Hellman, A. B., Adams, L. M., & Hansen, M. (2017). Spatiotemporal regulation of autophagy during *Caenorhabditis elegans* aging. *eLife*, *6*, 1–23. https://doi.org/10.7554/eLife.18459.

Chapin, H. C., Okada, M., Merz, A. J., & Miller, D. L. (2015). Tissue-specific autophagy responses to aging and stress in *C. elegans*. *Aging*, 7, 419–434. http://doi.org/10.18632/aging.100765.

Chen, Y., Scarcelli, V., & Legouis, R. (2017). Approaches for studying autophagy in *Caenorhabditis elegans*. *Cell*, *6*, 1–19. https://doi.org/10.3390/cells6030027.

Dikic, I., & Elazar, Z. (2018). Mechanism and medical implications of mammalian autophagy. *Nature Reviews. Molecular Cell Biology*, *19*, 349–364. https://doi.org/10.1038/s41580-018-0003-4.

Djeddi, A., Michelet, X., Culetto, E., Alberti, A., Barois, N., & Legouis, R. (2012). Induction of autophagy in ESCRT mutants is an adaptive response for cell survival in *C. elegans*. *Journal of Cell Science*, *125*, 685–694. https://doi.org/10.1242/jcs.091702.

Dooley, H. C., Razi, M., Polson, H. E. J., Girardin, S. E., Wilson, M. I., & Tooze, S. A. (2014). WIPI2 links LC3 conjugation with PI3P, autophagosome formation, and pathogen clearance by recruiting Atg 12-5-16L1. *Molecular Cell*, *17*, 238–252. https://doi.org/10.1016/j.molcel.2014.05.021.

Fujita, N., Itoh, T., Omori, H., Fukuda, M., Noda, T., & Yoshimori, T. (2008). The Atg16L complex specifies the site of LC3 lipidation for membrane biogenesis in autophagy. *Molecular Biology of the Cell*, *19*, 2092–2100. https://doi.org/10.1091/mbc.E07-12-1257.

Girard, L. R., Fiedler, T. J., Harris, T. W., Carvalho, F., Antoshechkin, I., Han, M., et al. (2007). WormBook: The online review of *Caenorhabditis elegans* biology. *Nucleic Acids Research*, *35*, D472–D475. https://doi.org/10.1093/nar/gkl894.

Guo, B., Huang, X., Zhang, P., Qi, L., Liang, Q., Zhang, X., et al. (2014). Genome-wide screen identifies signaling pathways that regulate autophagy during *Caenorhabditis elegans* development. *EMBO Reports*, *15*, 705–713. https://doi.org/10.1002/embr.201338310.

Hibshman, J. D., Leuthner, T. C., Shoben, C., Mello, D. F., Sherwood, D. R., Meyer, J. N., et al. (2018). Non-selective autophagy reduces mitochondrial content during starvation in *Caenorhabditis elegans*. *American Journal of Physiology. Cell Physiology*, *315*, C781–C792. https://doi.org/10.1152/ajpcell.00109.2018.

Ichimura, Y., Imamura, Y., Emoto, K., Umeda, M., Noda, T., & Ohsumi, Y. (2004). In vivo and in vitro reconstitution of Atg8 conjugation essential for autophagy. *Journal of Biological Chemistry*, *279*, 40584–40592. https://doi.org/10.1074/jbc.M405860200.

Jia, K., Hart, A. C., & Levine, B. (2007). Autophagy genes protect against disease caused by polyglutamine expansion proteins in *Caenorhabditis elegans*. *Autophagy*, *3*, 21–25. https://doi.org/10.4161/auto.3528.

Kabeya, Y. (2000). LC3, a mammalian homologue of yeast Apg8p, is localized in autophagosome membranes after processing. *The EMBO Journal*, *19*, 5720–5728. https://doi.org/10.1093/emboj/19.21.5720.

Kang, C., You, N. J., & Avery, L. (2007). Dual roles of autophagy in the survival of *Caenorhabditis elegans* during starvation. *Genes and Development*, *21*, 2161–2171. https://doi.org/10.1101/gad.1573107.

Klionsky, D. J., Klionsky, D. J., Abdelmohsen, K., Abe, A., Abedin, M. J., Abeliovich, H., et al. (2016). Guidelines for the use and interpretation of assays for monitoring autophagy (3rd edition). *Autophagy*, *12*, 1–222. https://doi.org/10.1080/15548627.2015.1100356.

Kozlowski, L., Garvis, S., Bedet, C., & Palladino, F. (2014). The *Caenorhabditis elegans* HP1 family protein HPL-2 maintains ER homeostasis through the UPR and hormesis. *Proceedings of the National Academy of Sciences of the United States of America*, *111*, 5956–5961. https://doi.org/10.1073/pnas.1321698111.

Lapierre, L. R., De Magalhaes Filho, C. D., McQuary, P. R., Chu, C. C., Visvikis, O., Chang, J. T., et al. (2013). The TFEB orthologue HLH-30 regulates autophagy and modulates longevity in *Caenorhabditis elegans*. *Nature Communications*, *4*, 2267. https://doi.org/10.1038/ncomms3267.

Lu, Q., Yang, P., Huang, X., Hu, W., Guo, B., Wu, F., et al. (2011). The WD40 repeat PtdIns(3)P-binding protein EPG-6 regulates progression of omegasomes to autophagosomes. *Developmental Cell*, *21*, 343–357. https://doi.org/10.1016/j.devcel.2011.06.024.

Meléndez, A., Tallóczy, Z., Seaman, M., Eskelinen, E. L., Hall, D. H., & Levine, B. (2003). Autophagy genes are essential for dauer development and life-span extension in *C. elegans*. *Science*, *301*, 1387–1391. https://doi.org/10.1126/science.1087782.

Mercer, T. J., Gubas, A., & Tooze, S. A. (2018). A molecular perspective of mammalian autophagosome biogenesis. *Journal of Biological Chemistry*, *293*, 5386–5395. https://doi.org/10.1074/jbc.R117.810366.

Mizushima, N., & Yoshimori, T. (2007). How to interpret LC3 immunoblotting. *Autophagy*, *3*, 542–545. https://doi.org/10.4161/auto.4600.

Mizushima, N., Yoshimori, T., & Levine, B. (2010). Methods in mammalian autophagy research. *Cell*, *140*, 313–326. https://doi.org/10.1016/j.cell.2010.01.028.

Nakatogawa, H., Ichimura, Y., & Ohsumi, Y. (2007). Atg8, a ubiquitin-like protein required for autophagosome formation, mediates membrane tethering and hemifusion. *Cell*, *130*, 165–178. https://doi.org/10.1016/j.cell.2007.05.021.

Nakatogawa, H., Suzuki, K., Kamada, Y., & Ohsumi, Y. (2009). Dynamics and diversity in autophagy mechanisms: Lessons from yeast. *Nature Reviews. Molecular Cell Biology*, *10*, 458–467. https://doi.org/10.1038/nrm2708.

Noda, T. (2017). Autophagy in the context of the cellular membrane-trafficking system: The enigma of Atg9 vesicles. *Biochemical Society Transactions*, *45*, 1323–1331. https://doi.org/10.1042/BST20170128.

O'Rourke, E. J., & Ruvkun, G. (2013). MXL-3 and HLH-30 transcriptionally link lipolysis and autophagy to nutrient availability. *Nature Cell Biology*, *15*(6), 668–676. https://doi.org/10.1038/ncb2741.

Orsi, A., Razi, M., Dooley, H. C., Robinson, D., Weston, A. E., Collinson, L. M., et al. (2012). Dynamic and transient interactions of Atg9 with autophagosomes, but not membrane integration, are required for autophagy. *Molecular Biology of the Cell*, *23*, 1860–1873. https://doi.org/10.1091/mbc.E11-09-0746.

Polson, H. E. J., De Lartigue, J., Rigden, D. J., Reedijk, M., Urbé, S., Clague, M. J., et al. (2010). Mammalian Atg18 (WIPI2) localizes to omegasome-anchored phagophores and positively regulates LC3 lipidation. *Autophagy*, *6*, 506–522. https://doi.org/10.4161/auto.6.4.11863.

Possik, E., Jalali, Z., Nouët, Y., Yan, M., Gingras, M. C., Schmeisser, K., et al. (2014). Folliculin regulates Ampk-dependent autophagy and metabolic stress survival. *PLoS Genetics*, *10*, e1004273–e1004273. https://doi.org/10.1371/journal.pgen.1004273.

Reggiori, F., Tucker, K. A., Stromhaug, P. E., & Klionsky, D. J. (2004). The Atg1-Atg13 complex regulates Atg9 and Atg23 retrieval transport from the pre-autophagosomal structure. *Developmental Cell*, *6*, 79–90. https://doi.org/10.1016/S1534-5807(03)00402-7.

Saitsu, H., Nishimura, T., Muramatsu, K., Kodera, H., Kumada, S., Sugai, K., et al. (2013). De novo mutations in the autophagy gene WDR45 cause static encephalopathy of childhood with neurodegeneration in adulthood. *Nature Genetics*, *45*, 445–449. https://doi.org/10.1038/ng.2562.

Settembre, C., De Cegli, R., Mansueto, G., Saha, P. K., Vetrini, F., Visvikis, O., et al. (2013). TFEB controls cellular lipid metabolism through a starvation-induced autoregulatory loop. *Nature Cell Biology*, *15*(6), 647–658. https://doi.org/10.1038/ncb2718.

Settembre, C., Di Malta, C., Polito, V. A., Arencibia, M. G., Vetrini, F., Erdin, S., et al. (2011). TFEB links autophagy to lysosomal biogenesis. *Science*, *332*(6036), 1429–1433. https://doi.org/10.1126/science.1204592.

Shintani, T., & Klionsky, D. J. (2004). Cargo proteins facilitate the formation of transport vesicles in the cytoplasm to vacuole targeting pathway. *Journal of Biological Chemistry*, *279*, 29889–29894. https://doi.org/10.1074/jbc.M404399200.

Silvestrini, M. J., Johnson, J. R., Kumar, A. V., Thakurta, T. G., Blais, K., Neill, Z. A., et al. (2018). Nuclear export inhibition enhances HLH-30/TFEB activity, autophagy, and lifespan. *Cell Reports*, *23*(7), 1915–1921. https://doi.org/10.1016/j.celrep.2018.04.063.

Singh, R., & Cuervo, A. M. (2011). Autophagy in the cellular energetic balance. *Cell Metabolism*, *13*, 495–504. https://doi.org/10.1016/j.cmet.2011.04.004.

Tian, Y., Li, Z., Hu, W., Ren, H., Tian, E., Zhao, Y., et al. (2010). *C. elegans* screen identifies autophagy genes specific to multicellular organisms. *Cell*, *141*, 1042–1055. https://doi.org/10.1016/j.cell.2010.04.034.

Weidberg, H., Shpilka, T., Shvets, E., Abada, A., Shimron, F., & Elazar, Z. (2011). LC3 and GATE-16 N termini mediate membrane fusion processes required for autophagosome biogenesis. *Developmental Cell*, *20*, 444–454. https://doi.org/10.1016/j.devcel.2011.02.006.

Wilkinson, D. S., Jariwala, J. S., Anderson, E., Mitra, K., Meisenhelder, J., Chang, J. T., et al. (2015). Phosphorylation of LC3 by the hippo kinases STK3/STK4 is essential for autophagy. *Molecular Cell*, *57*, 55–68. https://doi.org/10.1016/j.molcel.2014.11.019.

Wu, Y., Cheng, S., Zhao, H., Zou, W., Yoshina, S., Mitani, S., et al. (2014). PI3P phosphatase activity is required for autophagosome maturation and autolysosome formation. *EMBO Reports*, *15*, 973–981. http://doi.org/10.15252/embr.201438618.

Wu, F., Li, Y., Wang, F., Noda, N. N., & Zhang, H. (2012). Differential function of the two Atg4 homologues in the aggrephagy pathway in *Caenorhabditis elegans*. *Journal of Biological Chemistry*, *287*, 29457–29467. https://doi.org/10.1074/jbc.M112.365676.

Xie, Z., & Klionsky, D. J. (2007). Autophagosome formation: Core machinery and adaptations. *Nature Cell Biology*, *9*, 1102–1109. https://doi.org/10.1038/ncb1007-1102.

CHAPTER FIFTEEN

Monitoring stress-induced autophagic engulfment and degradation of the 26S proteasome in mammalian cells

Victoria Cohen-Kaplan[a,*,†], Ido Livneh[a,*,†], Yong Tae Kwon[b], Aaron Ciechanover[a,b]

[a]Technion Integrated Cancer Center (TICC), The Rappaport Faculty of Medicine and Research Institute, Technion-Israel Institute of Technology, Haifa, Israel
[b]Department of Biomedical Sciences, Protein Metabolism Medical Research Center, College of Medicine, Seoul National University, Seoul, South Korea
*Corresponding authors: e-mail address: vita_kaplan@hotmail.com; i.livneh@gmail.com

Contents

Abstract

Almost 70 years after the discovery of the lysosome, and about four decades following the unraveling of ubiquitin as a specific "mark of death," the field of protein turnover—the numerous processes it regulates, the pathologies resulting from its dysregulation,

Methods in Enzymology, Volume 619
ISSN 0076-6879
https://doi.org/10.1016/bs.mie.2018.12.022
© 2019 Elsevier Inc.
All rights reserved.

and the drugs that have been developed to target them—is still growing exponentially. Accordingly, the need for new technologies and methods is ever growing. One interesting question in the field is the mechanism(s) by which the "predators become prey". We have reported recently that the 26S proteasome, the catalytic arm of the ubiquitin system, is degraded by the autophagy–lysosome machinery, in a process requiring specific ubiquitination of the proteasome, and subsequent recognition by the shuttle protein p62/SQSTM1. Studying the modification(s), recognition sites, engulfment, and breakdown of the 26S proteasome via such "proteaphagy" has required the use of microscopy, subcellular fractionation, 'classical biochemistry', and proteomics. In this chapter, we present the essentials of these protocols, with emphasis on the refinements we have introduced in order for them to better suit the particular study of proteaphagy.

1. Introduction

The eukaryotic cell harbors two major proteolytic systems, the ubiquitin–proteasome and the autophagy–lysosome systems. The notion that cellular proteins are in a dynamic state is not a recent one, but though evidence supporting it emerged in the late 1930s, it was not fully accepted by the scientific community even after Christian de Duve has discovered the lysosome (Ciechanover, 2005; de Duve, Gianetto, Appelmans, & Wattiaux, 1953; Gianetto & de Duve, 1955; Hogness, Cohn, & Monod, 1955; Schoenheimer, 1942).

Since the discovery of the ubiquitin system in the late 1970s and early 1980s, the field of protein degradation has enjoyed rapid expansion. The mechanism of specific targeting of proteins and its energy dependence has been unraveled, and ubiquitin has been described as a *bona fide* degradation-targeting signal (Ciechanover, 1994; Glickman & Ciechanover, 2002; Hershko & Ciechanover, 1992; Hershko, Ciechanover, & Rose, 1981).

As for lysosomal degradation, it also requires metabolic energy—not for specific targeting of its substrates, but rather for maintaining the low intralysosomal pH required for optimal activity of the hydrolases (Schneider, 1981). Interestingly, for certain modes of lysosomal degradation, recognition of a specific targeting signal (KFERQ) within the coding sequence of the substrate proteins is required (Dice, Chiang, Spencer, & Backer, 1986). Importantly, the lysosome, besides being involved in targeting extracellular proteins, was also shown—initially by Christian de Duve—to degrade intracellular proteins via a mechanism that de Duve termed autophagy (de Duve, 1963; Klionsky, 2008). Autophagy was later found to operate both under basal and stress conditions, and its underlying mechanisms have been deciphered (Suzuki & Ohsumi, 2007; Thumm et al., 1994; Tsukada & Ohsumi, 1993).

The ubiquitin system is "universal" across all eukarya, and that its proper function is essential for the normal development and the basic activities of all eukaryotic cells and organisms. Similarly, autophagy plays important roles in enabling the cell to cope with stress. Study of the ubiquitin and autophagy pathways required the development of appropriate methodologies to dissect their details. Many of the methods in use nowadays are similar in principle to those that have been used in the fields of cell biology and biochemistry for decades. While many methods have been improved dramatically (e.g., microscopy), and new areas have been developed (e.g., proteomics), they did not eliminate the necessity for "classic" biochemical tools. Despite all of these technological advances, the selectivity with which we are able to isolate subcellular organelles, and our ability to reconstruct organellar and biochemical pathways in cell-free systems, has remained limited. These limitations have at times held back our ability to accurately understand such processes, especially as related to their intracellular topology (Fabre et al., 2013, 2014).

Nevertheless, significant progress has been made in our understanding of both autophagy and the ubiquitin–proteasome system (UPS), as well as regarding many aspects that the two share in common—such as mechanisms of action, key mediators, signaling molecules, and cellular functions. Both the UPS and autophagy were shown to recognize ubiquitin as a degradation signal, and initially, different modes of ubiquitination were thought to signal differentially to the two pathways (Cohen-Kaplan, Livneh, Avni, Cohen-Rosenzweig, & Ciechanover, 2016). The canonical ubiquitin proteasomal signal was identified as a K48-based polyubiquitin chain (Chau et al., 1989). Autophagy was suggested to recognize mostly K63-based ubiquitin chains (Tan et al., 2008), but targeting of membrane proteins to endosomes appeared to require monoubiquitination (Goh & Sorkin, 2013). Mono-ubiquitination was considered insufficient for proteasomal targeting (Thrower, Hoffman, Rechsteiner, & Pickart, 2000).

With time, it has become clear that neither the proteasome nor the autophagic machinery shows exclusivity for any ubiquitin chain type. Thus for example, it has been shown that K63- and K11-based chains can be recognized by the proteasome (Min, Mevissen, De Luca, Komander, & Lindon, 2015; Paraskevopoulos et al., 2014; Saeki et al., 2009), whereas K48- and K27-based chains were shown to be recognized by the autophagic machinery (Ikeda & Kerppola, 2008; Jin & Cui, 2017; Zhang, Xu, Scotti, Chen, & Tontonoz, 2013). In addition, monoubiquitination has been clearly shown as a *bona fide* proteasomal recognition signal, though at times it was difficult to separate it from multiple monoubiquitinations (Boutet, Disatnik, Chan, Iori, & Rando, 2007; Braten et al., 2016; Dimova et al., 2012; Kravtsova-Ivantsiv, Cohen, & Ciechanover, 2009; Liani et al., 2004; Livneh, Kravtsova-Ivantsiv, Braten, Kwon, & Ciechanover, 2017; Yin, Gui, Du, Frohman, & Zheng, 2010). Generally, we currently understand that the ubiquitin code is much more diverse and complex than originally assumed, and it requires for specificity other factors or distinct cellular localizations. Important in this respect are the points where the autophagosomal and the ubiquitin–proteasome systems meet one another—as in the case of sequential activation of the two systems following induction of stress, shared receptors for ubiquitinated proteins (e.g., Sequestosome 1 (SQSTM1)/p62), or the degradation of key components of each system by the other (Gao et al., 2010; Liu et al., 2016; Song et al., 2016; Zhao, Zhai, Gygi, & Goldberg, 2015). Such complexity calls for cautious evaluation and interpretation of data, as well as for the refinement of existing methods and the development of new ones.

An important question related to regulation of the ubiquitin system has been the degradation of its own components (de Bie & Ciechanover, 2011; Weissman, Shabek, & Ciechanover, 2011). In this chapter, we use mostly confocal fluorescence microscopy to monitor autophagosomal uptake of the 26S proteasome, with emphasis given to upregulation of the process following stress. In the next part of this chapter, we describe a method for isolation and purification of autophagic vesicles, which may be later employed for the study of their content. It should be emphasized that the autophagosomal uptake of the proteasome requires its prior ubiquitination on specific sites by an as yet to be identified ubiquitin ligase (E3) (Cohen-kaplan, Ciechanover, & Livneh, 2017). The ubiquitin moiety is required for recognition by SQSTM1/p62 via its ubiquitin-associating (UBA) domain, while the anchorage of the cargo to the growing autophagosomal vesicle is mediated via the LC3-interaction region (LIR) of SQSTM1/p62 (Cohen-Kaplan, Ciechanover, & Livneh, 2016).

2. Monitoring the autophagosomal uptake of the proteasome by immunofluorescence

In order to monitor autophagosomal uptake of the proteasome, we use human cervical carcinoma HeLa cells stably expressing the fusion protein GFP-LC3B, which is used widely as an autophagosomal marker (Shvets, Abada, Weidberg, & Elazar, 2011; Yoshii & Mizushima, 2017).

LC3 (microtubule-associated proteins 1A/1B light chain 3B) is the only known autophagic receptor for several shuttle proteins, which carry the cargo proteins destined for degradation (Shvets et al., 2011; Weidberg et al., 2010). LC3, a ubiquitin-like protein, is synthesized as pro-LC3, which is cleaved at the site of its mature C-terminus by the cysteine protease ATG4B, forming LC3-I, a cytosolic form of LC3 which is then conjugated with phosphatidylethanolamine (PE). LC3-PE (also called the lipidated form of LC3 or LC3-II) is directly incorporated into the expanding autophagosomal membrane, where it will eventually serve as receptor for the cargo bound to the shuttle proteins (Hanada et al., 2007; Kabeya et al., 2000; Slobodkin & Elazar, 2013; Tanida, Ueno, & Kominami, 2004).

In selective autophagy, shuttle proteins have an important role in the delivery of cargo to autophagosomes. It was shown that ubiquitin marks proteins not only for proteasomal but also for autophagosome–lysosomal degradation. Shuttle proteins, such as SQSTM1/p62 and Neighbor of BRCA1 gene 1 (NBR1), simultaneously bind to ubiquitinated proteins through their UBA domain, and to the autophagosomal receptor LC3 through the LIR motif (Bjørkøy et al., 2005; Kirkin et al., 2009). Since SQSTM1/p62 is engulfed by the forming autophagosome along with its cargo, it is also used as an autophagosomal marker (Klionsky et al., 2016).

Recently, we have shown that amino acid starvation increases specifically the ubiquitination of certain proteasomal subunits, resulting in selective autophagosomal uptake of the complex, which is mediated by SQSTM-1/p62 (Cohen-Kaplan, Livneh, Avni, Fabre, et al., 2016).

2.1 Induction of autophagosome formation by amino acid starvation

Autophagosomes are known to form in response to a variety of stimuli, including amino acid starvation, which is probably the most common autophagy-inducing stimulus as well as the best understood mechanistically (Onodera & Ohsumi, 2005). Therefore, in order to upregulate autophagy, we have subjected the cells to 4 h of amino acid deprivation.

Autophagosome formation is regulated in mammals by the mTOR pathway (formerly "mammalian," and currently "mechanistic" target of rapamycin) (Heitman, Movva, & Hall, 1991; Laplante & Sabatini, 2013). The mTOR protein is a kinase that is a key component of two complexes, mTORC1 and mTORC2. Many of the steps downstream of these complexes are regulated via phosphorylation of key mediators. With regard to autophagy, most of the literature is concerned with the mTORC1 complex, which is active under nutrient-rich conditions and negatively regulates autophagy. Active mTORC1 directly phosphorylates ATG13 and the ULK1 complex, thereby suppressing their activity, which would otherwise promote autophagy induction. Upon amino acid deprivation, mTOR dissociates from the mTORC1 complex (which under a normal supply of nutrients is assembled on the lysosomal surface), as well as from the ULK1–ATG13–FIP200 complex, resulting in rapid dephosphorylation of ULK1 and ATG13. Consequently, activated ULK1 phosphorylates both FIP200 and ATG13, leading to the activation of downstream autophagy effector proteins and the initiation of autophagosome formation (Ganley et al., 2009; Hosokawa et al., 2009; Jung et al., 2009; Zachari & Ganley, 2017).

After autophagosomes have been formed fully sealed vesicles, they fuse with lysosomes to form autolysosomes, thereby exposing their contents to the lysosomal hydrolytic enzymes. These enzymes are optimally active at pH ~4.5–5.0 (Ohkuma & Poole, 1978). Membranal lysosomal proton pumps (V-ATPases) maintain the low pH by active transport of H^+ ions from the cytosol into the lysosome (Schneider, 1981). The activity of lysosomal hydrolases can be inhibited by increasing the internal lysosomal acidic pH to ~7.2 (Cooper & Hausman, 2013; Takenouchi et al., 2014). Therefore, to prevent lysosomal degradation of autophagosomal cargo after fusion, and thereby improve the ability to detect proteins and organelles taken up by autophagosomes following amino acid starvation, researchers often treat cells with alkalinizing agents. Several lysosomotropic alkalinizing reagents are available, including ammonium chloride, chloroquine, and bafilomycin A1, and while all increase the lysosomal internal pH, they act via different mechanisms.

One commonly used reagent is chloroquine, a weak base that readily penetrates the lysosomal membrane when only monoprotonated, but, when it is converted into its diprotonated form in the lysosome interior, it is prevented from passing back through the lysosomal membrane and thus trapped within the organelle. This diprotonated chloroquine derivative significantly increases the lysosomal pH by a mechanism which is not completely understood, resulting in inactivation of the lysosomal acid hydrolases (Redmann et al., 2017).

2.1.1 Equipment

- 35 mm cell culture plates.
- 150 mm cell culture plates.
- Sterile glass cover slips (22 × 22 mm, 0.17 mm thickness).

2.1.2 Buffers and reagents

- High glucose (4.5 g/L) Dulbecco's Modified Eagle's medium (DMEM).
- Penicillin/streptomycin stock solution: 10,000 U/mL and 10 mg/mL, respectively.
- Sodium pyruvate stock solution: 100 m*M*.
- L-Glutamine stock solution: 200 m*M*.
- Fetal bovine serum (FBS).
- Washing buffer—sterile phosphate-buffered saline (PBS): 137 m*M* NaCl, 2.7 m*M* KCl, 10 m*M* Na_2HPO_4, 1.8 m*M* KH_2PO_4, 0.9 m*M* $CaCl_2$; pH 7.4.
- Amino acid starvation medium: Earle's Balanced Salts Solution with sodium bicarbonate (EBSS).
- Lysosome inhibitor: Chloroquine (CQ) 100 m*M* (stock solution).

2.1.3 Procedure

1. Culture GFP-LC3B-expressing HeLa cells in high glucose DMEM, supplemented with Penicillin/Streptomycin (100 U/mL and 0.1 mg/mL, respectively), sodium pyruvate (1 m*M*), L-glutamine (2 m*M*), and 10% (v/v) FBS in a humidified incubator containing 5% CO_2 at 37°C. Seed cells (i) for immunofluorescence staining on sterile glass cover slips (22 × 22 mm) in 35 mm plates and (ii) for autophagosome purification on 150 mm plates (~20 plates).
2. Grow cells to approximately 80%–90% confluence.
3. Wash cells twice with sterile PBS supplemented with $CaCl_2$, and aspirate the final wash entirely.
4. Add EBSS starvation medium or complete medium (to control cells), each supplemented with 100 μ*M* CQ, and incubate for 4 h at 37°C.
5. Detect autophagosomes using confocal fluorescence microscopy (see Section 2.2.2).

2.1.4 Notes

1. A population of cells stably expressing GFP-LC3B shows heterogeneity in the expression level of the transfected gene. In order to achieve a more homogeneous GFP-LC3B-expressing population, cells should be sorted using flow cytometry.
2. Washing buffer PBS should contain $CaCl_2$ in order to preserve the normal structure of cell–cell contacts, which are Ca^{2+}-dependent.

2.2 Use fluorescence microscopy to follow stress-induced autophagosome formation and uptake of ubiquitinated proteasomes

Fluorescent labeling of cell components, followed by confocal visualization, constitutes a strong tool for studying cell structure as well as protein localization, interactions, and dynamics. Fluorescent labeling of organelles and proteins can be achieved using synthetic fluorescent probes, immunostaining with fluorophore-conjugated antibodies, or via the expression of intrinsically fluorescent fusion proteins.

Since the lipidated form of LC3 (LC3-II) is directly incorporated into the expanding autophagosomal membrane, we use cells expressing the fusion protein GFP-LC3B for monitoring autophagosome formation and subsequent autophagosome–lysosome fusion following amino acid starvation. For the detection of the proteasome and ubiquitinated proteins within autophagosomes/autolysosomes, we subject cells to immunofluorescent staining with appropriate antibodies.

Since the 26S proteasome is composed of two main subcomplexes, the catalytic 20S core particle (CP) and the 19S regulatory particle (RP) (Finley, Chen, & Walters, 2016; Livneh, Cohen-Kaplan, Cohen-Rosenzweig, Avni, & Ciechanover, 2016), an important question is whether both are ingested as part of intact 26S proteasomes and therefore colocalize to the same autophagosomes, or whether the 26S complex is disassembled first and then the two subparticles are taken up independently. To address this question, immunofluorescent co-staining of the 19S and 20S should be performed and their localization to LC3-positive vesicles assessed. It should be noted that while colocalization suggests coentry, it does not rule out a scenario by which they are taken up into the same vesicle independently.

In order to protect GFP-LC3B from fluorescence quenching in the lysosome due to the acidity of the organelle, all experiments should be carried out in the presence of chloroquine, which neutralizes the lysosomal acidic pH and also allows the detection of GFP-LC3B in autolysosomes (Lőw et al., 2013).

2.2.1 Equipment

- Glass microscope slides.
- Zeiss LSM 700 confocal microscope.
- ZEN software (Zeiss) for visualization and analysis.

2.2.2 Buffers and reagents

- Washing buffer—sterile PBS: 137 mM NaCl, 2.7 mM KCl, 10 mM Na_2HPO_4, 1.8 mM KH_2PO_4, 0.9 mM $CaCl_2$; pH 7.4.
- 4% buffered paraformaldehyde (PFA) solution in PBS supplemented with 0.9 mM $CaCl_2$, pH 7.4 (titrated using NaOH).
- Blocking buffer: 10% (v/v) normal goat serum in PBS.
- Permeabilization buffer: 0.5% Triton X-100 in PBS.
- Primary antibodies for immunostaining: anti-20S proteasome subunit α6, anti-19S proteasome subunit RPN3, anti-SQSTM1/p62, anti-lysosome-associated membrane glycoprotein protein-1 (LAMP-1), anti-K48 ubiquitin chain, anti-K63 ubiquitin chain.
- Secondary antibodies: Alexa Fluor 568/633 nm-conjugated anti-IgG (Invitrogen).
- Fluorescence mounting medium supplemented with 4′,6-diamidino-2-phenylindole dihydrochloride (DAPI) (for nuclear staining).

2.2.3 Procedure

1. Wash cells twice with sterile PBS supplemented with $CaCl_2$.
2. Fix cells in 4% buffered PFA solution for 20 min at room temperature, or overnight at 4°C.
3. Wash cells three times with PBS.
4. Permeabilize cells by incubating for 5 min with permeabilization buffer.
5. Wash cells three times with PBS.
6. Incubate cells in blocking buffer for 1 h.
7. Dilute primary antibodies in blocking buffer according to manufacturers' instructions. For double staining, two different antibodies (of different host species) can be used simultaneously.
8. Incubate cells with diluted primary antibodies at room temperature for 2–3 h, or overnight at 4°C while gently shaking.
9. Wash cells three times with PBS (5 min each).
10. Dilute fluorophore-conjugated secondary antibodies in blocking buffer according to manufacturers' instructions. In case of double staining, two antibodies (targeting immunoglobulins derived from different host species) conjugated to different fluorescent dyes can be used simultaneously.
11. Incubate cells with diluted secondary antibodies at room temperature for 1 h.
12. Wash cells five times with PBS (5 min each).
13. Mount samples: place a small drop of mounting medium containing DAPI on a glass slide and then invert the cover slips on the slide with

the cells facing the medium drop. Incubate for a few seconds (or per manufacturer's instructions). Remove gently excess liquid by pressing the cover slip against the slide and then seal the cover slip with clear nail polish or a commercial sealant.

14. Store slides in the dark at 4°C. Acquire images within 24 h using a Zeiss LSM 700 or other confocal microscope with suitable filter set(s).
15. Acquire images using high-magnification, oil-immersion objectives, optimal resolution, and pinhole diameter. These will vary according to the type of the objective used [we use a setup of 1024 × 1024 pixels with the pinhole diameter adjusted to 1 μm and either: (i) 40 ×, NA = 1.3, oil-immersion; or (ii) 63 ×, NA = 1.4, oil-immersion objective].
16. Perform two-channel colocalization analysis and determine the Manders' Overlap Coefficient of colocalization (Manders, Verbeek, & Aten, 1993) using the ZEN software (Zeiss).

2.2.4 Notes

1. All the immunostaining incubations involving fluorescent dyes should be carried out in the dark.
2. Acquisition of all image data should be done using the same parameters (laser intensity, gain, pinhole size, etc.) and image magnification.

Fig. 1Ai (middle column) shows increased autophagosome formation following amino acid starvation. The number of GFP-LC3B (green)-positive vesicles in starved cells (bottom image) is much higher than in control cells (upper image). The same field was analyzed for autophagosomal uptake of the proteasome, using immunofluorescent staining of the 20S proteasome α6 subunit (red), and detection of GFP-LC3 (green) (Fig. 1Ai, right column). White arrows point to proteasomes within autophagosomes. As can be seen, the proteasome amount in autophagosomes of starved cells far exceeds its amount in control cells.

Fig. 1Aii shows higher magnifications of images in the right column of Panel A(i) (box). The colocalization of GFP-LC3B with the proteasome was assessed using Manders overlap coefficients and is presented in Fig. 1Aiii.

Fig. 1B shows that amino acid starvation induces autophagosomal uptake of both the 20S and 19S proteasomal subcomplexes. As can be seen, the 20S α6 and 19S RPN3 subunits colocalize in the same autophagosomes.

Fig. 2 shows colocalization of GFP-LC3B, SQSTM1/p62, and LAMP-1 (a lysosomal marker) in a large number of autolysosomes (yellow). The fusion of autophagosomes with lysosomes (autolysosome formation) was detected in the cytoplasm of control and starved cells even following chloroquine treatment.

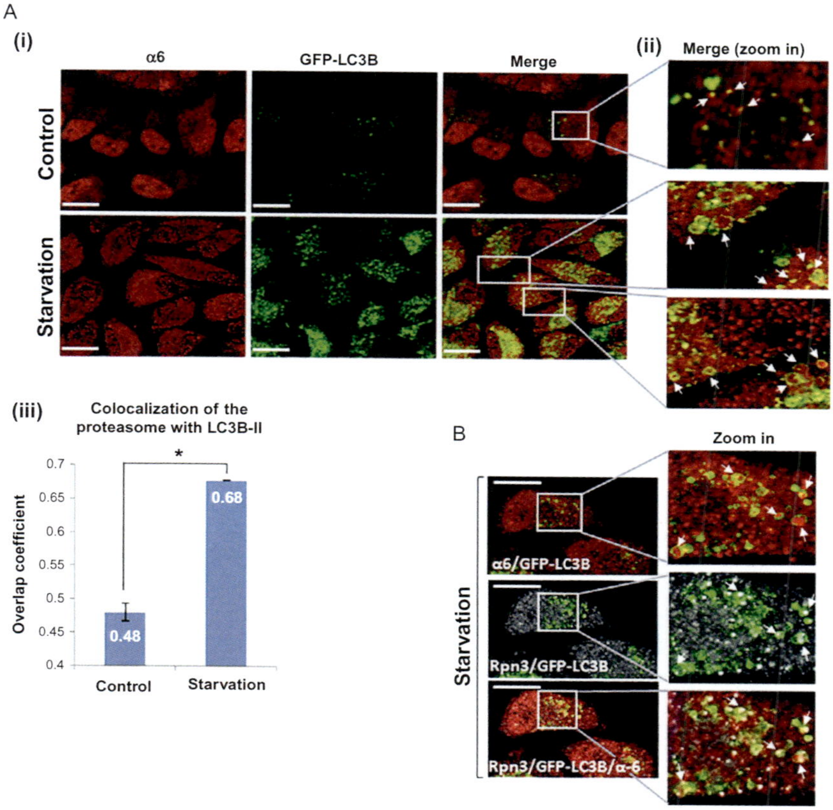

Fig. 1 Detection of autophagosome formation and autophagic uptake of the proteasome following amino acid starvation. (A) Amino acid starvation induces autophagosome formation and autophagosomal uptake of the 26S proteasome (i). Following 4 h of incubation in complete (*upper panel*) or amino acid-depleted (*lower panel*) medium, HeLa cells stably expressing GFP-LC3B were fixed in 4% of PFA and stained with anti-20S proteasome α6 antibody (*red*). Scale bars are 20 μm. (ii) High magnification of regions inside the box. *White arrows* point to autophagosomes with detected proteasomes. (iii) Colocalization between GFP-LC3B and 20S proteasome α6. Quantification was calculated according to Manders. $^*P < 0.0000012$. (B) Detection of 20S and 19S proteasomal subcomplexes within autophagosomes following amino acid starvation. HeLa cells stably expressing GFP-LC3 were subjected to immunofluorescence staining with anti-20S proteasome α6 (*upper image*) and anti-19S proteasome RPN3 (*gray color*; *middle image*) antibodies, following 4 h of amino acid starvation. Scale bars are 20 μm. The *right column* shows higher magnifications of images in the *left column* (*box*). *White arrows* point to autophagosomes containing both 20S CP (α6)] and 19S RP (RPN3), as is indicated in the merged image (bottom panel). *Adapted from Cohen-Kaplan, V., Livneh, I., Avni, N., Fabre, B., et al. (2016). p62- and ubiquitin-dependent stress-induced autophagy of the mammalian 26S proteasome.* Proceedings of the National Academy of Sciences of the United States of America, 113*(47), E7490–E7499. http://doi.org/10.1073/pnas.1615455113.* PNAS, *with permission.*

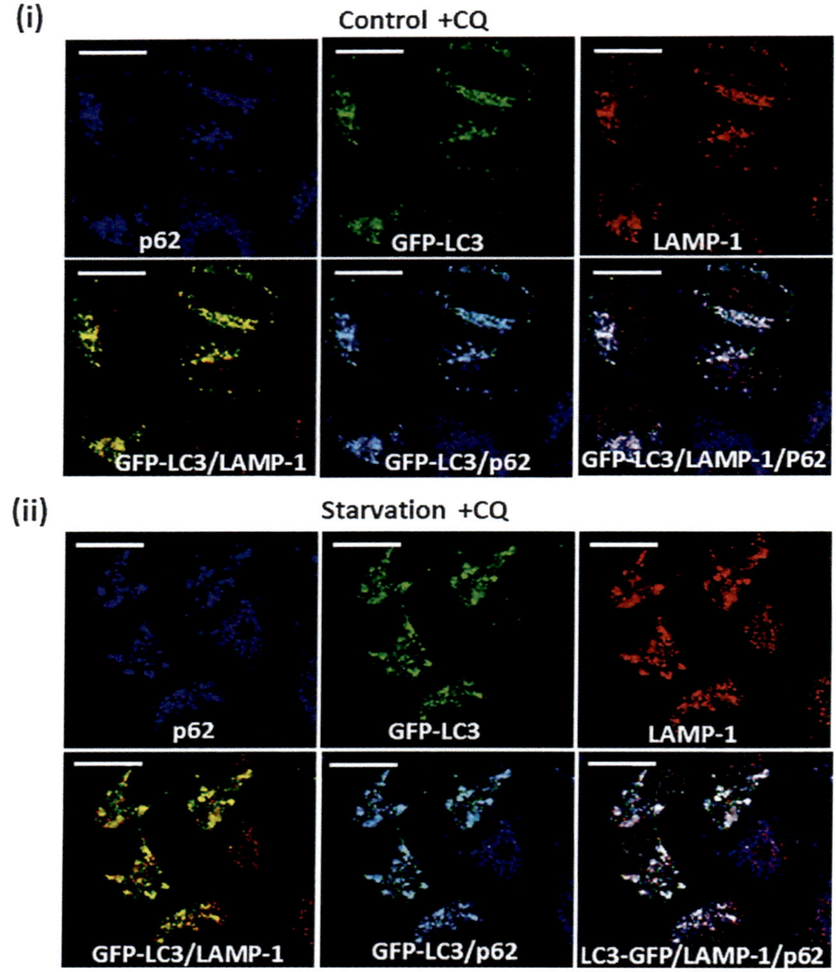

Fig. 2 Detection of colocalization of autophagosomes and lysosomes following chloroquine treatment. LC3B and SQSTM1/p62 colocalize with LAMP-1 in chloroquine-treated cells. HeLa cells expressing GFP-LC3B (*green*) were incubated in complete (i) or amino acid-depleted (ii) medium for 4 h. Following incubation, cells were fixed in 4% of PFA, and stained with anti-SQSTM1/p62 (*blue*) and anti-LAMP-1 (*red*) antibodies. Scale bars are 20 μm. *Adapted from Cohen-Kaplan, V., Livneh, I., Avni, N., Fabre, B., et al. (2016). p62- and ubiquitin-dependent stress-induced autophagy of the mammalian 26S proteasome.* Proceedings of the National Academy of Sciences of the United States of America, 113*(47), E7490–E7499. http://doi.org/10.1073/pnas.1615455113.* PNAS, *with permission.*

Fig. 3A shows autophagosomal uptake of proteins conjugated to K63- or K48-ubiquitin chains. As can be seen, conjugates of both chain types were found to be co-localized with LC3B. However, the number of K48-positive autophagosomes was much lower than that of K63-positive ones.

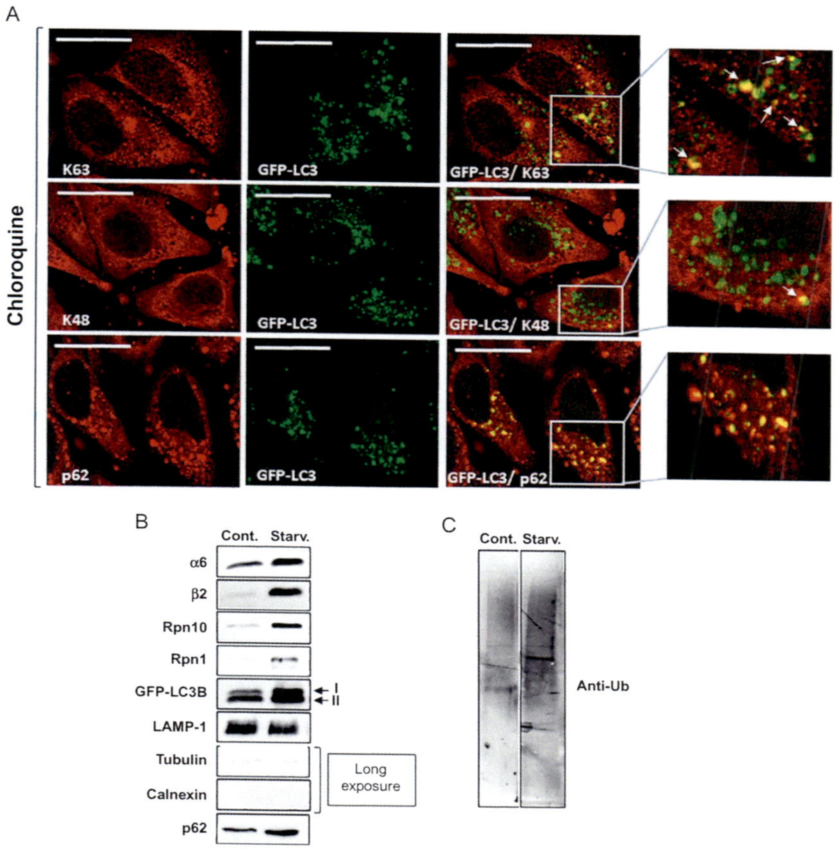

Fig. 3 Detection of ubiquitinated proteins within autophagosomes. (A) Ubiquitinated proteins colocalize with autophagosomes. HeLa cells stably expressing GFP-LC3B were treated with chloroquine for 4h and subjected to immunofluorescent staining with anti-K63, anti-K48 and anti-SQSTM1/p62 antibodies. Scale bars are 20 μm. The *right column* shows high magnification of regions inside the *box*. *White arrows* point to autophagosomes containing ubiquitinated proteins. (B) Amino acid starvation induces autophagosomal uptake of the proteasome. HeLa cells stably expressing GFP-LC3B were incubated for 4h in complete (Cont.) or EBSS (Starv.) medium supplemented with chloroquine. Autophagosome/autolysosome vesicles were purified, lysed, and resolved via SDS-PAGE followed by immunoblotting with anti-20S proteasome subunits α6 and β2, with anti-19S proteasome subunits RPN10 and RPN1, and with anti-LAMP1 (lysosome), anti-tubulin (cytosol), anti-calnexin (ER), and anti-p62 antibodies. (C) Amino acid starvation increases autophagosomal uptake of ubiquitinated proteins. Autophagosome/autolysosome vesicles isolated from control and starved cells (Fig. 1C) were resolved via SDS/PAGE and blotted with anti-Ub conjugates antibody. *Adapted from Cohen-Kaplan, V., Livneh, I., Avni, N., Fabre, B., et al. (2016). p62- and ubiquitin-dependent stress-induced autophagy of the mammalian 26S proteasome.* Proceedings of the National Academy of Sciences of the United States of America, 113*(47), E7490–E7499. http://doi.org/10.1073/pnas.1615455113.* PNAS, *with permission.*

3. Purification of autophagosomes and autolysosomes from GFP-LC3B-expressing HeLa cells

Purification of autophagosomes from cultured cells provides researchers with a powerful tool to analyze them biochemically, and to shed light on the underlying mechanisms of several key processes, such as autophagy and protein quality control. Interestingly, much of our knowledge concerning autophagy comes from microscopy. While microscopy allows one to follow the dynamics of the processes and visualization of a few proteins, purification of autophagosomes allows for global analyses of their contents and the changes occurring with time under different conditions. Needless to say, the two approaches are complementary.

The method for autophagosome purification described here is based on a previously published method (Seglen & Brinchmann, 2010), which describes autophagosome purification from rat hepatocytes. We have modified it in order to adapt it to our system—cultured HeLa cells stably expressing the GFP-LC3B fusion protein. Importantly, our method purifies both autophagosomes and autolysosomes, i.e., autophagosomes fused with lysosomes. Due to the continuous nature of autophagic flux, some of the engulfed cargo is localized within autolysosomes, in addition to that found in autophagosomes, and we therefore collect both vesicles. In order to prevent the degradation of autolysosomal contents, we conduct all experiments in the presence of chloroquine, which was shown to reduce autophagosome–lysosome fusion, though not entirely (Rajan et al., 2015) (Fig. 2).

3.1 Cell harvesting and disruption

In order to prevent rupture of the plasma membrane, cells should be harvested by trypsinization, rather than mechanically by scraping them off the plate. Cell lysis is carried out in a hypotonic buffer, as "regular" isotonic lysis buffers usually contain detergents (e.g., RIPA), which solubilize and disrupt the different cellular membranes, including that of autophagosomes, autolysosomes, and lysosomes. Hypotonic (osmotic) lysis, on the other hand, is effective for selective plasma membrane lysis and used widely for cellular fractionation. Incubation of cells in hypotonic buffer results in their swelling, to a point where the plasma membrane is weakened, and Dounce homogenization results in complete rupture of the cells without causing severe damage to organelles. Nevertheless, since lysosomes/autolysosomes and other organelles are partially sensitive to the hypotonic environment, in order to stabilize their membranes and avoid swelling and rupture, the hypotonic buffer is supplemented with 0.25 *M* sucrose (Koenig, 1974).

3.1.1 Equipment

- Benchtop cooled centrifuge.
- Glass Dounce homogenizer.

3.1.2 Buffers and reagents

- Sterile PBS: 137 m*M* NaCl, 2.7 m*M* KCl, 10 m*M* Na_2HPO_4, 1.8 m*M* KH_2PO_4; pH 7.4.
- Trypsin–ethylenediaminetetraacetic acid (EDTA) Solution B: 0.25% and 0.05%, respectively.
- PBS supplemented with 2% dialyzed FBS.
- Hypotonic buffer (HB)-I: 0.25 *M* sucrose, 1 m*M* EDTA, 10 m*M* 4-(2-hydroxyethyl)-1-piperazineethanesulfonic acid (HEPES); pH 7.3, freshly supplemented with protease inhibitor cocktail (Roche) and 5 m*M* ATP.
- Hypotonic buffer (HB)-II: 0.25 *M* sucrose, 1 m*M* EDTA, 10 m*M* HEPES; pH 7.3, freshly supplemented with protease inhibitor cocktail (Roche), 5 m*M* ATP, 10 m*M* iodoacetamide (IAA), 5 m*M* *N*-ethylmaleimide (NEM).
- DUB inhibitor stock solution ×10: 100 m*M* IAA, 50 m*M* NEM in HB-I. Aqueous solutions of IAA and NEM are unstable and are prepared freshly.

3.1.3 Procedure

1. Grow and starve cells as described under Section 2.1.3, Steps *1–5*.
2. Wash cells twice with sterile PBS.
3. Harvest cells by trypsinization (1 min at 37°C).
4. Collect detached cells from plates with PBS supplemented with 2% dialyzed FBS.
5. Centrifuge cells at 400 × *g* for 5 min, discard supernatant, and wash the pellet with ice-cold PBS. Repeat twice.
6. After discarding the second wash, centrifuge the nearly dry pellet, and thoroughly aspirate any remnants of PBS.
7. Resuspend cells in 2–3 mL of HB-I, and leave to swell on ice for 10–15 min.
8. Disrupt cells by homogenization with ~50 strokes of type B (tight) pestle of Dounce homogenizer.

9. Check cells under the microscope for 90% disruption, using 0.4% trypan blue as a marker for disrupted cells.
10. Add 1/10 volume of DUB inhibitor stock solution.
11. Centrifuge the cell homogenate at 400 × *g* for 2 min at 4 °C to pellet debris and leftover intact cells, and collect the supernatant.
12. Centrifuge the supernatant at 1500 × *g* for 2 min at 4 °C to pellet nuclei and plasma membranes, and collect the postnuclear supernatant (PNS).
13. Wash nuclear pellet with 1 mL of HB-II, and centrifuge at 1500 × *g* for 2 min at 4 °C.
14. Pool the supernatant from *Step 13* with the PNS from *Step 12*, add NaCl to a final concentration of 135 m*M*, and keep on ice.

3.1.4 Notes

- In order to neutralize trypsin activity which may otherwise damage the cells, and in order to improve cell yield, detached cells should be collected from plates with PBS supplemented with 2% dialyzed FBS.
- Cell fractionation is a fairly long process. Prolonged exposure of cellular organelles to the hypotonic environment can result in disruption of organelles as well (de Duve et al. 1955). Since lysosomes are sensitive to hypotonic lysis (Schröter et al., 1999; Zhao, Wang, & Zhang, 2005), we assumed that autolysosomes (autophagosomes fused with lysosome) can also undergo osmotic lysis. We noted that adding NaCl (to a final concentration of 135 m*M*) to the cell lysate immediately after collecting the PNS increases the yield of autophagosomes/autolysosomes.
- It was reported (Seglen & Brinchmann, 2010) that treatment of cell lysate with glycyl-L-phenylalanine 2-naphthylamide (GPN) is included in the protocol in order to induce rapid osmotic disruption of lysosomes that accumulate GPN-cleaved products (Jadot, Colmant, Wattiaux-De Coninck, & Wattiaux, 1984). Since the main goal of our study is to detect 26S proteasome and ubiquitinated proteins engulfed within *intact* autophagosomes and autolysosomes following amino acid starvation, we omitted the GPN treatment in order to prevent bursting, which may affect not only lysosomes but also autolysosomes (Lawrence & Brown, 1992; Rajan et al., 2015).
- To detect ubiquitinated proteins, inhibitors of DUBs (IAA and NEM) should be added to cell lysates. We noted that the addition of IAA and NEM to the hypotonic buffer at the first step of the procedure (cell swelling) prevents cell disruption. Therefore, DUB inhibitors should be added immediately after cell burst is detected (using 0.4% trypan blue staining).

3.2 Purification of autophagosomes/autolysosomes by discontinuous density gradients

Density gradient centrifugation is a widely used method for subcellular fractionation of organelles (Alberts et al., 2002). During ultracentrifugation, organelles float or sediment toward their isopycnic position within the gradient. Sucrose-based gradients are commonly used for resolving organelles according to their density. One important disadvantage of this method, however, is its osmotic pressure/density ratio, causing shrinking of organelles and subsequent reduction in density differences between certain organelles. Media based on iodinated matrices, like isotonic Nycodenz and OptiPrep, have partially replaced sucrose as they are characterized by lower osmotic force and viscosity compared with sucrose, and therefore are less damaging to organelles (Neves, Perez, Spencer, Melo, & Weller, 2009).

3.2.1 *Equipment*

- Ultracentrifuge.
- Tubes for ultracentrifugation.

3.2.2 *Buffers and reagents*

- HB ×2: 0.5 *M* sucrose, 2 m*M* EDTA, 20 m*M* HEPES; pH 7.3.
- Buffered medium for Nycodenz: 3 m*M* KCl, 0.3 m*M* EDTA, 5 m*M* Tris(hydroxymethyl)aminomethane hydrochloride (Tris–HCl); pH 7.5.
- Buffered diluents for Nycodenz: 217.8 m*M* sucrose, 3 m*M* KCl, 0.3 m*M* EDTA, 5 m*M* Tris–HCl; pH 7.5.
- Homogenization concentrate: 100 m*M* EDTA, 1 *M* HEPES; pH 7.3.
- HB-II: 0.25 *M* sucrose, 1 m*M* EDTA, 10 m*M* HEPES; pH 7.3, freshly supplemented with protease inhibitors cocktail (Roche), 5 m*M* ATP, 10 m*M* IAA, and 5 m*M* NEM.
- Percoll (Sigma-Aldrich).
- Nycodenz®(Histodenz™) (Sigma-Aldrich).
- OptiPrep™ Density Gradient Medium (Sigma-Aldrich)

3.2.3 *Preparation of iso-osmotic Nycodenz, buffered OptiPrep, and 33% Percoll solutions*

- Isotonic 36% Nycodenz solution is prepared by dissolving 36 g of powdered Nycodenz and bringing it to 100 mL with buffered medium. Iso-osmotic Nycodenz solutions of desired concentrations (22% and 9%) are prepared by diluting 36% Nycodenz solution with the sucrose-containing buffered diluent.

- Buffered 60% OptiPrep is prepared by diluting 100 parts of OptiPrep (60%, w/v) aqueous solution with 1 part of homogenization concentrate.
- Buffered 30% OptiPrep is prepared by diluting 1 part of 60% OptiPrep with 1 part of HB-II
- 33% Percoll solution is prepared by mixing 33 mL of Percoll with 50 mL of HB-II X2 and 17 mL of sterile doubly distilled water.

3.2.4 Procedure

1. To pellet mitochondria, prepare a discontinuous (two-step) Nycodenz gradient by injecting 1 part (v/v) 22.5% Nycodenz below 2.4 parts 9.5% Nycodenz.
2. Place 2 parts PNS on top of the discontinuous Nycodenz gradient. In the end, the gradient should consist of: top layer—2 parts PNS; middle layer—2.4 parts 9.5% Nycodenz; and bottom layer—1 part 22.5% Nycodenz.
3. Centrifuge the Nycodenz gradient for 1 h at 140,000 × *g* at 4°C.
4. Using a Pasteur pipette, collect the ring [AP/AL + ER (Fig. 4), containing autophagosomes (AP), autolysosomes (AL), endoplasmic reticulum (ER), Golgi, and endosomes] formed between the 9.5% and 22.5% Nycodenz layers, and transfer it to a new tube.
5. Dilute the collected fraction (AP/AL + ER fraction) with HB-II at a 1:1 ratio.
6. To separate autophagosomes/autolysosomes from other small vesicular material, prepare a discontinuous (two-step) Percoll-Nycodenz gradient by injecting 1 part 22.5% Nycodenz underneath 3 parts 33% Percoll.
7. Place 1.4 parts (relatively to the 1 part of 22.5% Nycodenz) of the diluted AP + ER fraction (*Step 5*) on top of the Percoll-Nycodenz gradient (*Step 6*). At the end, the gradient should consist of: top layer—1.4 parts diluted AP + ER fraction; middle layer—3 parts 33% Percoll; and bottom layer—1 part 22.5% Nycodenz.
8. Centrifuge the Percoll-Nycodenz gradient at 72,000 × *g* for 30 min at 4°C.
9. Collect the lowermost ring (between the Percoll and Nycodenz)—autophagosome/autolysosome fraction (AP/AL fraction) using a Pasteur pipette.
10. Dilute the collected AP/AL fraction with 0.7 volume of 60% buffered OptiPrep.
11. To remove Percoll silica from the AP/AL fraction, prepare an OptiPrep gradient: from bottom to top—place 5.66 parts diluted autophagosomal

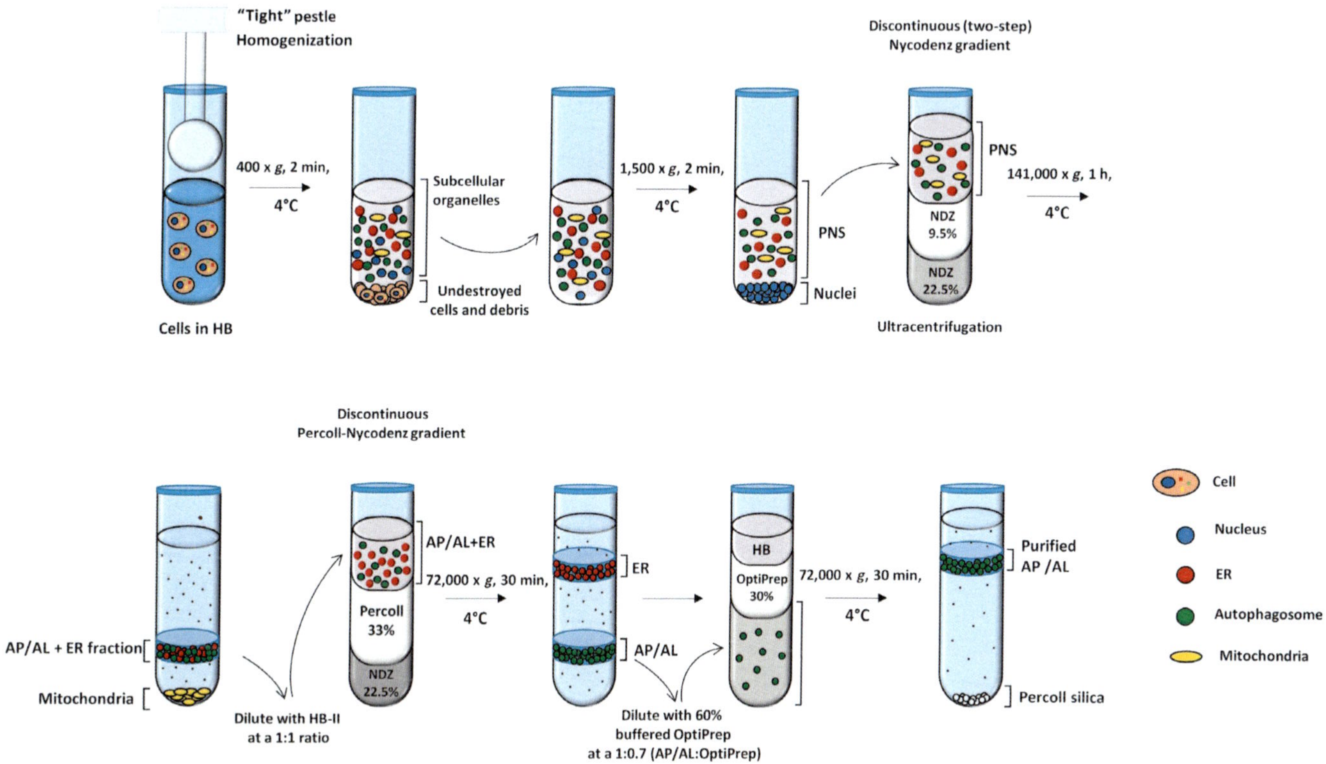

Fig. 4 Scheme for purification of autophagosomes/autolysosomes from HeLa cells. Details are described under Sections 3.1.3 and 3.2.4.

fraction at the bottom of the ultracentrifuge tube, and overlay with 1 part buffered 30% OptiPrep solution, and finally with 1.66 parts HB-II.

12. Centrifuge the gradient at 72,000 × *g* for 30 min at 4°C.
13. Collect the ring (autophagosomes/autolysosome; resting on the OptiPrep) in a minimal volume using a Pasteur pipette, and dilute it fourfold in HB-II supplemented with NaCl (to a final concentration of 135 m*M*).
14. Pellet autophagosomes/autolysosomes by centrifugation at 17,000 × *g* for 10 min at 4°C.
15. Wash the pellet three times with ice-cold PBS and centrifuge at 14,000 × *g* for 10 min between washes.
16. Add 30 μL of lysis buffer to the pellet and store at −80°C for further analysis.

3.2.5 Notes

- At *Step 3*, the samples can be left overnight in the centrifuge at 4°C.
- After centrifugation of the Percoll-Nycodenz gradient (*Step 8*), two rings can be observed. The upper ring contains small vesicular and membranous material, such as ER, Golgi, and endosomes, while the lower ring contains autophagosomes/autolysosomes. For easier collection of the autophagosomal/autolysosomal ring in *Step 9*, the upper part of the gradient (including the upper ring) can be collected and discarded prior to collecting the lower ring.

3.3 Detection of the proteasome (and possibly ubiquitinated proteins) in purified autophagosomal/autolysosomal fractions

The presence of the proteasome (and probably other ubiquitinated proteins) within autophagosomal/autolysosomal vesicles purified from cells grown in either complete or amino acid-depleted medium can be detected via Western blot (WB). In order to determine whether both the 20S and 19S proteasomal subcomplexes are ingested by autophagosomes during autophagy, the immunoblotting should be performed with antibodies targeting different proteasomal subunits of both subcomplexes. The purity of the autophagosome/autolysosome fraction can be confirmed by the lack of cytosolic and/or endoplasmic reticulum markers (e.g., tubulin and calnexin, respectively). Here, we discuss briefly the WB method, mainly by presenting the buffers and conditions we employ in our studies, while for more details readers are referred to Alegria-Schaffer (2014).

3.3.1 Equipment

- Standard SDS-PAGE equipment, including gel-running and transfer systems, and membrane/blot imaging device.
- 10% SDS-PAGE gels.
- ImageQuant, or an equivalent software for visualization and analysis.
- Polyvinylidene difluoride (PVDF) membranes.
- Spectrophotometer

3.3.2 Buffers and reagents

- Lysis buffer: 50 m*M* Tris–HCl; pH 7.4, 130 m*M* NaCl, and 1% Nonidet P-40, freshly supplemented with protease inhibitor cocktail (Roche), 5 m*M* ATP, 10 m*M* IAA, and 5 m*M* NEM.
- Bradford solution, or an equivalent kit/solution for protein concentration measurement.
- Running buffer (×10): 35 m*M* sodium dodecyl sulfate (SDS), 250 m*M* Tris Base, 1.92 *M* glycine.
- Transfer buffer: 25 m*M* Tris Base, 192 m*M* glycine, 20% methanol.
- Tris-buffered saline, Tween 20 (TBS-T): 150 m*M* NaCl, 20 m*M* Tris Base; pH 7.5, 0.05% Tween 20.
- Enhanced chemiluminescence (ECL) Western blotting solutions (available from various vendors).

3.3.3 Procedure

1. Add 50 μL of lysis buffer to the autophagosome/autolysosome pellet, and incubate for 15–30 min on ice.
2. Centrifuge at 14,000 × *g* for 15 min at 4°C. Collect supernatant and determine its protein concentration.
3. Resolve samples (up to 100 μg of total protein/lane) via SDS-PAGE, and transfer proteins onto PVDF membrane.
4. Blot the membrane by incubating it with the appropriate primary antibody, followed by sequential incubations in the presence of horseradish peroxidase (HRP)-conjugated secondary antibody, and a chemiluminescent substrate (ECL WB solutions).
5. Acquire image using an appropriate imager and software.

3.3.4 Notes

- In our research, we use HeLa cells stably expressing LC3 fused to GFP, the two forms of which—GFP-LC3-I and GFP-LC3-II—migrate at molecular mass of ∼43 and ∼41 kDa, respectively. Therefore, protein samples can be resolved on a 10% SDS-PAGE gel. For detection of endogenous LC3s, which migrate at molecular masses of ~15 and ~17 kDa, proteins should be resolved on either a 12.5% or a 15% SDS-PAGE gel.

Fig. 3B shows the presence of both the 20S and 19S proteasomal subcomplexes within isolated autophagosomal/autolysosomal vesicles. The 20S proteasome is probed with anti-α6 and β2, and the 19S is probed with anti-RPN10 and RPN1. Importantly, amino acid starvation increases the amount of both the 20S and 19S proteasomes in the vesicular fraction (Cohen-Kaplan, Livneh, Avni, Fabre, et al., 2016).

The same figure shows the purity of the isolated autophagosomal fraction, which is confirmed by a lack of cytosolic (tubulin) and endoplasmic reticulum (calnexin) markers.

Fig. 3C shows an increase in ubiquitin conjugates in purified autophagosomal/autolysosomal fraction following amino acid starvation.

4. Identification of ubiquitinated proteins and ubiquitin-modified sites on cellular proteins

Ubiquitin modification of proteins is altered during different pathophysiological conditions in both the amount of adducts and the specific ubiquitin-conjugation sites. Therefore, it is very important to identify the different proteins and sites altered by ubiquitin under these conditions. However, since only a small fraction of the cellular proteome is ubiquitinated at any moment, it is necessary to enrich the fraction of ubiquitinated proteins for high sensitivity tandem mass spectrometry.

One way to enrich for ubiquitin adducts is to use antibodies directed against the C-terminal amino acids (-G75G76-COOH) of ubiquitin bound to an internal lysine of the target protein (GG-K). The two Gly residues in the ubiquitin molecule are preceded by Arg, and the R-GG linkage is cleaved by trypsin during preparation of the peptides for mass spectrometry. The same treatment will also cleave C-terminally to the Arg or Lys residues flanking the ubiquitin-modified lysine (which is protected from tryptic cleavage). Therefore, it is possible to identify the protein from which each peptide is derived, provided that the molecular mass of the two additional Gly residues is added to the calculated molecular mass of the predicted peptides resulting from trypsin digestion.

We will review the procedure for enriching GG-modified peptides, while the sample preparations for mass spectrometry (MS), the subsequent separation of peptides, and analysis of the MS data are reviewed elsewhere (Burlingame, 2005).

4.1 Ubiquitination site detection using GG-modified peptide enrichment

4.1.1 Equipment

- 100 mm cell culture plates.
- Rotor for 15 mL tubes and centrifuge.
- End-over-end shaker.
- Horizontal shaker.

4.1.2 Buffers and reagents

- Urea lysis buffer: 8 *M* urea, 100 m*M* Tris–HCl; pH 7.6 supplemented with 10 m*M* IAA (DUB inhibitor).
- Immunoaffinity purification (IAP) buffer: 50 m*M* MOPS/NaOH; pH 7.2, 10 m*M* Na_2HPO_4 and 50 m*M* NaCl.
- Wash buffer: 500 m*M* NaCl, 3 m*M* KCl, 10 m*M* Na_2HPO_4, 2 m*M* KH_2PO_4, 0.1% octylglucoside; pH 7.4.
- Immobilized anti-K-ε-GG antibody. Available commercially (for example, from Cell Signaling Technologies).
- 0.2% Trifluoroacetic acid (TFA).

4.1.3 Procedure

1. Grow and starve cells as described under Section 2.1.3, *Steps 1–5*.
2. Wash cells twice with PBS at room temperature, and aspirate thoroughly.
3. Lyse cells on plate by adding 3 mL (per 100 mm plate) urea lysis buffer, incubate 3 min at room temperature, and scrape into a 15 mL plastic conical tube.
4. Sonicate (15 s on/15 s off) × 3, using a probe sonicator at 15 W intensity.
5. Centrifuge at 17,000 × *g* for 10 min at room temperature.
6. Collect supernatant. (Supernatant can be processed immediately or frozen in liquid N_2 and stored at −80°C.)
7. To prepare samples for MS, incubate 2 mg of protein with 2.8 m*M* DTT for 30 min at 60°C, followed by the addition of IAA to a final concentration of 8.8 m*M,* and incubation for additional 30 min at room temperature, while protected from light.

8. Dilute the solution fourfold with 20 m*M* HEPES pH 8.0 to a final concentration of 2 *M* urea.
9. Add modified Trypsin-TPCK at a 1:50 enzyme-to-substrate ratio and digest overnight at room temperature with mixing.
10. Desalt the tryptic peptides using Sep-Pak C18 column (Waters) and dry. Part of the peptide preparation also can be used for general proteomic analysis (rather than for identification of the specific GG-containing peptides).
11. Resuspend dry peptides generated in *Step* 7 in 700 μL buffer IAP ×1 (w/o detergent). Use a syringe with 30-G needle for thorough resuspension. Adjust pH to 7.0 by adding ~70 μL of NaOH 0.5 *M*. Verify by pH paper.
12. Centrifuge at 2000 × *g* for 5 min at 4°C and transfer supernatant to a new Eppendorf tube.
13. Transfer 40 μL of anti-K-ε-GG slurry in a new tube, and add to it the pH-adjusted lysate.
14. Incubate on an end-over-end shaker for 3 h, or overnight at 4°C.
15. Wash beads once with 500 μL IAP, and centrifuge at 3000 × *g* for 1 min at 4°C.
16. Wash beads twice with 700 μL wash buffer, and centrifuge at 3000 × *g* for 1 min at 4°C. During the second wash, incubate on an end-over-end shaker for 10 min at 4°C prior to centrifugation. Keep all fractions (washes) for MS analysis.
17. Elute GG-modified peptides by adding 100 μL of 0.2% TFA followed by an incubation on a horizontal shaker at 900 RPM for 10 min at room temperature. Centrifuge at 3000 × *g* for 1 min at 4°C, and transfer the supernatant to a new tube.
18. Repeat elution as in Step 15.
19. Pool the eluted fractions from *Steps 15–16* and freeze at −20°C.
20. Proceed to processing and analyzing the GG-modified peptides via MS as described under (Braten et al., 2016).

4.1.4 Notes

- Do not add urea lysis buffer onto cold plates, as urea precipitates at low temperatures.
- All buffers and solutions, including water, should be of MS-grade.
- We find that using 40 μL of the antibody-bead slurry is sufficient, compared with the double amount recommended by some manufacturers. The amount is probably assay-dependent, as well as cell type-dependent.

- For the inhibition of DUBs in samples to be analyzed by MS, we use IAA only and avoiding NEM. This is because the latter creates interference with the signal detected by the mass spectrometer.

Acknowledgments

HeLa GFP-LC3B cells were kindly provided by Dr. Zvulun Elazar from The Weizmann Institute of Science in Rehovot, Israel. Research in the laboratory of A.C. is supported by grants from the Dr. Miriam and Sheldon G. Adelson Medical Research Foundation (AMRF), the Israel Science Foundation (ISF), The German Israeli Foundation for Research and Development (GIF), and a special fund for research in the Technion established by Mr. Albert Sweet of California, USA. I.L. is supported by the Foulkes Fellowship. A.C. is an Israel Cancer Research Fund (ICRF) USA Professor.

References

Alberts, B., Johnson, A., Lewis, J., Raff, M., Roberts, K., & Walter, P. (2002). Fractionation of Cells. In B. Alberts (Ed.), *Molecular biology of the cell* (4th ed., pp. 445–450). New York, NY: Garland Science.

Alegria-Schaffer, A. (2014). Western blotting using chemiluminescent substrates. *Methods in enzymology, 541*, 251–259. https://doi.org/10.1016/B978-0-12-420119-4.00019-7.

Bjørkøy, G., Lamark, T., Brech, A., Outzen, H., Perander, M., Overvatn, A., et al. (2005). p62/SQSTM1 forms protein aggregates degraded by autophagy and has a protective effect on huntingtin-induced cell death. *The Journal of Cell Biology, 171*(4), 603–614. https://doi.org/10.1083/jcb.200507002.

Boutet, S. C., Disatnik, M. H., Chan, L. S., Iori, K., & Rando, T. A. (2007). Regulation of Pax3 by proteasomal degradation of monoubiquitinated protein in skeletal muscle progenitors. *Cell, 130*(2), 349–362. https://doi.org/10.1016/j.cell.2007.05.044.

Braten, O., Livneh, I., Ziv, T., Admon, A., Kehat, I., Caspi, L. H., et al. (2016). Numerous proteins with unique characteristics are degraded by the 26S proteasome following monoubiquitination. *Proceedings of the National Academy of Sciences of the United States of America, 113*(32), E4639–E4647. https://doi.org/10.1073/pnas.1608644113.

Burlingame, A. L. (2005). Biological mass spectrometry. *Methods in Enzymology, 402*, 1–478.

Chau, V., Tobias, J. W., Bachmair, A., Marriott, D., Ecker, D. J., Gonda, D. K., et al. (1989). A multiubiquitin chain is confined to specific lysine in a targeted short-lived protein. *Science (New York, N.Y.), 243*(4898), 1576–1583. https://doi.org/10.1126/science.2538923.

Ciechanover, A. (1994). The ubiquitin-proteasome proteolytic pathway. *Cell, 79*(1), 13–21. https://doi.org/10.1016/0092-8674(94)90396-4.

Ciechanover, A. (2005). Proteolysis: From the lysosome to ubiquitin and the proteasome. *Nature Reviews Molecular Cell Biology, 6*, 79–86. https://doi.org/10.1038/nrm1552.

Cohen-Kaplan, V., Ciechanover, A., & Livneh, I. (2016). p62 at the crossroad of the ubiquitin-proteasome system and autophagy. *Oncotarget, 7*(51), 83833–83834. https://doi.org/10.18632/oncotarget.13805.

Cohen-kaplan, V., Ciechanover, A., & Livneh, I. (2017). Stress-induced polyubiquitination of proteasomal ubiquitin receptors targets the proteolytic complex for autophagic degradation. *Autophagy, 13*(4), 1–2. https://doi.org/10.1080/15548627.2016.1278327.

Cohen-Kaplan, V., Livneh, I., Avni, N., Cohen-Rosenzweig, C., & Ciechanover, A. (2016). The ubiquitin-proteasome system and autophagy: Coordinated and independent activities. *The International Journal of Biochemistry & Cell Biology., 79*, 403–418. https://doi.org/10.1016/j.biocel.2016.07.019.

Cohen-Kaplan, V., Livneh, I., Avni, N., Fabre, B., Ziv, T., Kwon, Y. T., et al. (2016). p62- and ubiquitin-dependent stress-induced autophagy of the mammalian 26S proteasome. *Proceedings of the National Academy of Sciences of the United States of America, 113*(47), E7490–E7499. https://doi.org/10.1073/pnas.1615455113.

Cooper, G., & Hausman, R. (2013). Lysosomes. In G. Cooper (Ed.), *The cell: A molecular approach* (6th ed., pp. 412–420). Sinauer Associates, Inc.

de Bie, P., & Ciechanover, A. (2011). Ubiquitination of E3 ligases: Self-regulation of the ubiquitin system via proteolytic and non-proteolytic mechanisms. *Cell Death and Differentiation, 18*(9), 1393–1402. https://doi.org/10.1038/cdd.2011.16.

de Duve, C. (1963). General properties of lysosomes. In *Ciba foundation symposium on lysosomes*. London: J.A. Churchill Ltd., Ciba Found.

de Duve, C., Gianetto, R., Appelmans, F., & Wattiaux, R. (1953). Enzymic content of the mitochondria fraction. *Nature, 172*, 1143–1144.

de Duve, C., Pressman, B. C., Gianetto, R., Wattiaux, R., & Appelmans, F. (1955). Tissue fractionation studies. 6. Intracellular distribution patterns of enzymes in rat-liver tissue. *Biochemical Journal, 60*(4), 604–617. https://doi.org/10.1042/bj0600604.

Dice, J. F., Chiang, H. L., Spencer, E. P., & Backer, J. M. (1986). Regulation of catabolism of microinjected ribonuclease A. Identification of residues 7-11 as the essential pentapeptide. *Journal of Biological Chemistry, 261*(15), 6853–6859. https://doi.org/10.1186/1754-6834-6-131.

Dimova, N. V., Hathaway, N. A., Lee, B. H., Kirkpatrick, D. S., Berkowitz, M. L., Gygi, S. P., et al. (2012). APC/C-mediated multiple monoubiquitylation provides an alternative degradation signal for cyclin B1. *Nature Cell Biology, 14*(2), 168–176. https://doi.org/10.1038/ncb2425.

Fabre, B., Lambour, T., Delobel, J., Amalric, F., Monsarrat, B., Burlet-Schiltz, O., et al. (2013). Subcellular distribution and dynamics of active proteasome complexes unraveled by a workflow combining in vivo complex cross-linking and quantitative proteomics. *Molecular & Cellular Proteomics : MCP, 12*(3), 687–699. https://doi.org/10.1074/mcp.M112.023317.

Fabre, B., Lambour, T., Garrigues, L., Ducoux-Petit, M., Amalric, F., Monsarrat, B., et al. (2014). Label-free quantitative proteomics reveals the dynamics of proteasome complexes composition and stoichiometry in a wide range of human cell lines. *Journal of Proteome Research, 13*(6), 3027–3037. https://doi.org/10.1021/pr500193k.

Finley, D., Chen, X., & Walters, K. J. (2016). Gates, channels, and switches: Elements of the proteasome machine. *Trends in Biochemical Sciences, 41*(1), 77–93. https://doi.org/10.1016/j.tibs.2015.10.009.

Ganley, I. G., Lam, D. H., Wang, J., Ding, X., Chen, S., & Jiang, X. (2009). ULK1.ATG13. FIP200 complex mediates mTOR signaling and is essential for autophagy. *The Journal of Biological Chemistry, 284*, 12297–12305. https://doi.org/10.1074/jbc.M900573200.

Gao, Z., Gammoh, N., Wong, P.-M., Erdjument-Bromage, H., Tempst, P., & Jiang, X. (2010). Processing of autophagic protein LC3 by the 20S proteasome. *Autophagy, 6*(1), 126–137. Retrieved from http://www.ncbi.nlm.nih.gov/pubmed/20061800.

Gianetto, R., & de Duve, C. (1955). Tissue fractionation studies. 4. Comparative study of the binding of acid phosphatase, beta-glucuronidase and cathepsin by rat-liver particles. *The Biochemical Journal, 59*(3), 433–438.

Glickman, M. H., & Ciechanover, A. (2002). The ubiquitin-proteasome proteolytic pathway: Destruction for the sake of construction. *Physiological Reviews, 82*(2), 373–428. https://doi.org/10.1152/physrev.00027.2001.

Goh, L. K., & Sorkin, A. (2013). Endocytosis of receptor tyrosine kinases. *Cold Spring Harbor Perspectives in Biology, 5*(5), a017459. https://doi.org/10.1101/cshperspect.a017459.

Hanada, T., Noda, N. N., Satomi, Y., Ichimura, Y., Fujioka, Y., Takao, T., et al. (2007). The Atg12-Atg5 conjugate has a novel E3-like activity for protein lipidation in autophagy. *Journal of Biological Chemistry*, *282*(52), 37298–37302. https://doi.org/10.1074/jbc.C700195200.
Heitman, J., Movva, N. R., & Hall, M. N. (1991). Targets for cell cycle arrest by the immunosuppressant rapamycin in yeast. *Science (New York, N.Y.)*, *253*(5022), 905–909. Retrieved from http://www.ncbi.nlm.nih.gov/pubmed/1715094.
Hershko, A., Ciechanover, A., & Rose, I. A. (1981). Identification of the active amino acid residue of the polypeptide of ATP-dependent protein breakdown. *The Journal of biological chemistry*, *256*(4), 75–78.
Hershko, A., & Ciechanover, A. (1992). The ubiquitin system for protein degradation. *Annual Review of Biochemistry*, *61*(1), 761–807. https://doi.org/10.1146/annurev.bi.61.070192.003553.
Hogness, D. S., Cohn, M., & Monod, J. (1955). Studies on the induced synthesis of beta-galactosidase in *Escherichia coli*: The kinetics and mechanism of sulfur incorporation. *Biochimica et Biophysica Acta*, *16*(1), 99–116.
Hosokawa, N., Hara, T., Kaizuka, T., Kishi, C., Takamura, A., Miura, Y., et al. (2009). Nutrient-dependent mTORC1 association with the ULK1-Atg13-FIP200 complex required for autophagy. *Molecular Biology of the Cell*, *20*(7), 1981–1991. https://doi.org/10.1091/mbc.E08-12-1248.
Ikeda, H., & Kerppola, T. K. (2008). Lysosomal localization of ubiquitinated jun requires multiple determinants in a lysine-27–linked polyubiquitin conjugate. *Molecular Biology of the Cell*, *19*(11), 4588–4601. https://doi.org/10.1091/mbc.e08-05-0496.
Jadot, M., Colmant, C., Wattiaux-De Coninck, S., & Wattiaux, R. (1984). Intralysosomal hydrolysis of glycyl-L-phenylalanine 2-naphthylamide. *The Biochemical Journal*, *219*(3), 965–970. Retrieved from http://www.ncbi.nlm.nih.gov/pubmed/6743255.
Jin, S., & Cui, J. (2017). BST2 inhibits type I IFN (interferon) signaling by accelerating MAVS degradation through CALCOCO2-directed autophagy. *Autophagy*, *14*, 171–172. https://doi.org/10.1080/15548627.2017.1393590.
Jung, C. H., Jun, C. B., Ro, S.-H., Kim, Y.-M., Otto, N. M., Cao, J., et al. (2009). ULK-Atg13-FIP200 complexes mediate mTOR signaling to the autophagy machinery. *Molecular Biology of the Cell*, *20*(7), 1992–2003. https://doi.org/10.1091/mbc.e08-12-1249.
Kabeya, Y., Mizushima, N., Ueno, T., Yamamoto, A., Kirisako, T., Noda, T., et al. (2000). LC3, a mammalian homologue of yeast Apg8p, is localized in autophagosome membranes after processing. *The EMBO Journal*, *19*, 5720–5728. Retrieved from https://www.ncbi.nlm.nih.gov/pmc/articles/PMC305793/pdf/cdd565.pdf.
Kirkin, V., Lamark, T., Sou, Y.-S., Bjørkøy, G., Nunn, J. L., Bruun, J.-A., et al. (2009). A role for NBR1 in autophagosomal degradation of ubiquitinated substrates. *Molecular Cell*, *33*(4), 505–516. https://doi.org/10.1016/j.molcel.2009.01.020.
Klionsky, D. J. (2008). Autophagy revisited: A conversation with Christian de Duve. *Autophagy*, *4*(6), 740–743. https://doi.org/10.4161/auto.6398.
Klionsky, D. J., Abdelmohsen, K., Abe, A., Abedin, M. J., Abeliovich, H., Acevedo Arozena, A., et al. (2016). Guidelines for the use and interpretation of assays for monitoring autophagy (3rd edition). *Autophagy*, *12*(1), 1–222. https://doi.org/10.1080/15548627.2015.1100356.
Koenig, H. (1974). The isolation of lysosomes from brain. *Methods in Enzymology*, *31*, 457–477.
Kravtsova-Ivantsiv, Y., Cohen, S., & Ciechanover, A. (2009). Modification by single ubiquitin moieties rather than polyubiquitination is sufficient for proteasomal processing of the p105 NF-kappaB precursor. *Molecular Cell*, *33*(4), 496–504. https://doi.org/10.1016/j.molcel.2009.01.023 S1097-2765 (09)00067-7 [pii].

Laplante, M., & Sabatini, D. M. (2013). mTOR signaling in growth control and disease. *Cell*, *149*(2), 274–293. https://doi.org/10.1016/j.cell.2012.03.017.mTOR.

Lawrence, B. P., & Brown, W. J. (1992). Autophagic vacuoles rapidly fuse with pre-existing lysosomes in cultured hepatocytes. *Journal of Cell Science*, *102*(Pt 3), 515–526. Retrieved from http://www.ncbi.nlm.nih.gov/pubmed/1324248.

Liani, E., Eyal, A., Avraham, E., Shemer, R., Szargel, R., Berg, D., et al. (2004). Ubiquitylation of synphilin-1 and alpha-synuclein by SIAH and its presence in cellular inclusions and Lewy bodies imply a role in Parkinson's disease. *Proceedings of the National Academy of Sciences of the United States of America*, *101*(15), 5500–5505. https://doi.org/10.1073/pnas.0401081101.

Liu, W. J., Ye, L., Huang, W. F., Guo, L. J., Xu, Z. G., Wu, H. L., et al. (2016). p62 links the autophagy pathway and the ubiqutin–proteasome system upon ubiquitinated protein degradation. *Cellular & Molecular Biology Letters*, *21*(1), 29. https://doi.org/10.1186/s11658-016-0031-z.

Livneh, I., Cohen-Kaplan, V., Cohen-Rosenzweig, C., Avni, N., & Ciechanover, A. (2016). The life cycle of the 26S proteasome: From birth, through regulation and function, and onto its death. *Cell Research*, *26*(8), 869–885. https://doi.org/10.1038/cr.2016.86.

Livneh, I., Kravtsova-Ivantsiv, Y., Braten, O., Kwon, Y. T., & Ciechanover, A. (2017). Monoubiquitination joins polyubiquitination as an esteemed proteasomal targeting signal. *BioEssays*, *39*(6), 1700027. https://doi.org/10.1002/bies.201700027.

Lőw, P., Varga, Á., Pircs, K., Nagy, P., Szatmári, Z., Sass, M., et al. (2013). Impaired proteasomal degradation enhances autophagy via hypoxia signaling in Drosophila. *BMC Cell Biology*, *14*(1), 29. https://doi.org/10.1186/1471-2121-14-29.

Manders, E. M. M., Verbeek, F. J., & Aten, J. A. (1993). Measurement of co-localization of objects in dual-colour confocal images. *Journal of Microscopy*, *169*(3), 375–382. https://doi.org/10.1111/j.1365-2818.1993.tb03313.x.

Min, M., Mevissen, T. E. T., De Luca, M., Komander, D., & Lindon, C. (2015). Efficient APC/C substrate degradation in cells undergoing mitotic exit depends on K11 ubiquitin linkages. *Molecular Biology of the Cell*, *26*(24), 4325–4332. https://doi.org/10.1091/mbc.e15-02-0102.

Neves, J. S., Perez, S. A. C., Spencer, L. A., Melo, R. C. N., & Weller, P. F. (2009). Subcellular fractionation of human eosinophils: Isolation of functional specific granules on isoosmotic density gradients. *Journal of Immunological Methods*, *344*(1), 64–72. https://doi.org/10.1016/j.jim.2009.03.006.

Ohkuma, S., & Poole, B. (1978). Fluorescence probe measurement of the intralysosomal pH in living cells and the perturbation of pH by various agents. *Proceedings of the National Academy of Sciences of the United States of America*, *75*(7), 3327–3331. https://doi.org/10.1073/pnas.75.7.3327.

Onodera, J., & Ohsumi, Y. (2005). Autophagy is required for maintenance of amino acid levels and protein synthesis under nitrogen starvation. *Journal of Biological Chemistry*, *280*, 31582–31586. https://doi.org/10.1074/jbc.M506736200.

Paraskevopoulos, K., Kriegenburg, F., Tatham, M. H., Rösner, H. I., Medina, B., Larsen, I. B., et al. (2014). Dss 1 is a 26S proteasome ubiquitin receptor. *Molecular Cell*, *56*(3), 453–461. https://doi.org/10.1016/j.molcel.2014.09.008.

Rajan, R., Karbowniczek, M., Pugsley, H. R., Sabnani, M. K., Astrinidis, A., & La-Beck, N. M. (2015). Quantifying autophagosomes and autolysosomes in cells using imaging flow cytometry. *Cytometry Part A*, *87*(5), 451–458. https://doi.org/10.1002/cyto.a.22652.

Redmann, M., Benavides, G. A., Berryhill, T. F., Wani, W. Y., Ouyang, X., Johnson, M. S., et al. (2017). Inhibition of autophagy with bafilomycin and chloroquine decreases mitochondrial quality and bioenergetic function in primary neurons. *Redox Biology*, *11*, 73–81. https://doi.org/10.1016/j.redox.2016.11.004.

Saeki, Y., Kudo, T., Sone, T., Kikuchi, Y., Yokosawa, H., Toh-e, A., et al. (2009). Lysine 63-linked polyubiquitin chain may serve as a targeting signal for the 26S proteasome. *The EMBO Journal*, *28*(4), 359–371. https://doi.org/10.1038/emboj.2008.305.

Schneider, D. L. (1981). ATP-dependent acidification of intact and disrupted lysosomes. Evidence for an ATP-driven proton pump. *Journal of Biological Chemistry*, *256*(8), 3858–3864.

Schoenheimer, R. (1942). *The dynamic state of body constituents*. Cambridge, MA, USA: Harvard University Press.

Schröter, C. J., Braun, M., Englert, J., Beck, H., Schmid, H., & Kalbacher, H. (1999). A rapid method to separate endosomes from lysosomal contents using differential centrifugation and hypotonic lysis of lysosomes. *Journal of Immunological Methods*, *227*(1–2), 161–168. Retrieved from http://www.ncbi.nlm.nih.gov/pubmed/10485263.

Seglen, P. O., & Brinchmann, M. F. (2010). Purification of autophagosomes from rat hepatocytes. *Autophagy*, *6*(4), 542–547. https://doi.org/10.4161/auto.6.4.11272.

Shvets, E., Abada, A., Weidberg, H., & Elazar, Z. (2011). Dissecting the involvement of LC3B and GATE-16 in p62 recruitment into autophagosomes. *Autophagy*, *683*(7), 683–688. https://doi.org/10.4161/auto.7.7.15279.

Slobodkin, M. R., & Elazar, Z. (2013). The Atg8 family: Multifunctional ubiquitin-like key regulators of autophagy. *Essays in Biochemistry*, *55*, 51–64. https://doi.org/10.1042/bse0550051.

Song, P., Li, S., Wu, H., Gao, R., Rao, G., Wang, D., et al. (2016). Parkin promotes proteasomal degradation of p62: Implication of selective vulnerability of neuronal cells in the pathogenesis of Parkinson's disease. *Protein & Cell*, 7(2), 114–129. https://doi.org/10.1007/s13238-015-0230-9.

Suzuki, K., & Ohsumi, Y. (2007). Molecular machinery of autophagosome formation in yeast, Saccharomyces cerevisiae. *FEBS Letters*, *581*(11), 2156–2161. https://doi.org/10.1016/j.febslet.2007.01.096.

Takenouchi, T., Sekiyama, K., Tsukimoto, M., Iwamaru, Y., Fujita, M., Sugama, S., et al. (2014). Chapter 14. Role of autophagy in P2X7 receptor-mediated maturation and unconventional secretion of IL-1β in microglia. *Autophagy*, *6*, 211–222. https://doi.org/10.1016/B978-0-12-801032-7.00014-9.

Tan, J. M. M., Wong, E. S. P., Kirkpatrick, D. S., Pletnikova, O., Ko, H. S., Tay, S.-P., et al. (2008). Lysine 63-linked ubiquitination promotes the formation and autophagic clearance of protein inclusions associated with neurodegenerative diseases. *Human Molecular Genetics*, *17*(3), 431–439. https://doi.org/10.1093/hmg/ddm320.

Tanida, I., Ueno, T., & Kominami, E. (2004). Human light chain 3/MAP1LC3B is cleaved at its carboxyl-terminal met 121 to expose Gly 120 for lipidation and targeting to autophagosomal membranes. *Journal of Biological Chemistry*, *279*(46), 47704–47710. https://doi.org/10.1074/jbc.M407016200.

Thrower, J. S., Hoffman, L., Rechsteiner, M., & Pickart, C. M. (2000). Recognition of the polyubiquitin proteolytic signal. *The EMBO Journal*, *19*(1), 94–102. https://doi.org/10.1093/emboj/19.1.94.

Thumm, M., Egner, R., Koch, B., Schlumpberger, M., Straub, M., Veenhuis, M., et al. (1994). Isolation of autophagocytosis mutants of Saccharomyces cerevisiae. *FEBS Letters*, *349*(2), 275–280. https://doi.org/10.1016/0014-5793(94)00672-5.

Tsukada, M., & Ohsumi, Y. (1993). Isolation and characterization of autophagy-defective mutants of Saccharomyces cerevisiae. *FEBS Letter*, *333*(1), 169–174.

Weidberg, H., Shvets, E., Shpilka, T., Shimron, F., Shinder, V., & Elazar, Z. (2010). LC3 and GATE-16/GABARAP subfamilies are both essential yet act differently in autophagosome biogenesis. *The EMBO Journal*, *29*, 1792–1802. https://doi.org/10.1038/emboj.2010.74.

Weissman, A. M., Shabek, N., & Ciechanover, A. (2011). The predator becomes the prey: Regulating the ubiquitin system by ubiquitylation and degradation. *Nature Reviews Molecular Cell Biology*, *12*(9), 605–620. https://doi.org/10.1038/nrm3173.

Yin, H., Gui, Y., Du, G., Frohman, M. A., & Zheng, X.-L. (2010). Dependence of phospholipase D1 multi-monoubiquitination on its enzymatic activity and palmitoylation. *The Journal of Biological Chemistry*, *285*(18), 13580–13588. https://doi.org/10.1074/jbc.M109.046359.

Yoshii, S. R., & Mizushima, N. (2017). Monitoring and measuring autophagy. *International Journal of Molecular Sciences*, *18*(9), 1865. https://doi.org/10.3390/ijms18091865.

Zachari, M., & Ganley, I. G. (2017). The mammalian ULK1 complex and autophagy initiation. *Essays in Biochemistry*, *61*(6), 585–596. https://doi.org/10.1042/EBC20170021.

Zhang, L., Xu, M., Scotti, E., Chen, Z. J., & Tontonoz, P. (2013). Both K63 and K48 ubiquitin linkages signal lysosomal degradation of the LDL receptor. *Journal of Lipid Research*, *54*(5), 1410–1420. https://doi.org/10.1194/jlr.M035774.

Zhao, H., Wang, X., & Zhang, G.-J. (2005). Lysosome destabilization by cytosolic extracts, putative involvement of Ca^{2+}/phospholipase C. *FEBS Letters*, *579*(6), 1551–1556. https://doi.org/10.1016/j.febslet.2005.01.061.

Zhao, J., Zhai, B., Gygi, S. P., & Goldberg, A. L. (2015). mTOR inhibition activates overall protein degradation by the ubiquitin proteasome system as well as by autophagy. *Proceedings of the National Academy of Sciences of the United States of America*, *112*(52), 15790–15797. https://doi.org/10.1073/pnas.1521919112.

FSC
www.fsc.org
MIX
Papier | Fördert gute Waldnutzung
FSC® C083411

Kolibri 360 GmbH
Ferdinand-Jühlke-Straße 7
99095 Erfurt, Deutschland
produktsicherheit@kolibri360.de